Communications in Computer and Information Science 2692

Series Editors

Rationale

The CCIS series is devoted to the publication of proceedings of computer science conferences. Its aim is to efficiently disseminate original research results in informatics in printed and electronic form. While the focus is on publication of peer-reviewed full papers presenting mature work, inclusion of reviewed short papers reporting on work in progress is welcome, too. Besides globally relevant meetings with internationally representative program committees guaranteeing a strict peer-reviewing and paper selection process, conferences run by societies or of high regional or national relevance are also considered for publication.

Topics

The topical scope of CCIS spans the entire spectrum of informatics ranging from foundational topics in the theory of computing to information and communications science and technology and a broad variety of interdisciplinary application fields.

Information for Volume Editors and Authors

Publication in CCIS is free of charge. No royalties are paid, however, we offer registered conference participants temporary free access to the online version of the conference proceedings on SpringerLink (http://link.springer.com) by means of an http referrer from the conference website and/or a number of complimentary printed copies, as specified in the official acceptance email of the event.

CCIS proceedings can be published in time for distribution at conferences or as post-proceedings, and delivered in the form of printed books and/or electronically as USBs and/or e-content licenses for accessing proceedings at SpringerLink. Furthermore, CCIS proceedings are included in the CCIS electronic book series hosted in the SpringerLink digital library at http://link.springer.com/bookseries/7899. Conferences publishing in CCIS are allowed to use our online conference service (Meteor) for managing the whole proceedings lifecycle (from submission and reviewing to preparing for publication) free of charge.

Publication process

The language of publication is exclusively English. Authors publishing in CCIS have to sign the Springer CCIS copyright transfer form, however, they are free to use their material published in CCIS for substantially changed, more elaborate subsequent publications elsewhere. For the preparation of the camera-ready papers/files, authors have to strictly adhere to the Springer CCIS Authors' Instructions and are strongly encouraged to use the CCIS LaTeX style files or templates.

Abstracting/Indexing

CCIS is abstracted/indexed in DBLP, Google Scholar, EI-Compendex, Mathematical Reviews, SCImago, Scopus. CCIS volumes are also submitted for the inclusion in ISI Proceedings.

How to start

To start the evaluation of your proposal for inclusion in the CCIS series, please send an e-mail to ccis@springer.com

Wilson J. Sarmiento · Eduard L. Sierra-Ballén
Editors

Advances in Computing

19th Colombian Conference on Computing, CCC 2025
Cajica, Colombia, September 10–12, 2025
Revised Selected Papers, Part II

Springer

Editors
Wilson J. Sarmiento
Universidad Militar Nueva Granada
Bogotá, Colombia

Eduard L. Sierra-Ballén
Universidad Militar Nueva Granada
Bogotá, Colombia

ISSN 1865-0929 ISSN 1865-0937 (electronic)
Communications in Computer and Information Science
ISBN 978-3-032-28446-4 ISBN 978-3-032-28447-1 (eBook)
https://doi.org/10.1007/978-3-032-28447-1

This Springer imprint is published by the registered company Springer Nature Switzerland AG
The registered company address is: Gewerbestrasse 11, 6330 Cham, Switzerland

Preface

The Colombian Conference on Computing (CCC) is an annual event organized by the Colombian Computing Society, aimed at promoting scientific discussion in the fields of computer science and informatics in Colombia. This conference brings together researchers, students, and professionals from both Colombia and abroad. The conference serves as an ideal venue for constructive discussions on advancements in computing, with the academic community playing a central role. It also invites professionals from both the public and private sectors to actively participate in the exchange of knowledge and experiences. Through its distinguished national and international keynote speakers, presenters, and workshops, the event seeks to strengthen the development and positioning of computing in the country and the Latin American region.

The 19th edition of the Colombian Congress of Computing (CCC 2025) was held in person from September 10 to 12, 2025, at the Nueva Granada Campus of the Universidad Militar Nueva Granada. Located at Kilometer 2 on the Cajicá-Zipaquirá road in Cundinamarca, the campus provided a beautiful setting in the heart of Colombia, a region rich in natural, cultural, and historical wonders.

The CCC 2025 received a total of 124 submissions across all tracks, including 74 long papers, 39 short papers and workshops, and 11 submissions for the Master's and Doctoral Symposium. Following a rigorous double-blind peer-review process, where each paper was evaluated by at least three reviewers, 40 full papers, 10 short papers/workshops, and 4 symposium papers were selected for publication, representing an acceptance rate of 43.5%. Additionally, the conference featured further contributions that are not included in these proceedings, specifically 6 Demos and Showcases and 21 posters.

The technical program was organized into several thematic sections. The area of **Artificial Intelligence and Machine Learning** included 11 accepted papers, followed by the **Education and Learning Technologies** section with nine works, and **Cybersecurity, Data Management, and Computational Thinking** with five contributions. The fields of **Software Engineering and System Architectures**, **Image Processing and Computer Vision**, and **Human-Computer Interaction and Simulation** each featured four papers, while the domain of **Biomedical Signal Processing** was represented by three technical works. Additionally, the proceedings contain 10 contributions from the **Short Papers and Workshops** track—which integrates the workshops on Applications of Computing in Medicine and Women in Computing—along with four papers selected for the **Master's and Doctoral Symposium**.

Special recognition was given to the best paper of the conference: "New Kid in the Classroom: Exploring Student Perceptions of AI Coding Assistants," authored by Sergio Rojas-Galeano, for its significant contribution to the field of AI in education.

We would like to express our gratitude to the Program Committee members and reviewers for their dedicated effort in ensuring the scientific quality of the congress. We also thank the authors for their contributions and the *Universidad Militar Nueva Granada* for hosting and financial support of this successful event. Finally, we thank the Colombian

Computing Society and Springer for their continued support in the publication of these proceedings.

September 2025

Wilson J. Sarmiento
Eduard L. Sierra-Ballén

Organization

General Chair

Wilson J. Sarmiento	Universidad Militar Nueva Granada, Colombia

Program Committee Chairs

Wilson J. Sarmiento	Universidad Militar Nueva Granada, Colombia
Eduard Sierra-Ballén	Universidad Militar Nueva Granada, Colombia

Steering Committee

Eduard Sierra-Ballén	Campus Nueva Granada, Cajicá, Universidad Militar Nueva Granada, Colombia
Jorge Jaramillo	Campus Calle 100, Bogotá, Universidad Militar Nueva Granada, Colombia
Wilfrido Arteaga	Campus Nueva Granada, Cajicá, Universidad Militar Nueva Granada, Colombia
Oscar Avilés	Campus Calle 100, Bogotá, Universidad Militar Nueva Granada, Colombia
Johany Armando Carreño	Colombian Computing Society, Colombia

Organizing Committee

General Organization

Eduard Sierra-Ballén	Universidad Militar Nueva Granada, Colombia
Jorge Jaramillo	Universidad Militar Nueva Granada Colombia
Wilson J. Sarmiento	Universidad Militar Nueva Granada, Colombia
Edith Pulido Herrera	Universidad Militar Nueva Granada, Colombia
Nathaly Higuera	Universidad Militar Nueva Granada, Colombia

Graphic Design

Carlos Bahamón	Universidad Militar Nueva Granada, Colombia
Carlos Aranzazu	Universidad Militar Nueva Granada, Colombia

Webmaster

Cristian Plazas	Universidad Militar Nueva Granada, Colombia

Community Managers

Maria Paula Pedraza	Universidad Militar Nueva Granada, Colombia
Andrea Paniagua	Colombian Computing Society, Colombia

Student Volunteers

Multimedia Engineering Program
Universidad Militar Nueva Granada, Colombia
Andrés Felipe Alcázar Silva
Monserrath Amado Torres
Lizbeth Katherine Arevalo Fernandez
Anjhi Tatiana Bonilla Bernal
Carolina Castro
Daniel Colmenares Bolivar
Kevin Daniel Contreras Baron
Heidy Natalia Garzón González
Jesús Mateo Llanos Toro
Stefany Lizeth López Buitrago
Ricardo López Murcia
Sergio Alberto Mendez Burgos
Mateo Alejandro Moya Mendoza
Diego Fernando Poveda Rojas
María Paula Ramos Ruiz
César Fernando Rodríguez Prada
Ana Maria Sierra Montaño
Juan Diego Velásquez Pérez

Program Committee

Silvana Aciar	Universidad Nacional de San Juan, Argentina
Carlos Astengo	Tecnológico de Monterrey, Mexico
Fredy Bernal	Universidad Militar Nueva Granada, Colombia

Felipe Besoain	Universidad de Talca, Chile
Marta Camacho	Institución Universitaria Colegio Mayor del Cauca, Colombia
Alexander Ceron	Universidad Militar Nueva Granada, Colombia
Julio Chavarro	Universidad Tecnológica de Pereira, Colombia
Cesar A. Collazos	Universidad del Cauca, Colombia
Rosanna Costaguta	Universidad Nacional de Santiago del Estero, Argentina
Victor Adriel Oliveira	University of Applied Sciences, St. Pölten, Austria
Gloria Mercedes Diaz	Instituto Tecnológico Metropolitano, Colombia
Nestor Duque Méndez	Universidad Nacional de Colombia, Colombia
Alejandro Fernandez	Universidad Nacional de La Plata, Argentina
Juan Manuel Gonzalez	Benemérita Universidad Autónoma de Puebla, Mexico
Toni Granollers	Universitat de Lleida, Spain
Emilcy Hernández-Leal	Universidad de Medellín, Colombia
Huizilopoztli Luna-Garcia	Universidad Autónoma de Zacatecas, Mexico
Valeria Martins	Universidade Presbiteriana Mackenzie, Brazil
Luciana Nedel	Universidade Federal do Rio Grande do Sul, Brazil
Manuel Ortega	University of Castilla - La Mancha, Spain
Wilson Sarmiento	Universidad Militar Nueva Granada, Colombia
Eduard Sierra-Ballen	Universidad Militar Nueva Granada, Colombia
Marco Javier Suárez	Universidad Pedagógica y Tecnológica de Colombia, Colombia

Additional Reviewers

Aciar, Silvana	UNSJ, Argentina
Aguilar Vera	Raúl, UADY, Mexico
Aliaga-Vergara	Camila, UTALCA, Chile
Amato, Cibelle	UPM, Brazil
Andres, Rodriguez	UNLP, Argentina
Arceo Olague, José	UAZ, Mexico
Aristizábal Q., Luz	UNAL, Colombia
Ballesteros, Dora	UMNG, Colombia
Bengochea, Luis	UAlcalá, Spain
Bermon, Leonardo	UNAL, Colombia
Bocanegra, Jose	UDFJC, Colombia
Bruno, Mario	UPLA, Chile

Calleros, Claudia Blanca	UV, Mexico
Cano, Sandra	PUCV, Chile
Castillo, Luis	UCaldas, Colombia
Celaya-Padilla, Jose	UAZ, Mexico
Challiol, Cecilia	UNLP, Argentina
Cornide Reyes	Hector, UDA, Chile
Coronel, Luis	UFPSO, Colombia
Cortes Rico, Laura	Cornell, USA
Diaz, David	FEN-UChile, Chile
Díaz-Arancibia, Jaime	UFRO, Chile
Duque, Leonardo	ITM, Colombia
Eliseo, Maria Amelia	UPM, Brazil
Espino-Salinas, Carlos	UAZ, Mexico
Galván, Carlos E.	UAZ, Mexico
García-Alonso, José	UEx, Spain
Gatica, Gustavo	UNAB, Chile
Giraldo, William	Uniquindío, Colombia
González, Enrique	PUJ, Colombia
Guillen, Edward	UMNG, Colombia
Henao Gallo, Oscar	UTP, Colombia
Hernández-García, Ruber	UCM, Chile
Jaramillo, Cesar	UTP, Colombia
Jimenez, Robinson	UMNG, Colombia
Lagos, Juan	UFRO, Chile
Lescano, Germán	UNSE, Argentina
Levano, Marcos	UCT, Chile
Lopez, María	Unicauca, Colombia
Marin, Beatriz	UNAB, Colombia
Mas-Sansó, Ramon	UIB, Spain
Mendoza, Ricardo	TecNM, Mexico
Meneses, Claudio	UCN, Chile
Mnedez, Yeny	UMayor, Chile
Mon, Alicia	UNLAM, Argentina
Moreira, Fernando	UPT, Portugal
Muñoz, Luis Freddy	FUP, Colombia
Olivares, Cristian	UAH, Chile
Orozco, Mauricio	UNAL, Colombia
Osorio, Mariutsi	ITM, Colombia
Osorio Germán	UNAL, Colombia
Palacios, Pablo	UDP, Chile
Pantoja, Wilson	Unicauca, Colombia
Pulido, Edith	UMNG, Colombia

Quintero, Christian	UMNG, Colombia
Quiñones, Daniela	PUCV, Chile
Ramirez, Leonardo	UMNG, Colombia
Revelo Sánchez, Oscar	Udenar, Colombia
Rico, Dewar	UFPSO, Colombia
Rodriguez, Jorge	UNCOMA, Argentina
Rodriguez, Arles	UNAL, Colombia
Rodriguez, Paula	ITM, Colombia
Rojas-Mora, Julio	UCT, Chile
Sanchez, Robinson	UTP, Colombia
Serrano, Jairo	UTB, Colombia
Silva Sprock, Antonio	UCV, Venezuela
Silveira, Ismar	UPM, Brazil
Solis, Andrés	Unicomfacauca, Colombia
Solis Robles, Roberto	UAZ, Mexico
Sturm, Christian	THI, Germany
Tabares, Valentina	UNAL, Colombia
Torres, Pablo	UTPL, Ecuador
Torres, Diego	UNLP, Argentina
Torres-Avilés, Rodrigo	UBB, Chile
Uribe, Ana	UNAL, Colombia
Vidal-Silva, Cristian	UTALCA, Chile
Villareal, Angela	ITM, Colombia
Villegas, Maria	Uniquindío, Colombia
Zuñiga, René	Unicauca, Colombia

Contents

E. Image Processing and Computer Vision

Automated Maritime Surveillance: Deep Learning-Based Boat Detection in the Inner Bay of Cartagena, Colombia

Cristian Jimenez Barrera[1], Oscar Eduardo Olaya Vera[2](✉), Edwin Puertas[1,2], Juan C. Martinez-Santos[1,2], and Jairo E. Serrano[1,2]

[1] Armada Nacional de Colombia, Cartagena, Colombia
cristian.jimenez@armada.mil.co, {epuerta,jcmartinezs,jserrano}@utb.edu.co
[2] Universidad Tecnológica de Bolívar, Cartagena, Colombia
oolaya@utb.edu.co

Abstract. The article discusses the development and training of a *Deep Learning*-based model for the detection and tracking of tourist vessels in the inner bay of Cartagena de Indias (Colombia). Initially, external *datasets* were tested without satisfactory results, leading to the creation of a specific dataset for this project. Images were captured using an electro-optical camera in a strategic location and manually labeled for training the YOLOv8 model. The system was validated on low-performance computers to verify its effectiveness in resource-constrained environments. Adjustments were made to the model to improve accuracy under adverse conditions, achieving a detection rate of 95%. This research highlights the importance of building customized *datasets* and the synergy between academic knowledge and operational applications in maritime security.

Keywords: Boat Detection · Computer Vision · Deep Learning · Image Processing · Neural Network Modeling and Training · Dataset Annotation · Maritime Security

1 Introduction

Maritime tourism transportation in the inner bay of Cartagena has a significant economic and operational impact. Multiple vessels transit daily to destinations such as the Rosario Islands and Barú, generating a constant flow of maritime traffic [1]. Despite efforts by the General Maritime Directorate (Dimar) and Coast Guard units to supervise, no automated system is available to enable continuous monitoring and registration of vessels, making it challenging to obtain precise statistics and implement effective control measures.

In response to this issue, this study proposes the development of an automated vessel detection and tracking system based on computer vision and deep neural networks [13]. The YOLOv8 architecture is employed to train a detection model using a dataset captured from the bay, allowing the system to adapt to real

W. J. Sarmiento and E. L. Sierra-Ballén (Eds.): CCC 2025, CCIS 2692, pp. 3–14, 2027.
https://doi.org/10.1007/978-3-032-28447-1_1

operating conditions. The authors analyzed the model's performance in practical tests and evaluated its applicability in low-performance computing environments [17].

The development of this system not only represents a technological advancement in maritime monitoring but also provides a decision-support tool for security and traffic management. The implementation of advanced computer vision techniques will enhance the efficiency of surveillance systems in the region.

2 Background

Over more than two decades, various naval institutions have led the design, development, and manufacturing of fire control and surveillance systems for naval units. During this period, multiple systems integrating electro-optical sensors—such as daylight vision cameras, infrared thermal cameras, and laser rangefinders—have been implemented to enhance situational awareness and support tasks in weapon assignment and fire control in surface operations [14].

Since around 2005, several of these institutions have ventured into advanced image processing for the detection, recognition, and identification of targets of interest in maritime environments, utilizing tools such as OpenCV and applying contrast analysis techniques for target tracking. These solutions have been developed on FPGA-based platforms and through software written in C++, which represented a milestone in advancing institutional technological autonomy.

Starting in the 2010s, optronic systems began to incorporate image processing capabilities directly within the hardware. Although these solutions present certain limitations, they have proven highly useful for operators, improving the effectiveness of surveillance and control missions.

With the rise of artificial intelligence tools starting in 2020, image processing has evolved significantly, enabling the development of more robust and accessible solutions through widely documented commercial software [3]. In this new context, the integration of machine learning and deep learning techniques has facilitated the creation of more precise and efficient models for automated detection tasks.

In parallel, the training of operational personnel in machine learning and deep learning, fueled by the synergy between naval institutions and academia, has contributed to reducing knowledge gaps, strengthening technological independence, and promoting innovative solutions with dual applications. This collaboration has advanced the development of vessel detection systems that not only enhance operational capability but also positively impact regional maritime security.

3 Design and Implementation

The vessel detection system was designed with the objectives of generating a dataset of representative images from Cartagena Bay and training a Deep Learning model adapted to real operating conditions, as well as the technical characteristics of the Electro-Optical Sensors (EOT) used by the National Navy of Colombia [15].

3.1 Optronic Location

To obtain appropriate images for model training, the Escuela Naval Almirante Padilla (ENAP) was selected. A reference camera, CONTROP'S QUAD iSea-40, was used (see Fig. 1).

Fig. 1. EOT Location in the Inner Bay of Cartagena de Indias

3.2 Image Processing

To obtain images from the EOT, an analog composite video signal is received and integrated with a video converter, which is responsible for digitizing the signal. This module transforms the analog signal into a digital format through a composite video-to-USB converter, enabling frame capture for processing in YOLOv8 [8] (See Fig. 2).

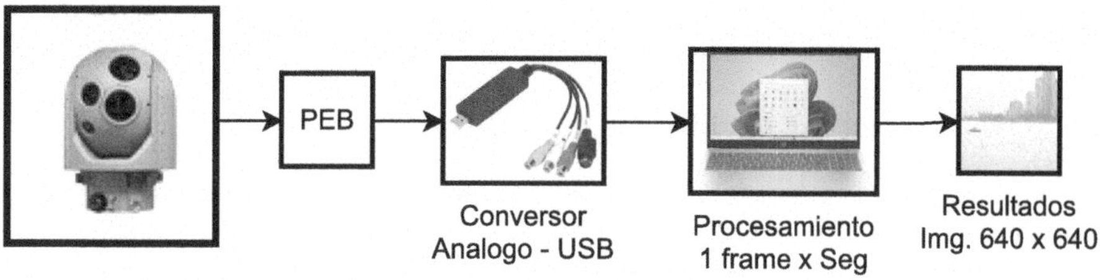

Fig. 2. Video processing modules

Images were extracted at a rate of one frame per second and resized to 640×640 pixels, the recommended size for training YOLOv8. Subsequently, they were organized into a dataset with the following structure:

- **Training**: 358 images.
- **Validation**: 300 images.

- **Testing**: 299 images.

Labeling was performed using the LabelImg tool, generating files in .txt format that document the bounding box coordinates of each boat [6]. A YAML file was then generated for the YOLOv8 training configuration:

```
train: ./data/images/train
val: ./data/images/val
test: ./data/images/test
nc: 1
names: ['boat']
```

This file enables the correct loading of images and labels into the model.

3.3 Model Implementation

The training was conducted using the Python programming language with the Ultralytics library and GPUs provided by Google Colaboratory (Colab) [18]. For validation, the system was integrated with a military daytime camera, and visual aids were programmed for the operator, including:

- Central cursor for tracking.
- Bounding box around the detected boat.
- Audible alert upon active detection.

Following training, the best.pt file was generated, enabling the development of a detection program for these vessels [9].

The following describes the operation of the developed system for assisted boat detection using GPS data. The system integrates computer vision techniques with positioning data acquisition and processing to provide monitoring capabilities.

3.4 Overall System Architecture

The system comprises several interconnected modules that manage data acquisition (video and GPS), signal processing, object detection, and information presentation to the user through a graphical interface. Figure 3 presents a general diagram of the proposed system architecture.

The system was developed using the Python programming language, leveraging various specialized libraries for each task. Table 1 lists the main libraries used and their function within the system. An Arduino Nano was used for GPS data acquisition, with data transmitted via RS-232 [12].

The following details the operation of each of the main modules:

GPS Data Acquisition: The GPS data acquisition module ('gps.py') reads information from a GPS receiver connected to the system via a serial port.

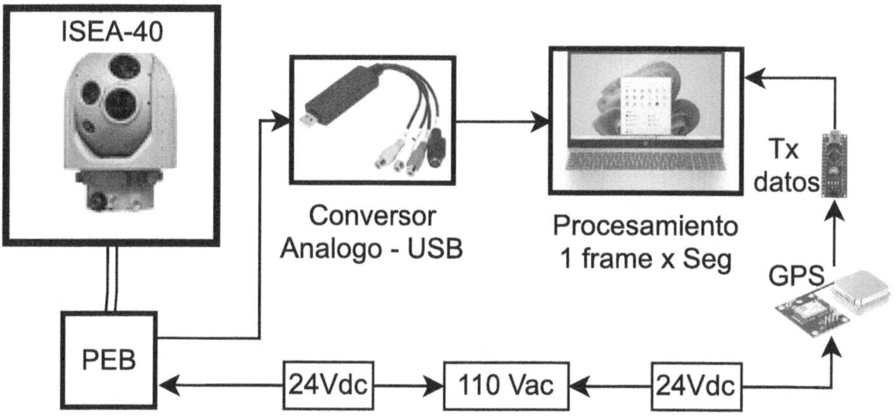

Fig. 3. System architecture with integrated GPS

Communication follows the NMEA 0183 standard. Specifically, *GPRMC* sentences are processed, containing essential navigation data such as UTC, latitude, longitude, speed over ground (SOG), and date.

The processed GPS information is displayed in the graphical user interface, providing the operator with the current location and platform speed. A monitoring mechanism was implemented to detect GPS signal loss, providing notifications when data is not being updated [5].

The GPS sensor is integrated into the system to provide a geospatial reference point, enabling future developments to estimate the precise position of detected vessels. The methodology takes into account the field of view (FOV) geometry of the camera, its tilt angle, and the resolution of the captured image [2].

By converting detected object coordinates from the image reference frame to a geographic coordinate system, a precise, real-time estimation of the vessel's location can be obtained. This process involves spatial projection calculations, considering the angular offset of the camera, the aspect ratio of the sensor, and the estimated distance between the capture device and the detected target [16].

Implementing this technique will enable future versions of the model to not only identify vessels in the image but also assign latitude-longitude coordinates, optimizing automated monitoring and tracking in maritime security applications [4].

The Object Detection Module (detection.py) manages video capture from a connected camera and applies an object detection model based on the YOLO (You Only Look Once) architecture. A trained model (best.pt) is used to identify boats.

Detection processing is applied to each frame captured by the camera, with the model outputting the object's location (bounding box) and a confidence score for each detected object. A confidence threshold is applied to filter out less reliable detections. For each confirmed detection, the system calculates the centroid coordinates of the object (Fig. 4).

Fig. 4. Example of boat detection with bounding box.

Information regarding the number of detected boats, including details such as confidence level and centroid location, is displayed in the graphical interface. Additionally, an audible signal—a beep—is implemented to alert the user to detections with significant confidence levels [7].

The system operates in two modes: "Detection," where video analysis identifies boats, and "Camera Only," where video transmission runs without the detection process.

The user interface ('interface.py'), built with the Tkinter library, provides a visual medium for interacting with the system and viewing relevant information. The interface is divided into panels that display real-time video transmission, boat detection details (including count and specific detection information), and acquired GPS data (latitude, longitude, time, date, and speed).

The interface includes basic controls, such as a button to stop system execution, which terminates video capture and GPS reading processes and closes the application. It also features the previously mentioned mode selector ("Detection" and "Camera Only") [11].

3.5 Execution Flow

Table 1. Python libraries used in the system.

Library	Primary Function
OpenCV	Image and video processing, camera capture.
Tkinter	Creation of the graphical user interface (GUI).
Ultralytics	Implementation of YOLO for object detection.
Pyserial	Communication with the GPS module via the serial port.
PIL (Pillow)	Image manipulation for the GUI.
threading	Concurrent execution of tasks (GPS reading and detection).

System execution begins with the 'main.py' script, which launches the graphical interface. The interface, in turn, initiates two concurrent processes: one thread for continuous GPS reading and the primary function for object detection in the video stream. Both processes run in a loop until the user chooses to stop the system by pressing the corresponding button in the GUI, activating an orderly shutdown mechanism to release resources, such as the camera and serial port.

4 Results

The model was trained using images of size 640×640 for a total of 45 epochs. The graphs and metrics indicate strong performance in terms of precision and *recall*.

Figure 5 presents the normalized confusion matrix of the model, showing an accuracy rate of 99% for the *Track* class.

Table 2 displays the key performance metrics of the model.

Table 2. Summary of Model Metrics

Metric	Value
mAP@0.5	0.978
Precision	0.94
Recall	0.93

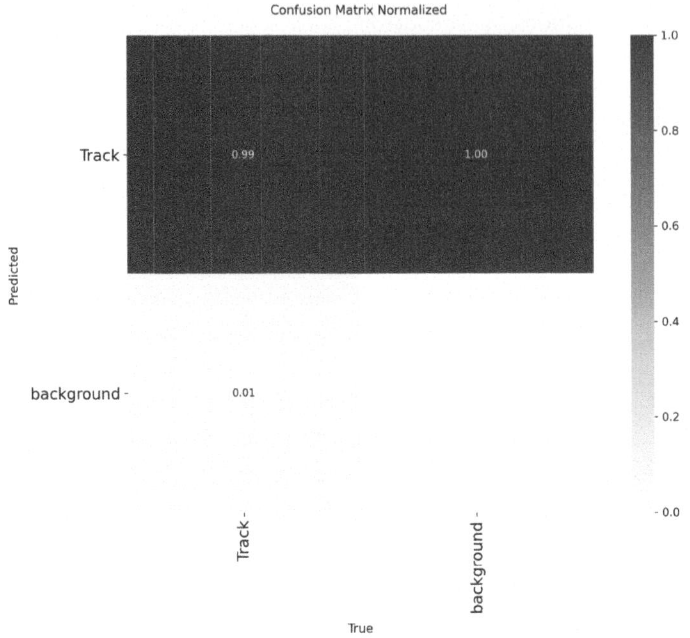

Fig. 5. Confusion matrix of the trained model.

Figure 6 illustrates the evolution of the model through the Precision-Recall and Confidence-Recall curves, providing a detailed performance analysis:

- Precision-Recall Curve: Shows the balance between precision and recall at various thresholds, reflecting the model's ability to correctly identify objects without compromising the detection of difficult cases.
- Confidence-Recall Curve: Analyzes the sensitivity of the model concerning the confidence threshold, helping determine the optimal point to minimize false negatives while maintaining overall detection efficiency.

These metrics are crucial for refining system configuration, ensuring robust detection adapted to real-world operational conditions [10].

After analyzing the detection training results, Fig. 7 shows the outcome of GPS integration alongside vessel detection using trained models.

The YOLOv8 model trained for vessel detection in Cartagena Bay demonstrated strong performance, achieving an *mAP@0.5* of 0.978 and a *recall* of 0.94, indicating high capability in correctly identifying objects of interest. False negatives were minimized through dataset refinement, hyperparameter adjustments, and confidence threshold optimization, thereby enhancing precision in complex scenarios. The confusion matrix indicates effective detection of the *Track* class. However, improvements can still be made in differentiating the *background* class using data balancing techniques and data augmentation. The results confirm

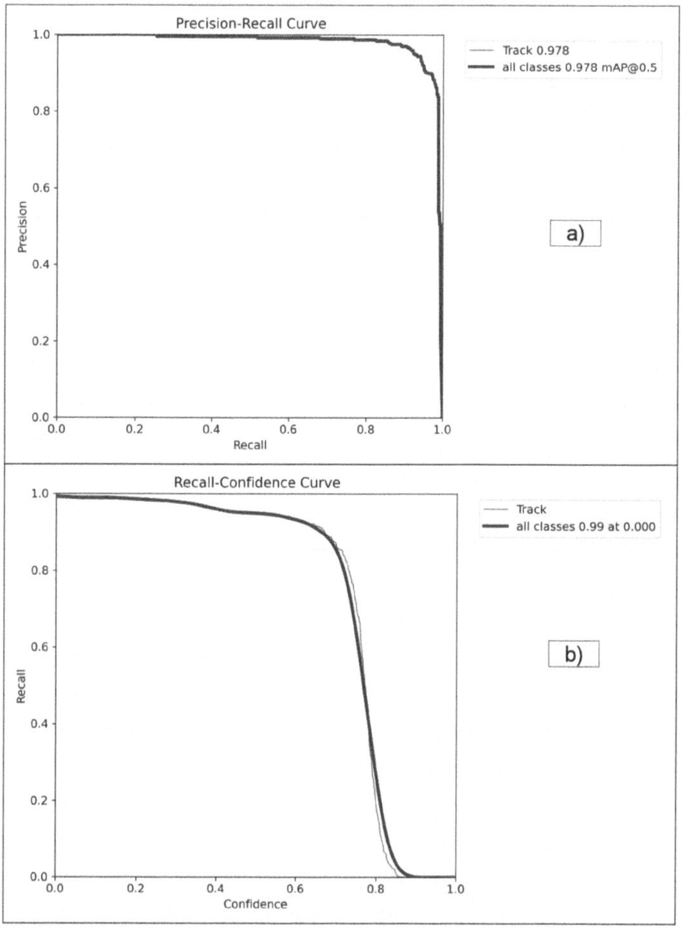

Fig. 6. Curves of: a) *Precision Vs Recall* and b) *Recall vs Confidence.*

the feasibility of the model for maritime security applications and suggest that fine-tuning the post-processing pipeline could further optimize its real-world performance.

The performance of the vessel detection model is evaluated using key metrics that reflect its accuracy and identification capability. The mAP@0.5 (0.978) indicates a high level of reliability in detecting objects with at least 50 % overlap with reality. The precision (0.94) represents the proportion of correct detections relative to the total predictions made, while the recall (0.93) measures the model's ability to correctly identify all vessels present in the image. These values demonstrate strong performance, minimizing false negatives and optimizing detection in maritime environments.

The Precision vs Recall and Confidence vs Recall curves help visualize the model's behavior under different confidence thresholds. The first shows how precision balances with detection capability in difficult cases, while the second illustrates how the model can adjust detection sensitivity to minimize false negatives without compromising reliability. These results confirm the feasibility of the Deep Learning-based approach for maritime security and suggest future improvements in background differentiation and the integration of multiple sensors to enhance system robustness under adverse conditions.

Fig. 7. Result of vessel detection.

5 Conclusions

The developed model for automatic vessel detection in Cartagena Bay has demonstrated high effectiveness, achieving an *mAP@0.5* of 0.978 and a *recall* of 0.94. The combination of advanced computer vision techniques and deep neural networks, along with a custom dataset, has enhanced vessel identification and tracking in real-world environments.

The system has been validated on low-performance computers, confirming its viability for operational applications with limited resources. Implementing an optimized confidence threshold has helped reduce false positives without compromising detection accuracy.

The GPS-based vessel detection system represents a significant advancement in maritime security. Combining deep learning techniques with geospatial data enables precise real-time identification of vessels, improving monitoring and nautical traffic management. The implementation of YOLOv8 has achieved a 95% detection rate, validating its effectiveness in operational environments with limited computational resources.

To further optimize the system, several strategic improvements are proposed. Dataset expansion, by incorporating images captured under different lighting, weather, and viewing angles, will enhance the model's ability to adapt to real-world scenarios. Additionally, advanced data augmentation techniques, such as brightness variations, rotations, and simulations of adverse meteorological conditions, will improve the model's generalization and reduce possible detection errors.

Another key improvement is the integration of multiple sensors, combining data from optical cameras with information from radar, infrared sensors, or LiDAR technology. This will optimize detection in low-visibility conditions, ensuring more reliable monitoring. Likewise, the exploration of new architectures will be considered, evaluating more recent models like YOLOv11, should they offer significant improvements over YOLOv8 in terms of performance and adaptability.

More precise geospatial estimation is another fundamental aspect to enhance the system's operational capabilities. Refining GPS utilization alongside computer vision techniques will allow for the calculation of exact vessel coordinates, facilitating monitoring and tracking. Finally, optimization for low-performance hardware is a priority, ensuring the model can run efficiently on resource-limited devices without compromising detection accuracy.

These developments will not only strengthen system robustness but also contribute to incident prevention and increased efficiency in maritime surveillance. The combination of these advancements will result in a more sophisticated and adaptable system, further solidifying its usefulness in maritime security and nautical traffic management applications.

Acknowledgments. This research was made possible through the strategic support of the **Centro de Desarrollo Tecnológico Naval (CEDNAV)**, which provided the infrastructure and technical resources necessary for system deployment and validation. We also acknowledge the academic guidance and expertise of the faculty at the **Universidad Tecnológica de Bolívar**, particularly in the areas of *Deep Learning* and computer vision, whose contributions were instrumental in refining the detection model. The efforts of the research team and collaborators in data collection, annotation, and experimental testing were essential to the successful completion of this work.

References

1. A.V.G.: Turismo y sostenibilidad ambiental y social a través del etnomarketing en isla grande (archipiélago de nuestra señora del rosario, colombia). Revista de Estudios Andaluces **42**, 158–176 (2021). https://doi.org/10.12795/rea.2021.i42.08
2. Chen, S., Gao, M., Shi, P., Zeng, X., Zhang, A.: Target ship recognition and tracking with data fusion based on bi-yolo and OC-sort algorithms for enhancing ship navigation assistance. J. Marine Sci. Eng. **13**(2), 366 (2025)
3. Kumar, A., Singh, R.: Deep learning models for digital image processing: a review. Artif. Intell. Rev. **56**(3), 1–25 (2023). https://doi.org/10.1007/s10462-023-10631-z

4. Li, M., Wang, Q., Liao, Y.: Target tracking using video surveillance for enabling machine vision services at the edge of marine transportation systems based on microwave remote sensing. J. Cloud Comput. **13**, Article 47 (2024). https://doi.org/10.1186/s13677-024-00604-0
5. Mahadevan, H.S., Kumar, A., Grundmann, R., Schwarze, A.: Assessment of situational awareness in relation to advanced navigation systems using ship handling simulators. Eng. Proc. **88**(1), 36 (2025). https://doi.org/10.3390/engproc2025088036
6. Mustapha, A.A., Saruchi, S.A., Supriyono, H., Solihin, M.I.: A hybrid deep learning model for waste detection and classification utilizing yolov8 and CNN. Eng. Proc. **84**(1), 82 (2025). https://doi.org/10.3390/engproc2025084082
7. Palladino, A., et al.: An application-agnostic automatic target recognition system using vision language models. Proc. AAAI Conf. Artif. Intell. **39**(28), 28878–28884 (2025). https://doi.org/10.1609/aaai.v39i28.35154
8. Prihandoko, P., Rumapea, S.A., Fawwaz, M.F.: Implementation of yolov8 in object recognition systems for public area security in kebun raya bogor. Ultimatics: Jurnal Teknik Informatika **17**(1), 1–10 (2025). https://doi.org/10.31937/ti.v17i1.4133
9. Priya, N.K., Sridevi, P., Rukmini, R., Manasa, G.L., Kumari, G.K.: Ai-enabled real time object detection and classification for retail application. Strad Res. **12** (2025). https://doi.org/10.5281/Zenodo.15250853
10. Quach, L.D., Quoc, K.N., Quynh, A.N., Ngoc, H.T.: Evaluating the effectiveness of yolo models in different sized object detection and feature-based classification of small objects. J. Adv. Inf. Technol. **14**(5), 907–917 (2023)
11. Sapra, D., Jadeja, A., Vhanesha, B., Shah, Z.: Real-time cartoonization: fusion of computer vision and artistic rendering in python. Indian J. Natural Sci. **14**(82), 69283 (2024)
12. Stick, S.B.A.: Smart blind assistant stick. J. Eng. Sci. Technol. Rev. **18**(1), 110–119 (2025)
13. Tsuda, M.E., Miller, N.A., Saito, R., Park, J., Oozeki, Y.: Automated Viirs boat detection based on machine learning and its application to monitoring fisheries in the east China sea. Rem. Sens. **15**(11), 2911 (2023). https://doi.org/10.3390/rs15112911
14. Vivas, N., Barrera, C.J., Paz, K.B., Amaya, D.G.: Stability analysis for a remotely controlled weapon station. Res Militaris **13**(3), 687–695 (2023). https://resmilitaris.net/issue-content/stability-analysis-for-a-remotely-controlled-weapon-station-2052
15. Yang, S., Xiao, W., Zhang, M., Guo, S., Zhao, J., Shen, F.: Image data augmentation for deep learning: a survey. arXiv preprint (2022). https://arxiv.org/abs/2204.08610
16. Zhang, S., Wu, R., Xu, K., Wang, J., Sun, W.: R-CNN-based ship detection from high resolution remote sensing imagery. Rem. Sens. **11**(6), 631 (2019). https://doi.org/10.3390/rs11060631
17. Zhou, Y., Wang, H., Wang, X.: A survey on image caption generation with deep learning. IEEE Access **9**, 9913–9930 (2021). https://doi.org/10.1109/ACCESS.2021.3051283
18. Zulfi, M.R., Murdika, U., Darajat, A.U.: Gender identification and population detection in a room using yolov8. Emitor: Jurnal Teknik Elektro. 77–85 (2025). https://journals2.ums.ac.id/index.php/emitor/article/download/5734/3210

Comparative Analysis of ConvNeXt vs. Vision Transformer for Cucumber Disease Detection with CutMix + Mixup

Andrea Menco Tovar(✉), Jairo E. Serrano, Edwin Puertas, and Juan Carlos Martinez-Santos

Universidad Tecnologica de Bolivar, Cartagena, Colombia
{amenco,jserrano,epuerta,jcmartinezs}@utb.edu.co

Abstract. Early and accurate detection of crop diseases is essential for optimizing agricultural production and minimizing economic losses. This study evaluates the performance of two advanced deep learning architectures, Vision Transformer (ViT) and ConvNeXt, applied to the classification of diseases in cucumber crops. Data augmentation techniques, specifically CutMix and Mixup, as well as a hybrid combination of both, were implemented to enhance model robustness and generalization. The dataset used consisted of 6,400 images categorized into eight classes representing both disease conditions and healthy states of cucumber plants. Experimental results demonstrate that ConvNeXt, when combined with the hybrid CutMix + Mixup strategy, achieved outstanding performance, with an accuracy close to 99.7%. Furthermore, the Friedman statistical test confirmed significant differences among configurations, supporting the effectiveness of the hybrid approach in improving model performance. This work provides evidence that integrating data augmentation techniques with modern deep learning architectures offers an effective solution for the automatic and timely identification of plant diseases in precision agriculture.

Keywords: Disease · Vision Transformer · ConvNeXt · Data Augmentation · Cucumber Crops

1 Introduction

Precision agriculture has become an essential tool for the early detection of crop diseases, the generation of plant stress maps, and the integration of diverse technologies aimed at improving efficiency, reducing costs, and optimizing resource use [15]. Crop diseases are a widespread issue that negatively impacts both yield and quality in agricultural production. Affected plants often exhibit characteristic signs or lesions on their leaves, stems, flowers, or fruits [1,12,13,16]. The cucumber is a widely cultivated climbing plant of high economic and nutritional value globally, known for its rapid growth, low-calorie content, and richness in water, vitamins, and essential minerals. However, due to various abiotic stressors,

W. J. Sarmiento and E. L. Sierra-Ballén (Eds.): CCC 2025, CCIS 2692, pp. 15–28, 2027.
https://doi.org/10.1007/978-3-032-28447-1_2

cucumber diseases significantly reduce yield and quality, resulting in substantial economic losses for farmers [17,18].

Traditionally, the identification of plant diseases and pests relies on direct visual inspection by agricultural or forestry experts or experienced farmers. However, this process is often subjective, time-consuming, labor-intensive, and prone to diagnostic errors, particularly when performed by less experienced personnel. Misdiagnoses may lead to excessive or inappropriate chemical applications, compromising crop quality and productivity, increasing environmental contamination, and causing avoidable economic losses [1,16,18]. Therefore, early and accurate detection of plant diseases is crucial for implementing effective control measures, minimizing losses, and promoting more sustainable and secure agricultural practices.

In recent years, computer vision approaches have shown great promise in addressing classification and detection tasks in this domain [6]. The development of automated recognition systems based on Convolutional Neural Networks (CNNs) and Vision Transformer (ViT) has become increasingly critical in modern agriculture, particularly for the rapid and accurate identification of diseases in crops such as cucumber [1,20]. These systems provide objective, consistent, and efficient assessments of plant health, overcoming the limitations of traditional methods that rely heavily on human expertise. Integrating advanced image processing and artificial intelligence techniques, including data augmentation, makes it possible to detect diseases at early stages, thereby optimizing the application of phytosanitary treatments and significantly reducing the economic and environmental impact associated with conventional crop management [13].

Although recent studies have explored automatic disease detection using deep learning, challenges remain in terms of accuracy, stability, and generalization when faced with visual variations in agricultural images. This study aims to evaluate and compare the effectiveness of advanced deep learning models, specifically ViT and ConvNeXt, combined with data augmentation techniques, such as CutMix and Mixup, for cucumber disease classification. The goal is to enhance automatic detection and support timely decision-making in agriculture, thereby increasing productivity and ensuring food security.

This paper is structured to provide a comprehensive analysis of deep learning methodologies and their practical implementation, along with a concise description of the algorithms employed. Section 2 presents the state of the art, focusing on recent research in data augmentation, ConvNeXt, and ViT. Section 3 describes the methodology, including dataset composition, preprocessing, techniques, and evaluation metrics. Section 4 details the experimental results, highlighting comparisons across augmentation strategies. Finally, Sect. 5 provides the findings of the study.

2 State of the Art

To enhance disease recognition in crops, Chen et al. [4] proposed Progressive Mixup Prompt Learning (PMPL). For data augmentation, they employed

Mixup, arguing that it enables the model to effectively learn critical rare features while reducing overfitting and improving robustness. Their approach achieved an average accuracy of 88.77% and a harmonic mean of 59.89%. Similarly, Cheng et al. [5] introduced a calibration method called Mix-Center Loss, which combines the Mixup augmentation technique with Center Loss-based deep metric learning through the design of a novel loss function. They used Mixup to generate mixup centers, thereby enriching feature sampling. The method demonstrated a positive impact on the calibration of both CNN and transformer-based architectures, improving generalization performance.

Continuing with Mixup-based data augmentation, Cao et al. [2] formulated an iterative attack based on sub-frequency spectrum mixup transformation. They introduced a frequency-domain mixup scheme in which they divided the discrete cosine transform spectra of the original and auxiliary images into four sub-spectra. Each pair was then combined using different weighting coefficients. By applying this strategy iteratively, they generated multiple transformations of the same input, diversifying input patterns and stabilizing gradient update directions during the adversarial attack. Additionally, Gan et al. [11] proposed an input transformation-based attack called Semantic Mixing Attack (SMA). They applied the Discrete Fourier Transform to secondary images sampled from other classes, extracted mid-low frequency components containing critical semantic features using a Butterworth filter, converted them back to the spatial domain, and locally combined them with the primary image. This approach enriched semantic feature diversity without significantly altering the visual appearance, thereby stabilizing gradient directions and improving the transferability of adversarial examples.

Regarding the use of CutMix, Zhang et al. [21] employed it as an intra-class data augmentation technique in the domain of side-channel attack traces. They grouped the original traces by label and generated new traces by combining segments from two traces sharing the same label using CutMix, thus ensuring consistency between synthesized traces and their labels. It produced a balanced augmented trace set that improved the training of Multilayer Perceptron and CNN models for key recovery, resulting in faster convergence and fewer required traces. Subsequently, Ferdinand et al. [9] integrated CutMix into their Rehearsal-CutMix model. During the compression phase, they generated mixed images by taking patches from old class examples stored in exemplar memory and combining them with new class examples using CutMix, adjusting the labels proportionally. It preserved prior class information and balanced the training set for knowledge distillation.

In the same vein, Deng et al. [3] employed CutMix adaptively based on model confidence. They computed the global confidence of each unlabeled image and, if it fell below the batch's average threshold, selected it for CutMix with a labeled image. It ensured that the cropped regions came from reliable data, improving the quality of pseudo-labels and reducing noise from low-confidence unlabeled samples. On the other hand, Feng et al. [8] applied Mixup and CutMix jointly for data augmentation during training. With a probability of 0.5 per batch, they randomly selected between Mixup, which linearly interpolates pairs of images and their labels, and CutMix, which pastes a patch from one image onto another, adjusting labels based on the combined area. After the transformation, labels changed from shape N×1 to N×241, promoting more diverse representations and improving model generalization. This combined approach achieved a top-1 accuracy of 86.28%, top-5 accuracy of 98.49%, and a weighted average of precision, recall, and F1-score of 0.86, along with an AUC of 0.9989.

In terms of ConvNeXt-based models in the literature, Hasan et al. [12] employed the pretrained ConvNeXt-Base architecture to extract visual features from images captured in rice fields. From the feature maps, they added a Convolutional Block Attention Module and several dense layers to perform a regression task predicting water height in alternate wetting and drying irrigation systems, achieving high performance. Likewise, Fu et al. [10] used ConvNeXt to extract deep features from thermal images of tool spindles. Based on these features, they trained a classification model to detect thermal errors. Images were processed through the first four ConvNeXt stages, followed by global pooling and a fully connected layer assigning each image to a thermal error label. The output probabilities from ConvNeXt were combined using a weighted integration scheme to obtain a continuous prediction of thermal error, achieving a mean absolute error of 0.7807 μm and an RMSE of 1.2461 μm.

Regarding the use of ViT models, Huang et al. [13] combined ConvNeXt and ViT. They utilized the former to extract local features from apple leaf images, leveraging the early ConvNeXt stages enhanced with attention mechanisms to capture fine details and suppress background noise. They then fused these feature maps with ViT to obtain a global representation. They concluded that ConvNeXt alone achieved 99.2% accuracy on the laboratory validation set but only 43.2% on the natural environment test set. In contrast, the EConv-ViT combination improved natural environment accuracy to 79.3%. Similarly, Valle-Medina et al. [20] detected and classified diseases in potato leaves using a deep learning model based on the Vision Transformer architecture. The dataset for training and testing was sourced from the PlantVillage platform, achieving accuracy and F1-score metrics of 99% and 98%, respectively.

3 Methodology

This section outlines the process followed for developing the proposed study, which we structured into three main components. The Dataset subsection describes the cucumber crop image database, detailing its composition and the partitioning into training, validation, and test sets. Next, the Preprocessing subsection explains the techniques applied to optimize the data, including resizing, normalization, and data augmentation. Finally, the Techniques and Models subsection presents the methods and architectures used to implement the experimental design.

3.1 *Dataset*

The dataset used in this study was the *Cucumber Disease Recognition Dataset* [17], which consists of 6,400 images organized into eight classes: anthracnose, bacterial wilt, belly rot, downy mildew, fruit rot caused by Pythium, gummy stem blight, healthy leaves, and fresh cucumber. These classes represent both disease conditions and wholesome states of cucumber crops, as illustrated in Fig. 1. Each class is balanced with 800 images, and we store them in separate subdirectories, which enables structured and efficient loading using standard supervised classification procedures. They collaborated with experts from an agricultural research institute to label cucumber diseases.

Fig. 1. Sample images from the cucumber dataset.

For the experimental phase, we divided the dataset into three non-overlapping subsets according to a stratified ratio: 70% for training, 15% for validation, and 15% for testing. This partitioning preserved the original class distribution across each subset. We applied data augmentation transformations exclusively to the training set. In contrast, the validation and test sets remained unaltered, aside from minimal transformations aimed at standardizing image size and scale. It ensured that the reported results accurately reflected the true generalization capability of each evaluated configuration.

3.2 *Preprocessing*

To ensure robust training and improve the generalization capability of the classification system, we implemented various preprocessing and data augmentation strategies. We uniformly resized all images to 224×224 pixels, with random transformations including rotations, horizontal flips, and adjustments to brightness and contrast applied exclusively during the training phase. Subsequently, we incorporated advanced augmentation techniques. Mixup generates synthetic samples by linearly interpolating two images and their corresponding labels, encouraging the model to focus on multiple discriminative regions. CutMix, on the other hand, replaces a patch of one image with a segment from another and adjusts the label proportionally to the mixed area. We also employed a hybrid strategy that combines both methods. These techniques are known to enhance generalization by mitigating overfitting, with parameters sampled from Beta distributions to preserve semantic consistency between classes.

We evaluated the augmentation strategies both independently and in combination across two architectures, ViT and ConvNeXt, resulting in six experimental configurations. To ensure the validity of the analysis, images used for validation and testing underwent no additional augmentations and were only resized and normalized. This setup allowed for a clear assessment of the impact of each augmentation technique on model performance.

3.3 *Technique*

This section presents the data augmentation techniques and models employed to conduct the experimental design.

Mixup

Blends two images and their labels with given probability Fig. 2. The Mixup hyperparameter defines the probability of applying the transformation, with a Value of 1.0 ensuring the mixing of all images and a value of 0.0 turning off the transformation. The purpose is to improve model robustness and reduce overfitting. For example, in retail product recognition systems, Mixup helps the model learn more robust features by blending images of different products, teaching it to identify items even when they're partially visible or obscured by other products on crowded store shelves. The Mixup ratio is a random value picked from a np.random.beta(32.0, 32.0) beta distribution, meaning each image contributes approximately 50%, with slight variations [19].

Fig. 2. Sample images of Mixup in operation

CutMix

Cuts a rectangular region from one image and pastes it onto another image with given probability Fig. 3. The CutMix hyperparameter defines the probability of applying the transformation, with CutMix=1.0 ensuring that all photos undergo this transformation and CutMix=0.0 turning it off completely. For example, with CutMix=0.5, each image has a 50% chance of having a region replaced with a patch from another image. Its purpose is to improve performance by simulating realistic occlusions yet preserving local feature integrity.

Fig. 3. Sample images of CutMix in operation

Vision Transformer

The ViT utilizes a structure based on traditional transformer blocks adapted for computer vision. Each block includes Multi-Head Self-Attention (MHSA) and fully connected feed-forward layers. We divided the images into fixed patches, which we processed as tokens. We added positional coding to retain spatial information. ViT employs techniques such as the use of LayerNorm normalization layers before self-attenuation operations and masking in pretextual training, allowing global relationships to be captured efficiently [7].

ConvNeXt

ConvNeXt is a modern CNN architecture proposed by researchers at Facebook AI Research FAIR. Its primary goal is to bring the performance of traditional CNNs to a competitive level with Vision Transformers while maintaining the CNN's structural advantages. It is based on the ResNet architecture but introduces ViT-inspired enhancements such as normalization, block design, and training strategies [14].

3.4 *Evaluation and Metrics*

We conducted a performance evaluation using the test subset, which was previously separated and remained unaltered during training. We employed the following metrics for multi-class classification tasks: overall accuracy, per-class precision, recall, and F1-score, each computed both individually and as macro-averages. Additionally, we generated confusion matrices to visualize the model's behavior across categories and to identify potential confusion between visually similar classes. It enabled a comprehensive assessment of each experimental configuration, facilitating direct comparisons between architectures and data augmentation techniques under equivalent validation conditions. Consistency among the various indicators was key to identifying the approaches with the highest generalization capacity and inference stability.

4 Results

The experimental design, summarized in Table 1, involved a systematic evaluation of two advanced deep learning architectures, ViT and ConvNeXt, in combination with three specific data augmentation techniques: CutMix, Mixup, and their hybrid approach, CutMix + Mixup. Each configuration was analyzed using accuracy, macro-averaged precision, macro-averaged recall, and macro-averaged F1-score, which provide a comprehensive assessment of classification performance. For the ViT model, both CutMix alone and CutMix combined with Mixup showed strong performance, with metrics approaching 97%. In contrast, Mixup alone yielded significantly lower results, around 50%, indicating that Mixup in isolation is not suitable for this architecture, particularly in the context of cucumber disease classification.

Table 1. Comparison of ViT and ConvNeXt performance using different augmentation techniques.

Model	Technique	Accuracy (%)	Precision (%)	Recall (%)	F1-score
ViT	CutMix	97.00	97.00	97.00	97.00
	Mixup	50.42	54.89	50.42	42.80
	CutMix + Mixup	97.29	97.32	97.29	97.28
ConvNeXt	CutMix	99.00	99.00	99.00	99.00
	Mixup	99.06	99.08	99.06	99.07
	CutMix + Mixup	**99.69**	**99.70**	**99.69**	**99.69**

In the case of the ConvNeXt model, results demonstrated outstanding and consistent performance across all experimental configurations. Notably, the configuration that combines CutMix and Mixup achieved near-perfect scores, reaching approximately 99.69% in accuracy, precision, recall, and F1 score. These findings establish ConvNeXt combined with the hybrid augmentation strategy as the optimal solution, significantly outperforming the ViT model. This experimental design supports the conclusion that integrating CutMix and Mixup techniques within the ConvNeXt model effectively enhances accuracy and robustness in image classification tasks.

The learning curve of the ConvNeXt model Fig. 4 trained with CutMix + Mixup shows a rapid reduction in training loss during the initial epochs, after which it stabilizes at approximately 0.50 to 0.55. In contrast, the validation loss decreases to low values between 0.03 and 0.05 and remains stable throughout the entire process. In parallel, training accuracy reaches moderate levels of around 70 – 75%, reflecting the label smoothing effect introduced by the mixing techniques. In contrast, validation accuracy consistently exceeds 99%, indicating the model's strong ability to classify validation samples under these augmentation conditions correctly. Moreover, the evolution of both loss and accuracy during training and validation shows stable behavior with no signs of overfitting. Throughout 50 epochs, the validation loss remained low and continued to decrease. In contrast, validation accuracy surpassed 99% from the early stages and exhibited no abrupt drops. This behavior suggests that CutMix and Mixup acted as effective regularizers, reducing variance without compromising generalization performance.

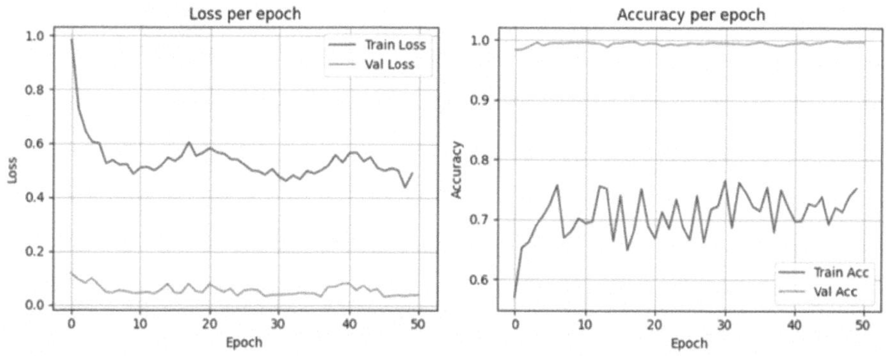

Fig. 4. Learning curve of the ConvNeXt model trained with CutMix + Mixup

The confusion matrix Fig. 5 presents the results of the experimental design using the ConvNeXt model with the combined data augmentation technique CutMix + Mixup for the classification of diseases and healthy conditions in cucumber plants. The model demonstrates overall strong performance, as most classes, including Anthracnose, Bacterial Wilt, Belly Rot, Downy Mildew, Fresh Cucumber, Healthy Leaves, and Fruit Rot caused by Pythium, achieved high accuracy, with 120 correct predictions out of 120 evaluated cases. We observed a slight misclassification in the Gummy Stem Blight class, where three samples were incorrectly classified as Anthracnose, resulting in 117 correct predictions out of 120. This outcome suggests a minor but specific challenge for the model in clearly distinguishing between these two particular classes.

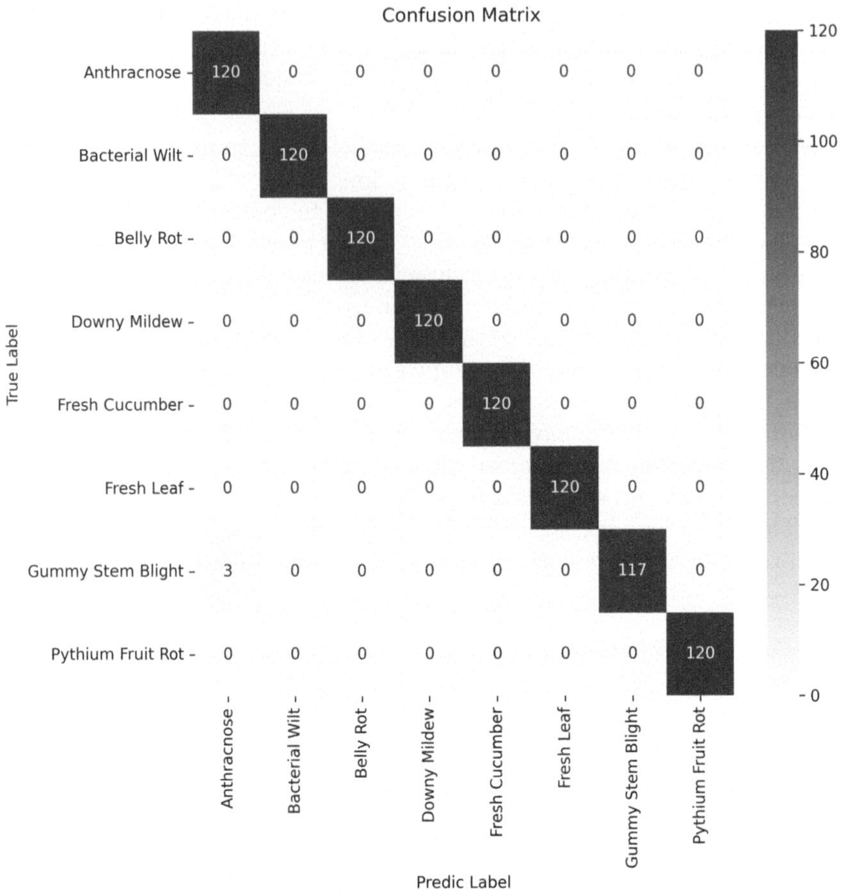

Fig. 5. Confusion matrix of the ConvNeXt model trained with CutMix + Mixup

Friedman Test

To determine whether statistically significant differences existed among the various experimental designs applied, we used the non-parametric Friedman test. This test is suitable in contexts where we compared multiple treatments across the same set of repeated measures without assuming the normality of the data. The choice of this statistical approach enabled the assessment of whether the observed performance differences were consistent, thereby providing a foundation for subsequent post-hoc analyses.

The results of the Friedman test Table 2 yielded a statistic of 20.00 with a p-value of 0.0012, indicating that statistically significant differences exist among the evaluated configurations. It suggests that model performance varies depending on the data augmentation technique applied. In particular, configurations that combined CutMix and Mixup demonstrated notably superior performance

Table 2. Results of the Friedman test.

Friedman statistic:	20.0000
p-value:	0.0012

compared to those that used only one of the techniques or showed lower results, such as ViT with Mixup. Based on this outcome, we conducted a post-hoc analysis using the Nemenyi test to identify which configurations differed significantly from one another.

Nemenyi Post-hoc Test

The Nemenyi post-hoc test was applied to perform pairwise comparisons between the evaluated configurations Fig. 6. The results indicated that the ConvNeXt configuration using the combined CutMix and Mixup augmentation achieved significantly superior performance compared to ViT with CutMix $p = 0.0300$ and, more notably, to ViT with Mixup $p = 0.0022$. Additionally, a significant difference was observed between ViT with Mixup and ConvNeXt with Mixup

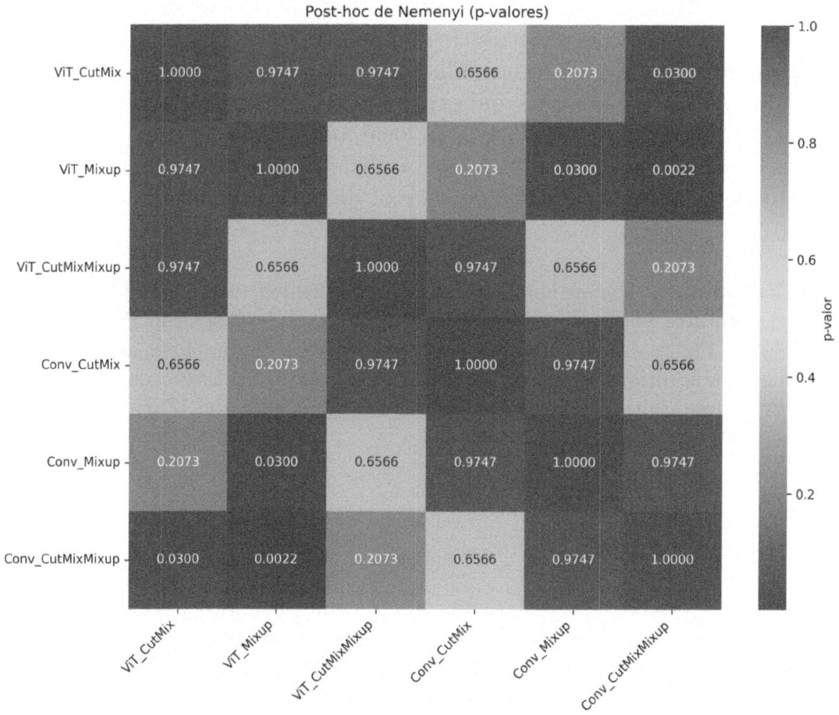

Fig. 6. Nemenyi Post-Hoc Test

$p = 0.0300$, reinforcing the conclusion that the ConvNeXt architecture is more robust and effective, even when individual augmentation techniques are applied. No statistically significant differences were found among configurations within the same model, suggesting stable performance when the architecture remains constant.

5 Conclusion

In conclusion, this study demonstrates the effectiveness of deep learning models—particularly ConvNeXt—when combined with CutMix and Mixup data augmentation techniques in a hybrid approach. This configuration delivers outstanding performance in the automatic classification of cucumber crop diseases, achieving an accuracy of nearly 99.7%. The hybrid strategy enhances the model's ability to generalize under visual variations, reducing overfitting and improving training stability. On the other hand, the Vision Transformer model showed competitive results when used with CutMix but revealed significant limitations when Mixup was applied in isolation, suggesting that specific augmentation techniques may not be equally effective across different architectures.

Among the strengths of this approach is the rigorous application of data augmentation methods that simulate real-world conditions, contributing to system robustness. However, a limitation of the study is that we conducted testing exclusively on a single dataset, which may affect the generalizability to other crops or environmental conditions. Therefore, it would be advisable to extend the analysis to different datasets and explore novel augmentation techniques and hybrid architectures that integrate both local and global features. Future work will focus on implementing real-time and in-field diagnostic systems to enhance early detection and improve the accuracy of disease recognition across various crops.

Acknowledgments. The authors express their gratitude to the Call 933 "Training in National Doctorates with a Territorial, Ethnic and Gender Focus in the Framework of the Mission Policy—2023" of the Ministry of Science, Technology and Innovation (Minciencia). In addition, we thank the team of the Artificial Intelligence Laboratory VerbaNex [1](https://github.com/VerbaNexAI, affiliated with the UTB, for their contributions to this project.)

References

1. Alamri, F.S., Sadad, T., Almasoud, A.S., Aurangzeb, R.A., Khan, A.: Mango disease detection using fused vision transformer with convnext architecture. Comput. Mater. Continua **83**, 1023–1039 (3 2025). https://doi.org/10.32604/CMC.2025.061890. https://sciencedirect.utb.elogim.com/org/science/article/pii/S1546221825002656

2. Cao, H., Sun, Q., Geng, R., Wang, X.: Subspectrum mixup-based adversarial attack and evading defenses by structure-enhanced gradient purification. Knowledge-Based Syst. **318**, 113357 (2025). https://doi.org/10.1016/J.KNOSYS.2025.113357. https://sciencedirect.utb.elogim.com/science/article/pii/S0950705125004046
3. Cao, Y., Luo, B., Chen, Y., Xu, L., Ding, C.: Confidence-aware mean teacher for semi-supervised metallographic image semantic segmentation. Computational Materials Science **249**, 113645 (2025). https://doi.org/10.1016/J.COMMATSCI.2024.113645. https://sciencedirect.utb.elogim.com/science/article/pii/S0927025624008668
4. Chen, H., Li, H., Zhao, J., Ruan, C., Huang, L.: Enhancing crop disease recognition via prompt learning-based progressive mixup and contrastive language-image pre-training dynamic calibration. Eng. Appl. Artif. Intell. **152**, 110805 (2025). https://doi.org/10.1016/J.ENGAPPAI.2025.110805. https://sciencedirect.utb.elogim.com/science/article/pii/S095219762500805X
5. Cheng, S., et al.: Improving model calibration in bone marrow cell classification through mixup and center loss fusion. Eng. Appl. Artif. Intell. **150**, 110620 (2025). https://doi.org/10.1016/J.ENGAPPAI.2025.110620. https://sciencedirect.utb.elogim.com/science/article/pii/S0952197625006207
6. Elharrouss, O., et al.: Vits as backbones: Leveraging vision transformers for feature extraction. Information Fusion **118**, 102951 (2025). https://doi.org/10.1016/J.INFFUS.2025.102951. https://sciencedirect.utb.elogim.com/science/article/pii/S1566253525000247
7. face, H.: Transformador de visión (vit). https://huggingface.co/docs/transformers/model_doc/vit
8. Feng, Y., Wang, Y., Wang, X., Bi, J., Xiao, Z., Luo, Y.: Large-scale image classification and nutrient estimation for Chinese dishes. J. Agri. Food Res. **19**, 101733 (2025). https://doi.org/10.1016/J.JAFR.2025.101733. https://sciencedirect.utb.elogim.com/science/article/pii/S2666154325001048
9. Ferdinand, Q., Clement, B., Papadakis, P., Oliveau, Q., Chenadec, G.L.: Feature expansion and enhanced compression for class incremental learning. Neurocomputing **634**, 129807 (2025). https://doi.org/10.1016/J.NEUCOM.2025.129807. https://sciencedirect.utb.elogim.com/science/article/pii/S0925231225004795
10. Fu, G., Lin, K., Lu, C., Wang, X., Wang, T.: Spindle thermal error regression prediction modeling based on convnext and weighted integration using thermal images. Expert Systems with Applications **274**, 127038 (2025). https://doi.org/10.1016/J.ESWA.2025.127038. https://sciencedirect.utb.elogim.com/science/article/pii/S0957417425006608
11. Gan, F., Wo, Y.: Improving the transferability of adversarial examples through semantic-mixup inputs. Knowledge-Based Syst. **316**, 113325 (2025). https://doi.org/10.1016/J.KNOSYS.2025.113325. https://linkinghub.elsevier.com/retrieve/pii/S0950705125003727
12. Hasan, A.R., Kundu, N.K., Hasan, S., Hoque, M.R., Shatabda, S.: Accurate water level monitoring in alternate wetting and drying rice cultivation using attention-based convnext architecture. Comput. Electr. Agri. **234**, 110216 (2025). https://doi.org/10.1016/J.COMPAG.2025.110216. https://sciencedirect.utb.elogim.com/science/article/pii/S0168169925003229
13. Huang, X., et al.: Econv-vit: a strongly generalized apple leaf disease classification model based on the fusion of convnext and transformer. Information Processing in Agriculture (2025). https://doi.org/10.1016/J.INPA.2025.03.001. https://sciencedirect.utb.elogim.com/science/article/pii/S2214317325000174

14. Liu, Z., Mao, H., Wu, C.Y., Feichtenhofer, C., Darrell, T., Xie, S.: A convnet for the 2020s. In: Proceedings of the IEEE Computer Society Conference on Computer Vision and Pattern Recognition, vol. 2022, 11966–11976 (2022). https://doi.org/10.1109/CVPR52688.2022.01167. https://arxiv.org/pdf/2201.03545
15. Natura, F.: ¿qué es agricultura sostenible y cómo se relaciona con el monitoreo climático? - fundación natura colombia (2024). https://natura.org.co/que-es-agricultura-sostenible
16. Sahin, M.E., Özkaya, U., Arisoy, C., İbrahim Coşar, H., Ulutaş, H.: Cucunetcnns: application of novel ensemble deep neural networks for classification of cucumber leaf disease. Ain Shams Eng. J. **16**, 103380 (2025). https://doi.org/10.1016/J.ASEJ.2025.103380. https://sciencedirect.utb.elogim.com/science/article/pii/S2090447925001212
17. Sultana, N., Shorif, S.B., Akter, M., Uddin, M.S.: Cucumber disease recognition dataset. Mendeley Data, 1 (2022). https://doi.org/10.17632/Y6D3Z6F8Z9.1
18. Tovar, A.M., Puertas, E., Martinez-Santos, J.C.: Implementation of convolutional neural networks for automated disease detection in cucumber crops. Commun. Comput. Inf. Sci. **2208 CCIS**, 156–167 (2024). https://doi.org/10.1007/978-3-031-75233-9_12
19. Ultralytics: Data augmentation using ultralytics yolo - ultralytics yolo docs. https://docs.ultralytics.com/guides/yolo-data-augmentation/#mixup-mixup
20. Valle-Medina, F.D., Castillo-Heredia, L.J., Correa-Peralta, M.A., Seraquive, J.E.G.: Clasificación de enfermedades en hojas de papa utilizando transformadores de visión. NOVASINERGIA REVISTA DIGITAL DE CIENCIA, INGENIERÍA Y TECNOLOGÍA **8**(1), 142–156 (2025). https://doi.org/10.37135/ns.01.15.06
21. Zhang, R., Mo, Y., Pan, Z., Zhang, H., Wei, Y., Wu, X.: Intra-class cutmix data augmentation based deep learning side channel attacks. Integration **102**, 102373 (2025). https://doi.org/10.1016/J.VLSI.2025.102373. https://sciencedirect.utb.elogim.com/science/article/pii/S0167926025000306

Detection and Classification of Biomass in Orthomosaics Generated by Aerial Photogrammetry Using UAVs

Katia M. Villarnobo-Gonzalez[1(✉)], Luis A. Ramirez-Bermudez[1], Ricardo López-López[1], Valente Vázquez-Velázquez[1], Donají X. Cruz-López[1], and Fernando Angeles-Uribe[2]

[1] Facultad de Ciencias, U. N. A. M., Cd. Mx., 04510 Mexico City, Mexico
km.villarnobo@ciencias.unam.mx
[2] Instituto de Astronomía, U. N. A. M., Cd. Mx., 04510 Mexico City, Mexico
https://www.fciencias.unam.mx/

Abstract. The monitoring of urban green areas is key for ecological conservation and sustainable planning. In hard-to-access zones, such as many vegetated areas within urban environments, the use of Unmanned Aerial Vehicle (UAVs also called drones) and computer vision techniques provides an efficient and non-invasive alternative for analyzing tree canopy coverage. This study was conducted in the Ajusco neighborhood, Coyoacán district (Mexico City, Mexico), an area with scattered vegetation amidst urban infrastructure. Agisoft Metashape was used to generate orthomosaics, followed by the application of digital image processing (DIP) techniques in Python, including bilateral filtering, segmentation in HSV color space, morphological operations, shape analysis, and the *Watershed* algorithm to separate dense regions. Dense vegetation was segmented, and biomass was estimated. Improvements such as color histogram-based subdivision and the integration of algorithms like *Random Forest* to refine classification are proposed.

Keywords: Biomass · Watershed · photogrammetry · computer vision · Mexico City

1 Introduction

The conservation and monitoring of urban green spaces have become a priority due to rapid urban growth and the need to build resilient and sustainable environments.

Urban vegetation, particularly trees, plays a crucial role in microclimate regulation, mitigating atmospheric pollutants, sequestering carbon, and improving psychological well-being [5]. In Mexico City, where green spaces are fragmented and face increasing pressure from urban land use, a detailed studies are essential for the ecological management of the urban environment.

W. J. Sarmiento and E. L. Sierra-Ballén (Eds.): CCC 2025, CCIS 2692, pp. 29–44, 2027.
https://doi.org/10.1007/978-3-032-28447-1_3

This project was developed in a representative area of the Ajusco neighborhood (see Fig. 1), located in the Coyoacán district, southern Mexico City, Mexico. This area combines residential zones with educational and commercial infrastructure, featuring a scattered yet significant distribution of urban trees situated on sidewalks, small parks, or undeveloped lots. The vegetation in this area provides crucial environmental services, as it is surrounded by major roadways and high-density housing zones [4].

In this context, digital image processing (DIP) plays a key role by enabling detailed extraction of green space information from visual data acquired via UAVs (drones). The implemented thecnics include: color segmentation, color space conversion (e.g., RGB to HSV), bilateral filtering, segmentation algorithms like *Watershed* allow for distinguishing individual objects within images, identifying tree canopies, calculating relevant metrics (area, biomass). These digital tools facilitate objective, repeatable, and efficient analysis, even in areas that are difficult to access or where field inspection is limited, thus helping environmental planning processes based on quantitative evidence [6].

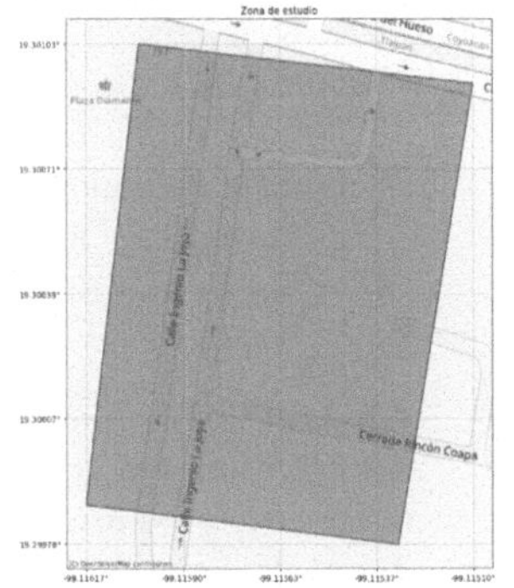

Fig. 1. Focus area.

This study not only generates valuable data on the number and distribution of trees, but also explores scalable automated analysis tools for other urban areas. By focusing on Ajusco neighborhood with its typical urban characteristics, but also with significant vegetation presence, this research advances the development of environmental surveillance techniques that are transferable to comparable settings across various metropolises.

2 Methodology and Theoretical Framework

2.1 Photogrammetric Processing

For photogrammetric processing and point cloud generation, we utilized a dataset comprising of 231 RGB images acquired by a UAV at an altitude of 36.4 m was used, with a spatial resolution of 1.23 cm / pixel, covering a total area of 0.0341 km^2.

The photogrammetric processing of the images yielded a digital elevation model (DEM) with a spatial of 2.47 cm/pixel and a point density of 0.164 $points/cm^2$. Additionaly, a georeferenced orthomosaic was obtained, which served as the cartographic base for subsequent analyzes.

The photogrammetric methodology exhibited inherent limitations, particularly in peripheral coverage quality. For this reason, the analysis focused on the central mapped zone, bounded by a polygon with vertices at the following geographic coordinates:

- Vertex 1: −99.116174, 19.299855
- Vertex 2: −99.115305, 19.299756
- Vertex 3: −99.115096, 19.300929
- Vertex 4: −99.116028, 19.301029

The acquired images contained areas with pronounced shadows, attributable to suboptimal illumination conditions during UAV operation. These shadows introduced voids in the point cloud density. Furthermore, distortions were identified at the edges of the model attributable to photogrammetric artifacts. Both technical challenges were mitigated using digital image processing techniques explained later.

2.2 Enhancement of Photogrammetric Images

To enhance the visual quality of the photogrammetric images, especially in shadowed areas caused by trees and structures, a gamma correction (power-law transformation) was applied. This technique non-linearly modifies the intensity distribution, enhancing details hidden in dark areas without overexposing brighter regions [3,6]. The gamma transformation follows the equation: $I_{adjusted} = I^{p}_{original}$, where p controls the degree of brightness adjustment in the image. A value of $p < 1$ amplifies intensities in dark areas, improving visibility in shaded zones, while $p > 1$ reduces overexposed areas. This technique not only enhances visual quality but also optimizes subsequent analysis stages, such as segmentation and vegetation biomass estimation, by revealing previously hidden details without distorting well-lit areas [5,12].

Transformation exponents p=0.3, 1, 1.7, 2, and 2.5 were evaluated to determine the optimal value for contrast and visibility enhancement in critical regions.

Subsequently, the resulting images were converted from RGB to HSV color space, and histogram equalization was applied exclusively to the V channel, achieving optimal redistribution of luminance values [6].

For optimal results, a p-value of 1.7 was selected. This setting provided the best compromise, boosting contrast in shadows while maintaining clarity in well-lit sections, which facilitated a clearer distinction between vegetation and shadows

2.3 Reconstruction of Missing Regions Using *inpainting*

Due to missing information in the processed orthophoto, the *inpainting* technique was applied. This method, also known as image reconstruction, fills deteriorated or missing zones coherently based on contextual information from adjacent regions. Its primary goal is to reconstruct these areas so they are imperceptible to the human eye, maintaining structural, color, and texture continuity in the image [1].

The procedure was applied to the optimized image ($p = 1.7$, previously selected). Initially, the image was converted to grayscale to simplify the detection of dark areas associated with gaps. Subsequently, a binary mask was generated using an inverted threshold, considering pixels with intensity below 10/255 as those to be filled.

To extend gap coverage and include edge regions, morphological dilation was applied using 3×3 pixel kernel.

Once the mask was obtained, the Navier-Stokes-based *inpainting* algorithm was applied with an influence radius of 3 pixels. This method was selected for its efficacy in reconstructing continuous structures by propagating edge information inward, simulating smooth feature diffusion. This allows edges and textures to be preserved without generating perceptible artifacts. The process was validated visually, analyzing structural continuity in the reconstructed areas, radiometric consistency with the surrounding environment, and the absence of artificial patterns or perceptible inconsistencies.

The final processing stage involved the removal of artificial edges introduced during the Navier-Stokes *inpainting* procedure. These edge artifacts, which typically originate in reconstructed regions, may exhibit subtle inconsistencies or visual artifacts. A structured crop based on the previously used binary mask was applied. For this, an element-wise multiplication operation was performed between the restored image and the original binary mask. Prior to this, dimensional compatibility of both arrays was ensured by converting the mask to a three-dimensional format and normalizing both matrices to the range $[0, 1]$ in `float32` format. The result was then converted back to `uint8` format for visualization and storage. This step ensured the final image maintained high visual fidelity, free from interference or residues associated with the reconstruction process, providing a cleaner and more accurate representation of the original scene.

2.4 Image Filtering

To improve the visual quality of the corrected image and mitigate residual artifacts generated by the *inpainting* process, a bilateral filter was applied. This nonlinear filtering method was chosen for its ability to smooth homogeneous regions without significantly affecting structural edges, which is particularly important in orthophotographs where transitions between objects like tree canopies, ground, and shadows represent critical information for analysis.

Unlike linear filters, bilateral filtering selectively smooths images by weighting pixel proximity and intensity similarity. In other words, a neighboring pixel influences the central pixel's value more if it is close and if its intensity value is similar. This dual weighting reduces noise in flat areas without blurring defined contours of objects in the image [6, 14].

For implementation, the following empirically adjusted parameters were used: a neighborhood diameter $d = 6$, a radiometric standard deviation of 100, and a spatial deviation of 150, achieving a balance between smoothing and edge preservation.

Subsequently, the absolute difference between the original and filtered images, amplified by a factor of 10, was calculated to visually highlight modified areas. This difference was converted to grayscale and then thresholded to generate a binary mask, where white regions indicated pixels significantly altered by filtering.

2.5 Segmentation

Segmentation is a process that divides a digital image into homogeneous regions based on one or more features [16]. To identify vegetation zones, color-based and shape-based segmentation techniques were implemented.

For color-based segmentation, the HSV image previously processed with bilateral filtering was used [3]. It is important to note that for the (H) channel, OpenCV does not use the standard range of 0° to 360° but rescales it to 0° to 179°, meaning each unit of H represents approximately 2° in reality.

In HSV space, the selected values covering various shades of green were defined within the intervals: H between 35 and 85, S between 110 and 125, and V between 20 and 255 (Fig. 2).

(a) RGB (b) Hue (c) Sat (d) Value

Fig. 2. Color segmentation in HSV space.

With these intervals, a binary mask identifying pixels within the specified range was generated. This mask was used to apply a `bitwise` operation on the original RGB image to visualize only regions classified as green.

Additionally, a histogram of the H value was calculated, considering only pixels within the mask, obtaining the tonal distribution of greens present in the image.

Finally, the original image, individual H, S, and V values, the binary mask, and the segmented image were visualized together to facilitate result interpretation.

2.6 Shape-Based Segmentation

Shape-based segmentation is classified as a method based on morphological or geometric features. Through connected component labeling and analysis of metrics like area and aspect ratio, specific structures such as tree canopies are filtered, discarding small or elongated elements [16]. This technique, complemented with morphological operators, allows delimiting relevant zones, in this case, forest coverage [13].

For shape-based segmentation, the V channel of the image was taken, and a binary threshold was applied, where all non-zero values were set to 255. This operation generates a binary mask highlighting regions of interest [12].

Once the mask was obtained, the connected components operator with 8-connectivity was applied, enabling the identification of contiguous regions in the image. Each of these regions was assigned a unique label, and statistics such as area and boundary coordinates were extracted [6].

Since the final goal of this analysis was to estimate vegetation biomass, processing was focused solely on large structures, likely corresponding to tree canopies. For this, a minimum area threshold of 50,000 pixels was set, generating a new tree mask containing only large green regions.

Subsequently, the contours of the filtered regions were highlighted, and *bounding boxes* were adjusted to each. From these shapes, geometric properties like aspect ratio (width/height) and contour area were calculated. These parameters were used as morphological filters to eliminate structures that, despite their size, did not exhibit arboreal morphology—such as roofs, courts, or paved surfaces—which often display highly rectangular shapes [8].

Based on these criteria, a new mask excluding objects with an aspect ratio between 0.7 and 2 or an area below 1,500 pixels was constructed. This shape-based segmentation refined the identification of tall vegetation, discarding artificial elements in the environment.

2.7 *Watershed*-Based Region Segmentation

Image segmentation based on the *Watershed* algorithm is a widely used technique to separate overlapping regions in complex images through morphological and distance transformations [6].

Initially, the mask was converted into a binary image using the Otsu thresholding method, which maximizes inter-class variance to find an optimal segmentation threshold [12]. Subsequently, a morphological opening with a 3×3 kernel was applied to eliminate small objects or noise while preserving the general structure of significant elements [13].

Next, the background was determined through dilation operations, while the foreground was extracted using the Euclidean distance transform, as regions farthest from edges indicate the center of objects [17]. The difference between these zones allowed the identification of an unknown region, which was later labeled using connected components.

The resulting markers were introduced into the *Watershed* algorithm, which acts as a controlled expansion of labels from initial markers, guided by local minima of the intensity gradient [9]. This process enabled the separation of adjacent regions and improved the delineation of structures.

Finally, the segmented regions were colored using pastel tones to facilitate visual interpretation. This morphological segmentation was key to detecting arboreal structures in the analyzed image (Fig. 3).

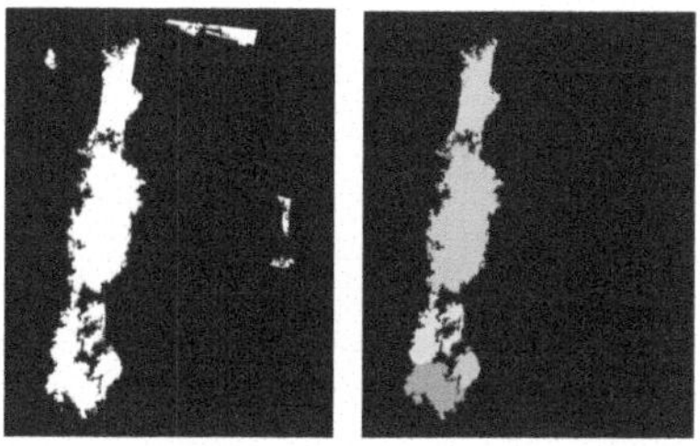

(a) Binary image (b) Segmented image

Fig. 3. Shape-based segmentation and *Watershed* segmentation result.

2.8 Biomass Estimation

Above-ground biomass estimation was performed based on the spatial segmentation of tree masses, previously delimited using morphological techniques and the *Watershed* algorithm. This approach relies on calculating the projected area visible in the orthorectified image and applying empirical allometric models [10,11].

First, the area occupied by each segmented region was quantified, considering the total number of pixels per object and the image's spatial resolution, expressed in meters per pixel. The projected surface area A (in m^2) of each region was obtained using the following expression:

$$A = N_{\text{px}} \cdot R^2 \tag{1}$$

where N_{px} represents the number of pixels in the identified region, and R corresponds to the image's spatial resolution (m/pixel), defined by the acquisition system (drone sensor and photogrammetry parameters).

To convert these areas into biomass estimates, allometric equations based on previously validated empirical models were used. These models estimate total above-ground biomass as a function of projected area through a power-law relationship:

$$\text{Biomass} = a \cdot (\text{area})^b \tag{2}$$

where biomass is expressed in kilograms (kg). The biomass of each tree species represented by the greatest cover within the corresponding segmentation patches was calculated. The average values of area and biomass for each species were obtained from the literature: Tree 2 (*Jacaranda* Montero [20]), Tree 3 (*Ficus elastica* [21]), Tree 4 (*Schinus molle* [24]), Tree 5 (*Fraxinus* [22]) and Tree 6 (*Quercus* [23]). Since a single average pair does not allow both coefficients to be uniquely determined, a typical value of $b \approx 0.9$ for urban trees was assumed [25], and a was then solved, yielding the following equations:

$$\text{Tree 2: } B_{\text{ind}} \approx 203.4 \cdot A^{0.9},$$
$$\text{Tree 3: } B_{\text{ind}} \approx 216 \cdot A^{0.9},$$
$$\text{Tree 4: } B_{\text{ind}} \approx 390 \cdot A^{0.9}$$
$$\text{Tree 5: } B_{\text{ind}} \approx 22.4 \cdot A^{0.9},$$
$$\text{Tree 6: } B_{\text{ind}} \approx 12.55 \cdot A^{0.9},$$

The Fig. 4 shows the complete process applied to identify tree areas.

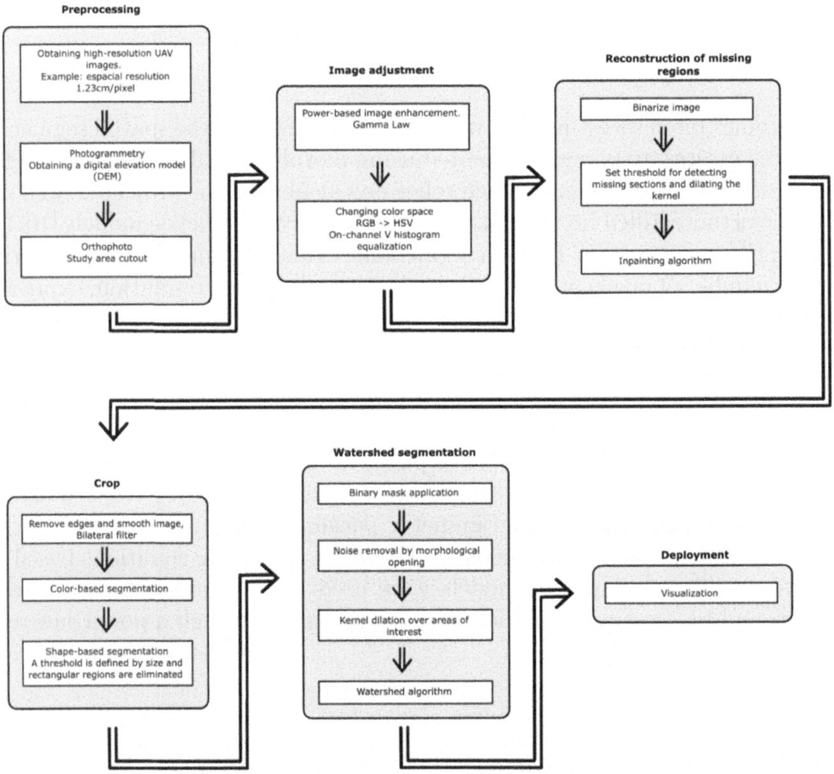

Fig. 4. Complete proccess diagram.

3 Results and Discussion

Figure 5 presents the visual results of the power transformation applied with values $p = 0.3$, 1, 1.7, 2.0, and 2.5. The image with $p = 0.3$ (Fig. 5(a) exhibits generalized overexposure, where bright areas lose detail, and shadows are minimally enhanced, resulting in an image with low structural contrast. The transformation with $p = 1$ (Fig. 5(b)), the original normalized image, was used as a reference.

Applying $p = 1.7$ (Fig. 5(e)) shows a notable increase in tonal differentiation of dark regions, highlighting contours of vegetation and structures in shadows more clearly without significant loss of information in illuminated areas. For $p = 2.0$ (Fig. 5(c)) and $p = 2.5$ (Fig. 5(d)), although contrast in dark areas is more pronounced, visibility in bright regions is reduced, as they tend to darken and blend, diminishing overall visual fidelity. Collectively, the analysis suggests $p = 1.7$ offers the best balance between detail enhancement and preservation of original tonal information.

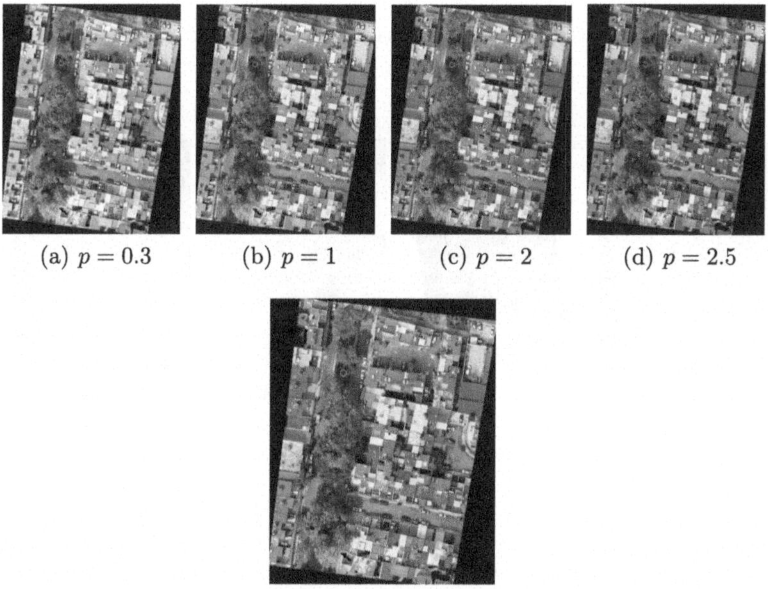

(a) $p = 0.3$ (b) $p = 1$ (c) $p = 2$ (d) $p = 2.5$

(e) $p = 1.7$

Fig. 5. Results of the power transformation applied with different values of p and equalization in HSV.

3.1 Image Reconstruction with Navier-Stokes

The application of the Navier-Stokes-based *inpainting* algorithm effectively reconstructed multiple regions with missing information in the orthophoto, primarily those associated with deep shadows cast by arboreal elements or elevated structures. Visually, significant improvement in structural continuity was evident, maintaining spatial coherence with the surroundings and avoiding perceptible artifacts [1].

Figure 6 shows the process results: the image in Fig. 5(e) taken as the original (Fig. 6(a)), the binary mask used to identify regions for reconstruction (Fig. 6(b)), and the final result after applying the algorithm (Fig. 6(c)). Navier-Stokes successfully interpolated information from edges into missing regions, simulating a natural flow of textures and intensities.

However, limitations were identified in particularly complex or extensive areas, where reconstruction failed to fully restore original texture or patterns. In these cases, filled regions exhibited some smoothness or loss of detail, especially at defined edges or areas with abrupt illumination gradients. These inconsistencies, though subtle, could affect performance in later stages of automatic segmentation or classification.

Additionally, as a side effect of the *inpainting* process, some reconstructed regions displayed artificial edges poorly integrated with their surroundings. To mitigate these effects, a subsequent crop guided by the original mask was applied, preserving only valid areas of the processed image and eliminating regions with artifacts (Fig. 6(d)).

(a) Original (b) Mask (c) Result (d) Crop

Fig. 6. Reconstruction with *inpainting* algorithm

Overall, the results suggest the Navier-Stokes method was suitable for improving visual continuity and filling moderate gaps but has limitations in high-complexity structural zones. Its use should be complemented with visual validations and additional post-processing treatments in critical contexts.

3.2 Bilateral Filtering

Differences between the image obtained after applying the Navier-Stokes algorithm and the filtered image show that most changes were concentrated in homogeneous regions, aligning with the bilateral filter's goal: smoothing textures without losing significant structural edges [14].

Areas where the filter's effect was most evident corresponded mainly to buildings. These regions, characterized by structures with defined edges but flat surfaces, showed higher sensitivity to the smoothing process, resulting in more uniform integration. However, slight loss of detail was observed in architectural elements like moldings or wall textures. Despite this, main contours—such as roof boundaries, sidewalks, or paths—were effectively preserved, fulfilling the filtering purpose without degrading essential structural information (See Fig. 7).

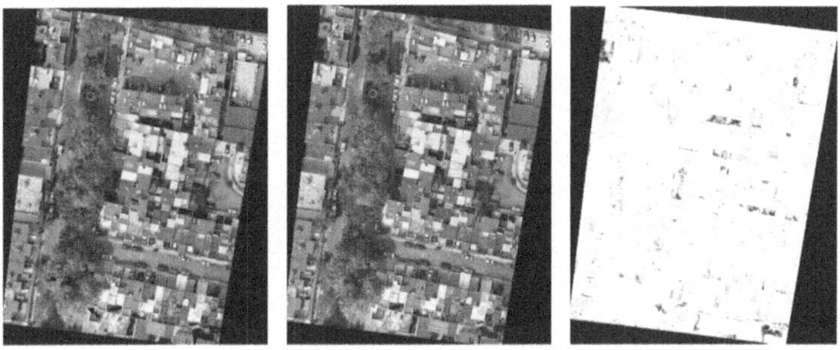

(a) Original (b) Bilateral filter (c) Difference

Fig. 7. Result after bilateral filtering of the reconstructed regions.

3.3 Color-Based Segmentation

Results indicate that color-based segmentation effectively identified vegetated regions, particularly tree canopies. The generated mask clearly highlights green areas, while unwanted elements like roofs, vehicles, or paved surfaces were successfully excluded due to their low saturation or tonal difference from green (Fig. 8).

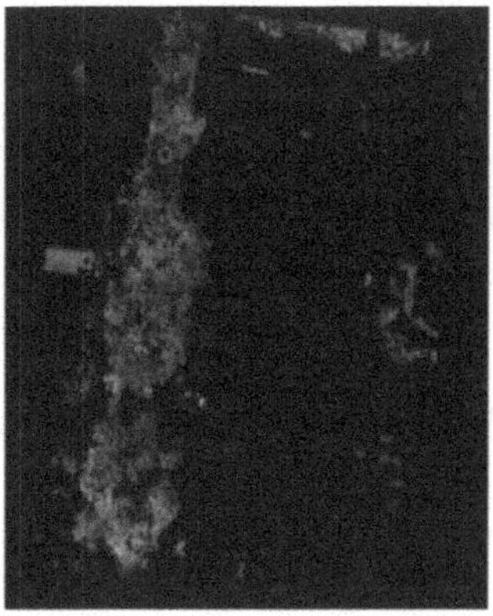

Fig. 8. Study area segmented by color. Regions with a high probability of vegetation are highlighted.

Figure 9 presents the histogram of the *Hue* value. Two peaks were identified: a primary one within the interval corresponding to green tones, validating that most segmented pixels indeed belong to this chromatic range. The second, smaller peak might be associated with internal variations in canopy color, such as shaded areas or species differences, and remains within the defined green range.

This spectral distribution shows that while segmentation predominantly corresponds to dense vegetation, it also captures internal color variations. This enriches subsequent morphological analysis without compromising precision in discriminating non-vegetative elements.

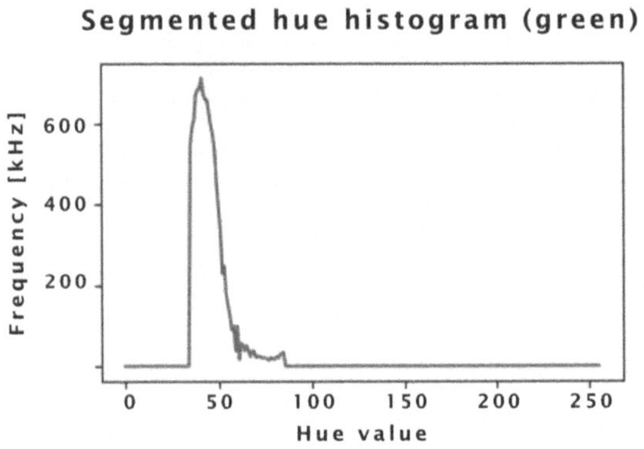

Fig. 9. Histogram of the *Hue* channel of the segmented image. Peaks are observed confirming the predominance of green tones.

3.4 Shape-Based Segmentation

Figure 10 shows the results of morphological shape-based segmentation. This process retained only regions meeting geometric criteria expected for tree canopies, such as minimum area and circular or irregular shape.

Elongated or rectangular structures, typically associated with buildings, roads, or other artificial elements in the original image, were eliminated. The remaining regions correspond to extensive vegetated masses, highly likely representing trees.

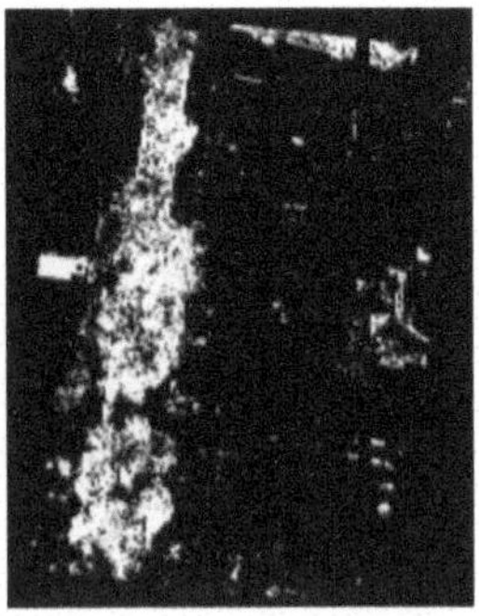

Fig. 10. Shape-Based segmentation.

3.5 Biomass Calculation

The segmentation applied to the processed image (Fig. 11) identified five main regions corresponding to arboreal zones. These areas were visualized using semi-transparent pastel colors, facilitating their differentiation and clearly showing their spatial distribution within the urban environment.

The segmented trees were individually labeled, and their respective projected areas and estimated above-ground biomass were calculated using the allometric models previously described. Results are presented in Table 1.

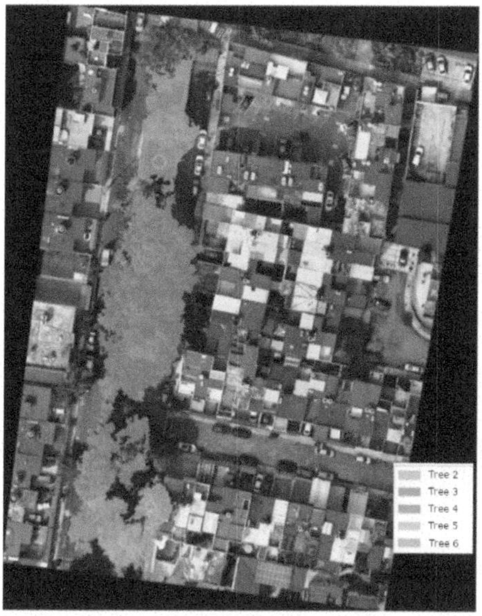

Fig. 11. Area segmented using the *Watershed* algorithm, with labeled regions corresponding to tree clusters.

Although segmented zones do not cover the entirety of visible tree coverage, they represent a significant proportion. Overestimation was observed in the region labeled "Tree 2" toward the image's lower part, while underestimation occurred in the upper part, possibly due to low resolution or partial cropping in that zone.

In contrast, other segmented regions showed more precise correspondence with actual tree distribution. However, some detected regions exhibit extensive areas that might not correspond to a single tree but to dense groupings or overlapping canopies. This possibility is inferred from both morphology and the extent of these zones.

To improve precision in individual tree estimation, subdividing these large regions into smaller subzones, analyzing the *Hue* channel histogram in each, and applying new segmentation based on internal color variations are suggested. This refinement would discriminate adjacent canopies and reduce overestimation errors.

Table 1. Projected area and estimated biomass for each segmented region.

Label	Area (cm^2)	Estimated Biomass (kg)
Tree 2	492	53.7
Tree 3	1130	121.1
Tree 4	299	65.95
Tree 5	110	34
Tree 6	296	19

3.6 Limitations

The algorithm developed in this study adequately identifies tree-covered areas; however, it is not capable of precisely segmenting individual tree crowns, which hinders species identification and directly affects biomass estimation. Furthermore, the algorithm's performance was compromised by the low quality of the aerial imagery captured by the drone, which negatively impacted the generation of the orthomosaic. The presence of artifacts and distortions in the images reduced the quality of the segmentation and limited the capacity for detailed analysis. For these reasons, the biomass estimation was performed only for the dominant species within the detected patches.

For future work, it is proposed to refine the algorithm by incorporating more advanced techniques, such as *Random Forest* [6], as well as to improve the planning and execution of drone flights. These improvements will help reduce errors in image processing, obtain more precise tree observations, progress towards more accurate individual tree segmentation and more detailed species-level classification, thereby increasing the accuracy of biomass estimation.

4 Conclusions

The analysis conducted in the Ajusco neighborhood, Coyoacán Municipality, demonstrates that using drone-captured images combined with digital processing techniques effectively identifies and quantifies tree coverage in urban environments. Segmentation using the *Watershed* algorithm enabled the delimitation of densely vegetated areas and biomass estimation, though refinement is needed in large regions where multiple tree canopies may be grouped.

Results confirm the potential of automated analysis for urban environmental monitoring, particularly in areas where vegetation is scattered and under pressure from intensive land use. Future integration of pattern recognition algorithms like *Random Forest* could significantly improve individual detection and strengthen green planning tools in densely populated cities.

References

1. Bertalmio, M., Sapiro, G., Caselles, V., Ballester, C.: Image inpainting. In: Proceedings of the 27th Annual Conference on Computer Graphics and Interactive Techniques (SIGGRAPH), pp. 417–424 (2000). https://doi.org/10.1145/344779.344972
2. Bowler, D.E., Buyung-Ali, L., Knight, T.M., Pullin, A.S.: Urban greening to cool towns and cities: a systematic review of the empirical evidence. Landsc. Urban Plan. **97**(3), 147–155 (2010). https://doi.org/10.1016/j.landurbplan.2010.05.004
3. Capilla, P., Artigas, J.M., Pujol, J.: Fundamentos de colorimetría, vol. 55. Universitat de València, Valencia (2002)
4. Congreso de la Ciudad de México: Informe de actividades del Congreso local sobre acciones en planteles educativos de Coyoacán. https://www.congresocdmx.gob.mx
5. Escobedo, F.J., Kroeger, T., Wagner, J.E.: Urban forests and pollution mitigation: analyzing ecosystem services and disservices. Environ. Pollut. **159**(8–9), 2078–2087 (2011). https://doi.org/10.1016/j.envpol.2011.01.010
6. Gonzalez, R.C., Woods, R.E.: Digital Image Processing, 4th edn. Pearson, New York (2018)
7. McPherson, G. E., Simpson, J. R., Peper, P. J., Maco, S. E., Xiao, Q.: Urban forestry and climate change. Cities Envir. **9**(1) (2016)
8. Jiang, Y., Li, C., Zhang, X.: Tree species classification based on UAV images using deep learning and metrics evaluation. Rem. Sens. **12**(12), 2002 (2020). https://doi.org/10.3390/rs12122002
9. Meyer, F.: Topographic distance and watershed lines. Signal Process. **38**(1), 113–125 (1994). https://doi.org/10.1016/0165-1684(94)90111-6
10. Nowak, D.J., Crane, D.E., Stevens, J.C.: Air pollution removal by urban trees and shrubs in the United States. Urban Forestry Urban Greening **4**(3–4), 115–123 (2006). https://doi.org/10.1016/j.ufug.2006.01.007
11. Nowak, D.J., Crane, D.E., Stevens, J.C.: Air pollution removal by urban trees and shrubs in the United States. Urban Forestry Urban Greening **4**(3–4), 115–123 (2013). https://doi.org/10.1016/j.ufug.2013.06.001
12. Otsu, N.: A threshold selection method from gray-level histograms. IEEE Trans. Syst. Man Cybern. **9**(1), 62–66 (1979). https://doi.org/10.1109/TSMC.1979.4310076
13. Soille, P.: Morphological Image Analysis: Principles and Applications. Springer, Berlin (2003)
14. Tomasi, C., Manduchi, R.: Bilateral filtering for gray and color images. In: Proceedings of the Sixth International Conference on Computer Vision (ICCV), pp. 839–846 (1998). https://doi.org/10.1109/ICCV.1998.710815
15. Tzoulas, K., et al.: Promoting ecosystem and human health in urban areas using Green Infrastructure: a literature review. Landsc. Urban Plan. **81**(3), 167–178 (2007). https://doi.org/10.1016/j.landurbplan.2007.02.003
16. Universidad de Sevilla: Capítulo 3 [Documento de proyecto académico]. https://biblus.us.es/bibing/proyectos/use/abreproy//fichero/10799%252Fmemoria
17. Vincent, L., Soille, P.: Watersheds in digital spaces: an efficient algorithm based on immersion simulations. IEEE Trans. Pattern Anal. Mach. Intell. **13**(6), 583–598 (1991). https://doi.org/10.1109/34.87344
18. Black Ice: The HSV color space. https://www.blackice.com/Help/Tools/Document%20Imaging%20OCX%20webhelp/WebHelp/The_HSV_color_space.html
19. Modelo de color: Detección de colores. https://omes-va.com/deteccion-de-colores/

20. Montero, M., Montagnini, F.: Modelos alométricos para la estimación de biomasa de diez especies nativas en plantaciones en la región Atlántica de Costa Rica. Recursos Naturales y Ambiente **45**, 112–119 (2008)
21. Tripathi, M., Joshi, H.: Allometric aboveground biomass models for northern Delhi ridge forest. Asian J. Res. Agr. Forestry **9**(4), 229–243 (2023)
22. Peper, P.J., Alzate, C.P., McNeil, J.W., Hashemi, J.: Allometric equations for urban ash trees (Fraxinus spp.) in Oakville, Southern Ontario, Canada. Urban Forestry Urban Greening, **13**(2), 175–183 (2014)
23. Canadell, J., Miquel, R., Andrés.: Pilar Biomass Equations for Quercus ilex L. in the Montseny Massif, Northeastern Spain. Forestry, **61**, 137-147 (1988). https://doi.org/10.1093/forestry/61.2.137
24. Begazo Curie, K., Jave Nakayo, J.: Estimación de la biomasa aérea total y por componente estructural en especies arbóreas de Lima. Perú. Revista Forestal del Perú **33**(1), 76–90 (2018)
25. Picard, N., Saint-André, L., Henry, M.: Manual for building tree allometric equations: application to tropical trees. For. Ecol. Manage. **262**(8), 1470–1483 (2012). https://doi.org/10.1016/j.foreco.2011.07.004

Weather Detection System Based on Vision Transformers to Enhance Robustness in Computer Vision Tasks

Diego Alberto Guevara Amaya[2(✉)], Jairo Enrique Serrano Castañeda[1], Juan C. Martínez-Santos[1], and Edwin Puertas[1]

[1] School of Digital Transformation, Universidad Tecnológica de Bolívar, Cartagena, Colombia

[2] Naval Technological Development Center, Colombian Navy, Cartagena, Colombia

guevarad@utb.edu.co

Abstract. This work presents an automatic classification system for environmental conditions in images, aimed at enhancing the robustness of downstream tasks such as object detection in varying environments. The system consists of two independently trained Vision Transformer (ViT) models using supervised learning with knowledge transfer. One model classifies atmospheric conditions (e.g., fog, rain, snow), while the other identifies the time of day (daytime, night, dawn_dusk). The training process is carried out in two stages: initially using the DAWN dataset, followed by fine-tuning with specialized subsets of the BDD100K dataset to improve generalization capabilities. The models achieve accuracies of up to 93% for lighting conditions and 89% for weather classification.

Keywords: Vision Transformer · Environmental Condition Classification · Adverse Weather · Adaptive Preprocessing · Computer Vision

1 Introduction

The variability in weather and lighting conditions affects the quality of images captured by visual systems. This variability poses a challenge for automatic analysis tasks, particularly in contexts such as object detection. Accurately identifying the conditions under which an image was captured enables the application of specific preprocessing techniques that stabilize visual features, improve dataset homogeneity, and contribute to the increased accuracy of subsequently trained models.

The proposed system functions as an environmental condition detector, similarly to WeatherNet [1], which employs a unified CNN-based framework (ResNet50) to identify various lighting conditions (e.g., dawn/dusk, day, night, glare) and atmospheric conditions (e.g., clear, rain, snow, fog) from street-level imagery. This approach aligns closely with our two-model architecture designed to classify both illumination and weather attributes independently.

W. J. Sarmiento and E. L. Sierra-Ballén (Eds.): CCC 2025, CCIS 2692, pp. 45–60, 2027.
https://doi.org/10.1007/978-3-032-28447-1_4

The system focuses on two key factors that directly affect image visibility, sharpness, and contrast:

- **Lighting conditions**, including: neutral lighting, low-light scenarios (e.g., night or shadow), backlighting, reflections and glare, and abrupt lighting transitions (such as when exiting a tunnel).
- **Atmospheric conditions**, such as: fog or haze, rain, snow, smoke, or dust.

These conditions degrade the captured image and alter its visual properties. Therefore, recognizing them allows for the selection of restoration techniques tailored to each situation. In computer vision systems applied to dynamic environments, such as autonomous driving or urban video surveillance, prior contextual identification becomes a key component to enhance the robustness and reliability of downstream detection tasks.

In this context, Vision Transformer (ViT) models offer a solid alternative due to their ability to capture global spatial relationships [2] without relying on convolutions. Unlike traditional CNN architectures [3], ViTs operate through attention mechanisms that model dependencies between distant regions in the image [4], enabling the identification of relevant global patterns under diverse visual conditions.

This work introduces an image-based environmental classification system composed of two independent Vision Transformer (ViT) models, trained through supervised learning [5] and transfer learning [6], as illustrated in Fig. 1. Both models are based on a pretrained architecture optimized on a 1000-class subset of ImageNet [5].

To adapt them to our domain, a fine-tuning process is required using a custom dataset containing specific classes associated with various types of environmental conditions. Therefore, the output section of the Vision Transformer is modified by replacing the original classification head with a new one, configured to match the appropriate number of classes according to the taxonomy of our problem.

Each model is designed to identify a specific contextual dimension:

- **Weather condition model**, trained on images labeled with attributes such as fog, rain, snow, among others.
- **Lighting condition model**, trained to recognize time-of-day categories (day, night, dawn/dusk), which influence the overall illumination of the scene.

Both models are trained on datasets with diverse visual attributes (DAWN and BDD100K), and their output can be integrated as a preliminary stage in object detection systems. This enables the application of adaptive preprocessing techniques based on the detected environmental context.

1.1 Related Work

Automatic detection of environmental conditions has been primarily addressed using convolutional neural network (CNN) architectures. A representative example is WeatherNet [1], which employs a unified ResNet50-based framework to

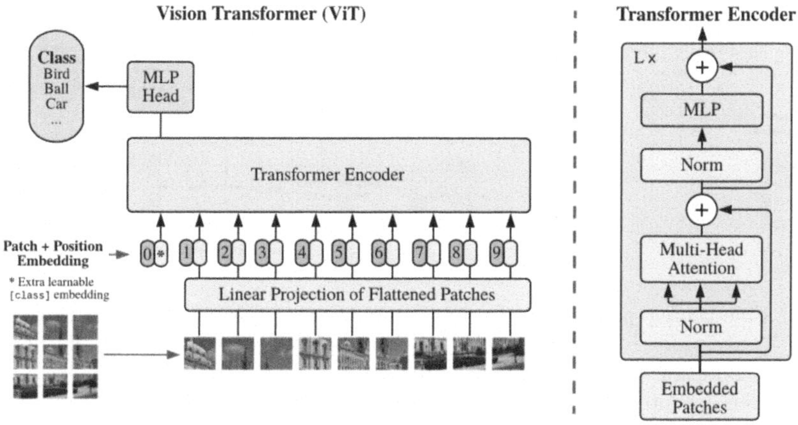

Fig. 1. Pretrained model adapts to a specific task. (taken from [7])

identify both atmospheric and illumination conditions from street-level imagery. While these approaches have shown competitive results, their generalization capability is constrained by the inductive bias of convolutions. In contrast, the introduction of Vision Transformers (ViT) by Dosovitskiy et al. [2] marked a turning point in computer vision, demonstrating that global attention mechanisms can capture long-range spatial dependencies and outperform CNNs in several cases when large-scale pretraining is available. Subsequent studies explored the transferability of these representations [6], highlighting their usefulness for classification tasks in specialized domains. In the specific context of weather and illumination conditions, recent works [8–10] have assessed the applicability of ViTs and their variants, reporting improvements under adverse scenarios and autonomous driving environments. Within this framework, the dual approach proposed in this work positions itself as an extension of these trends, by combining specialized models for atmospheric and illumination conditions to reinforce robustness in computer vision systems.

2 Methodology

To address the challenges of environmental condition recognition, the proposed system was designed following a sequential training strategy that leverages existing datasets and the transfer capabilities of Vision Transformers. As illustrated in Fig. 2, the approach begins with the adaptation of a base model to atmospheric conditions using the DAWN dataset and is subsequently extended with the BDD100K dataset to capture both illumination and weather variability. This two-stage process ensures progressive specialization, allowing the generation of models tailored to complementary environmental factors.

The proposed methodology follows a two-stage training pipeline. In the first stage, a Vision Transformer model (ViT_base_patch16_224) is adopted as the

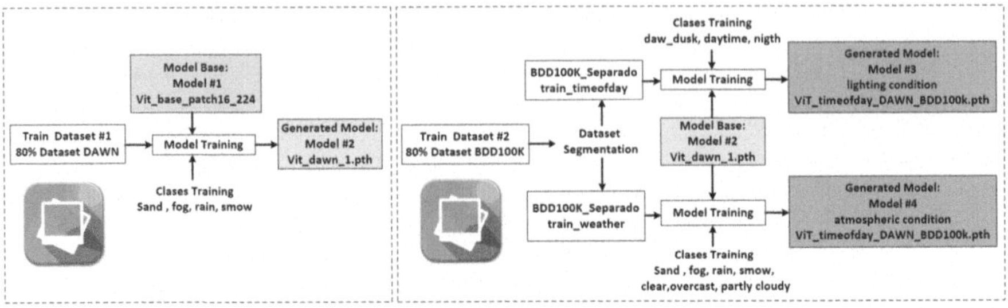

Fig. 2. Fine-tuning process of the `vit_base_patch16_224` model. The pre-trained Vision Transformer is adapted to downstream tasks, such as atmospheric and illumination condition classification, by updating its final layers with a labeled dataset while preserving internal representations learned from large-scale image recognition.

base architecture (Model #1). This model is trained with 80% of the DAWN dataset, which contains images under four atmospheric conditions: sand, fog, rain, and snow. The outcome of this stage is a domain-adapted model referred to as Model #2 (Vit_dawn_1.pth).

In the second stage, Model #2 is reused as the base for training with the BDD100K dataset. To address different environmental factors, the dataset is segmented into two subsets: (i) train_timeofday, with classes dawn/dusk, daytime, and night, and (ii) train_weather, with classes sand, fog, rain, snow, clear, overcast, and partly cloudy. Each subset is used to fine-tune the model toward a specific task.

This process produces two specialized models: Model #3, trained for lighting condition classification using the train_timeofday subset, and Model #4, trained for atmospheric condition classification using the train_weather subset. Both models are derived from the same base but are optimized for complementary tasks, enabling a more robust characterization of environmental conditions in diverse scenarios.

2.1 Dataset Selection and Preparation

Recent advances in artificial intelligence have highlighted the central role that data quality plays in the development of reliable models. As modern AI systems increasingly rely on large-scale datasets for training and evaluation, concerns have emerged not only about data availability, but also about its suitability for producing trustworthy outcomes. In their empirical study, Mohammed et al.(2022) [11], investigate how various data quality dimensions—including accuracy, completeness, and consistency—impact the performance of 19 widely used machine learning algorithms across classification, regression, and clustering tasks. Their findings demonstrate that polluted or low-quality data in either the training or test phases can significantly degrade model reliability, reinforcing the importance of rigorous data curation throughout the AI pipeline.

The performance of models is directly linked to the quality of the data used during training. In this context, data quality refers to its relevance, representativeness, and suitability for the research objective—covering all the features to be modeled and constituting a meaningful sample based on the study's goals.

Since a single dataset does not provide a comprehensive assessment of atmospheric and lighting conditions, two complementary datasets were used. The DAWN dataset offers images under weather conditions such as rain, fog, sand, and snow [12], while the BDD100K dataset provides a broader variety of weather conditions [13] *clear*, *fog*, *overcast*, *partly cloudy*, *rain*, *snow* as well as different lighting conditions *daytime*, *dawn_ dusk*, *night*.

The first dataset, DAWN, includes a folder for each weather condition. Each folder was split into 80% for training and 20% for testing, as shown in Fig. 3.

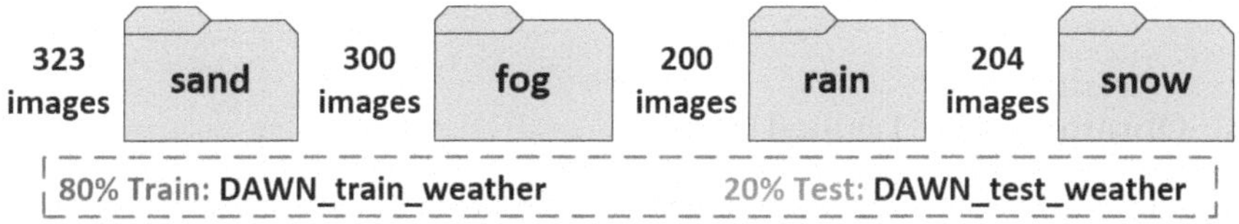

Fig. 3. Dataset DAWN split.

The second dataset, BDD100K, consists of two folders: one containing the images, subdivided into `train`, `val`, and `test` subsets; and another with corresponding annotations in `.json` files, organized under the same categories. Each `.json` file includes:

- Image ID
- Detected objects
- Image attributes

Image attributes are structured as follows:

- **weather**: classifies the atmospheric condition as *clear*, *overcast*, *partly cloudy*, *rainy*, *foggy*, *snowy*, or *undefined*.
- **scene**: describes the type of scene present in the image.
- **timeofday**: classifies the lighting condition as *daytime*, *dawn_ dusk*, *night*, or *undefined*.

Based on this information, all images from the `train` and `val` subsets of the **BDD100K** dataset were classified according to their `weather` and `timeofday` attributes, as shown in Fig. 4.

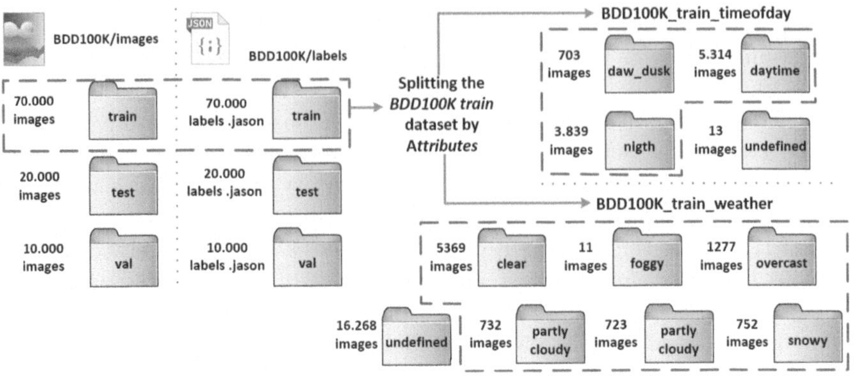

Fig. 4. Classification of images from the BDD100K train dataset according to attributes.

2.2 Obtaining the Trained Model

To build a general model capable of classifying weather conditions, the `vit_base_patch16_224` architecture was used a Vision Transformer pretrained on the ImageNet dataset [14]. This model is adapted to the new task through fine-tuning [15], which consists of reusing previously learned weights from a general classification task and refining them using a smaller, task-specific dataset. This approach retains general visual knowledge while specializing the model for a new domain.

Training is carried out using supervised learning, where the model receives input images with corresponding class labels and adjusts its internal parameters to minimize a loss function. In this case, the cross-entropy loss function [16] is used, suitable for multiclass classification. The training process includes the following stages:

DAWN Dataset Stage. In this stage, the model is trained using a dataset derived from DAWN, named `DAWN_train_weather`, with the goal of detecting four atmospheric conditions: *fog*, *rain*, *sand*, and *snow*, following this procedure:

1. **Dataset preparation** The dataset is organized into folders by class. Transformations include resizing to 224×224 pixels, tensor conversion, and normalization with appropriate mean and standard deviation. The dataset is then split stratified into 80% training and 20% validation sets, ensuring a balanced representation of each class.
2. **Data loading**
   `DataLoader` objects are created for the training and validation subsets. These load data in mini-batches of 32 images to improve computational efficiency, support GPU usage, and shuffle data during training.
3. **Model configuration**
   The `vit_base_patch16_224` architecture pretrained on ImageNet is used,

with its classification head adjusted to output four classes. The `AdamW` optimizer is set with a learning rate of 1×10^{-4}, appropriate for fine-tuning, and the loss function is set to `CrossEntropyLoss`.

4. **Training**
   The model is trained for five epochs using backpropagation, following the conventional deep learning training loop in which the entire dataset is processed multiple times to optimize the network parameters [17]. At each step, model output is computed, the classification loss is evaluated using `CrossEntropyLoss`, and the internal weights are updated to reduce prediction error. Accuracy is calculated at the end of each epoch, achieving a final training accuracy of 98.17%.
5. **Validation**
   After training, the model is evaluated on the validation set using `eval()` mode, without updating weights. This estimates generalization ability, yielding a final validation accuracy of 86.06%.

The resulting model, `Vit_dawn_1.pth`, is capable of detecting *fog*, *rain*, *sand*, and *snow* conditions. However, it may be retrained in the next stage for the following purposes:

- Improve accuracy in existing classes using a dataset with greater variability or quality.
- Expand the model's capability to detect new environmental conditions by incorporating additional classes.
- Extend functionality to include lighting condition detection, particularly timeofday (`timeofday`) attributes.

BDD100K Dataset Stage. In this stage, the previously trained model is retrained using datasets extracted from BDD100K, organized by visual attributes. The primary goal is to improve generalization by retraining the ViT model on a domain with higher variability.

This process improves accuracy for existing classes—also present in BDD100K by introducing more diverse examples in visual and contextual terms. It also broadens coverage by including new atmospheric conditions from `BDD100K_train_weather` and lighting conditions from `BDD100K_train_timeofday`.

Each model—focused on either weather or lighting conditions—is trained using the same methodological framework:

1. **Dataset preparation** A label set is defined with six classes: *clear*, *fog*, *overcast*, *partly cloudy*, *rain*, and *snow*. Each image is resized to 224×224 pixels [14], converted to tensor, and normalized using the parameters used during model pretraining. Stratified splitting ensures 80% training and 20% validation with balanced class distribution..

2. **Data loading** `DataLoader` objects are used for efficient loading of mini-batches of 32 images. This optimizes GPU usage, accelerates training, and ensures randomness in sample selection per epoch.
3. **Model configuration** The `vit_base_patch16_224` architecture pretrained on ImageNet is used. The final classification layer is modified to output six classes. The `CrossEntropyLoss` is used as the loss function, and `AdamW` optimizer is set with a learning rate of 1×10^{-4} [18], with no need for further hyperparameter tuning.
4. **Training and validation** The model is trained for five epochs using back-propagation. Loss is computed and weights are updated in each iteration. After each epoch, performance is evaluated on the validation set in `val()` mode. Epoch results are recorded for later analysis.

After this second stage, two specialized models are obtained: one for weather condition detection, with a final accuracy of 89.07% on the training set and 82.22% on the validation set; and another for lighting condition detection, achieving 95.73% training accuracy and 93.05% validation accuracy. Both models are retrained on subsets of the `BDD100K` dataset, improving their generalization and visual adaptability compared to the DAWN-only trained model.

Hyperparameter Selection and Sensitivity Analysis. The performance of deep learning models is strongly influenced by the choice of hyperparameters. To ensure the robustness of the proposed classifiers, we conducted a sensitivity analysis focusing on parameters that directly affect convergence, generalization, and computational efficiency. The analysis considered both the weather and lighting models.

The following hyperparameters were evaluated:

- Learning rate (LR): values of 1e-3, 5e-4, 1e-4, 5e-5 were tested using the AdamW optimizer to assess stability and convergence speed.
- Batch size: 16, 32, 64 to study its impact on gradient stability and GPU memory usage.
- Number of epochs: 5, 10, 20 with early stopping (patience = 3) to evaluate whether training beyond 5 epochs improved performance without overfitting.
- Weight decay: 0, 0.01, 0.05 to control overfitting and evaluate regularization strength.
- Dropout rate: 0.0, 0.1, 0.2 to assess its contribution to reducing variance in smaller classes.
- Learning rate scheduler: cosine decay, step decay, and constant schedules to analyze the effect of dynamic LR adjustment.

Each configuration was trained on a representative subset of the DAWN and BDD100K datasets, maintaining class balance. Performance was evaluated using accuracy, macro-F1 score, and per-class F1, particularly focusing on ambiguous categories such as fog, overcast, partly cloudy, and dawn/dusk.

The results of this analysis informed the final hyperparameter selection: LR = 1e-4, batch size = 32, dropout = 0.1, and weight decay = 0.01 with cosine learning rate scheduling. These values provided the best trade-off between accuracy, robustness, and computational cost.

2.3 System Evaluation

The evaluation stage aims to verify the performance of the proposed system under realistic conditions not seen during training. A rigorous protocol was designed to consider both the diversity of environmental conditions and the model's generalization capability on external data. This phase includes the construction of a combined evaluation set, integrating samples from the `DAWN` and `BDD100K` datasets that were previously reserved for external validation.

Using this dataset, the trained models were applied to classify images based on weather and lighting conditions, thereby assessing their inference capabilities in varied scenarios. The following subsections describe the process of building the evaluation set and implementing the classifier.

Construction of the Combined Evaluation Dataset. Images used for evaluation were never involved in any previous model training processes. For this reason, specific subsets were reserved exclusively for external validation during the dataset selection and preparation phase [19].

From the `DAWN` dataset, each weather condition folder contributed 20% of its images to form an independent test set named `DAWN_val_weather`.

Similarly, subsets were created from the **BDD100K** dataset using the `weather` and `timeofday` attributes available in the annotations, as shown in Fig. 5. This allowed the construction of the following derived datasets for final evaluation:

Since the weather-related datasets share common labels, a balanced sampling strategy was established. For each overlapping class between `DAWN_val_weather` and `BDD100K_val_weather`, the smaller available number of images was selected to avoid bias in evaluation due to dataset imbalance.

Additionally, for non-overlapping classes, a maximum of 100 images per class was selected to maintain controlled representation and prevent dominance by specific classes. The number of images per class was determined based on inter-dataset balancing criteria. Table 1 presents the number of images used per class, covering both weather and lighting conditions:

Implementation of the Environmental Condition Classifier. Once the combined evaluation set with class labels was constructed, each class was independently evaluated using the trained models, as illustrated in Fig. 6.

1. **Image preprocessing**
   The classification pipeline begins by loading each image according to its annotated class. Images are resized to 224×224 pixels, converted into tensor format,

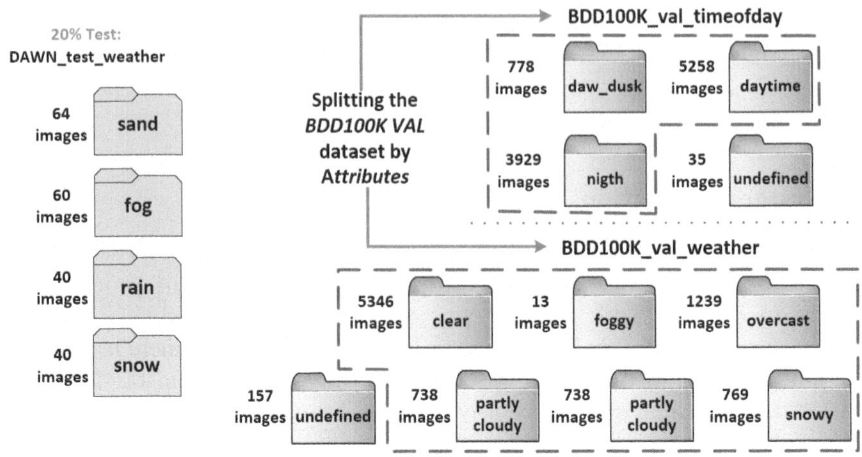

Fig. 5. Evaluation Dataset.

Table 1. Distribution of images per class in the evaluation dataset

Class	Number of images
Fog	26
Sand	100
Snow	82
Rain	80
Overcast	100
Partly cloudy	100
Sand	65
Clear	100
Undefined (weather)	130
Daytime	100
Night	100
Dawn/dusk	100

and normalized using the same mean and standard deviation values applied during model pretraining. These transformations ensure compatibility with the Vision Transformer (ViT) architecture.

2. **Inference and evaluation**
Inference is performed using two independently trained models: one dedicated to detecting lighting conditions and another focused on weather conditions. Predicted labels are compared against ground truth annotations, enabling the calculation of performance metrics such as class-wise precision, overall accuracy, and confusion matrices. To assess the model's behavior across multiple classes, a confusion matrix is employed. This matrix provides a comprehen-

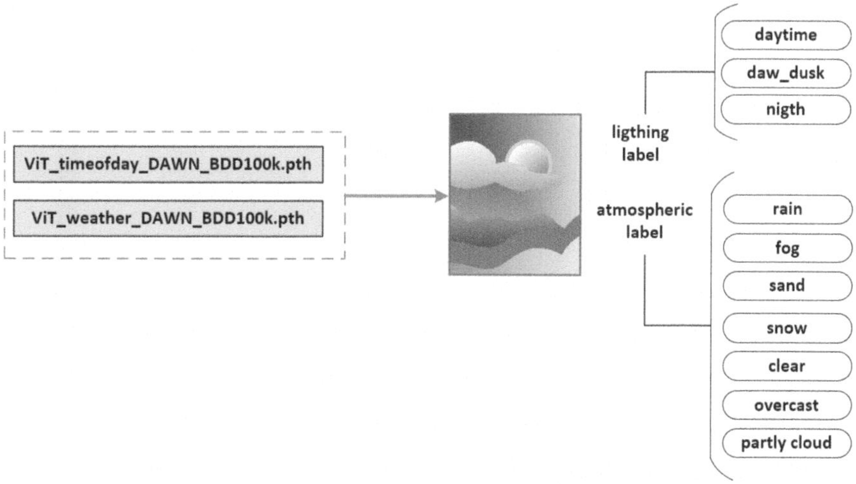

Fig. 6. Weather condition classifier.

sive summary of classification outcomes by organizing correct and incorrect predictions in a tabular format. Diagonal elements indicate true positives—correctly predicted samples—while off-diagonal elements represent misclassifications. This allows the identification of confusion between visually similar categories (e.g., `fog` and `overcast`) and highlights the model's strengths and weaknesses in distinguishing subtle environmental features. The confusion matrix is therefore a critical diagnostic tool in multiclass classification problems [20].

3 System Results

The system was evaluated on a dataset of labeled images that had not been used during training. Each image was analyzed by two specialized models: one designed to detect lighting conditions (*timeofday*) and another for classifying atmospheric conditions (*weather*). The results matrix shown in Fig. 7, summarizes the number of correct and incorrect classifications for each image group, structured by reference class and predicted labels.

The following observations can be made:

- The **lighting model** achieves high accuracy on clearly distinguishable classes such as `night` (96 out of 100 correctly classified) and `daytime` (99 correct). However, it shows greater confusion for the `dawn_dusk` class, which shares features with both previous categories.
- The **weather condition model** successfully identifies classes like `snow` (56/82 correct) and `rain` (51/80), although some misclassifications occur with visually similar conditions such as `overcast` and `fog`.

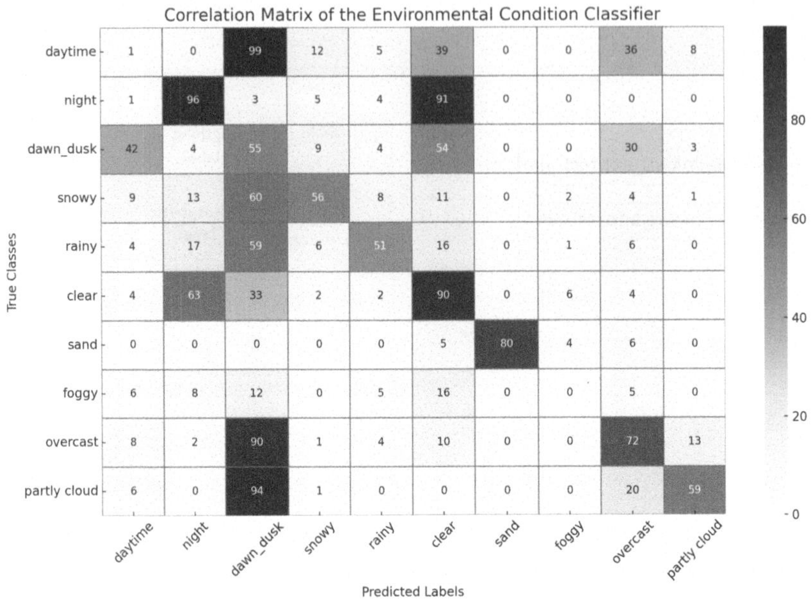

Fig. 7. Correlation matrix between true classes and predicted labels generated by the environmental condition classifier.

– In subsets like `overcast` and `partly cloudy`, the model's predictions are more scattered [8], suggesting potential visual ambiguity between these classes in the dataset.

These results confirm that the system is capable of correctly classifying most of the evaluated conditions. However, some overlapping visual features across classes affect overall performance.

Lighting Classification (timeofday): The model `ViT_timeofday_DAWN_BDD100k.pth` correctly classifies 99% of `daytime` images and 96% of `night` images. However, 94% of `partly cloudy` images are identified as `dawn_dusk`, indicating a visual bias toward transitional lighting conditions.

Weather Classification: The model `ViT_weather_DAWN_BDD100k.pth` shows high accuracy for `clear` (90%), `overcast` (72%), and `partly cloudy` (59%). The `foggy`, `rainy`, and `snowy` conditions exhibit more dispersed predictions, with some confusion involving `clear`, `overcast`, and `dawn_dusk` classes.

Representative examples of correctly and incorrectly classified images are shown in Fig. 8, demonstrating both the system's strengths in detecting well-defined conditions (e.g., `night`, `snow`) and the challenges posed by ambiguous classes like `dawn/dusk` and `partly cloudy`.

En la Fig. 8 Some representative examples of environmental and lighting condition detection, as classified by the specialized models, are shown. These images illustrate both correct and incorrect predictions, highlighting the challenges associated with borderline classes such as `dawn/dusk` and `partly cloudy`, while also demonstrating the system's ability to accurately identify well-defined conditions such as `night` or `snow`.

Fig. 8. Examples of classified images.

4 Discussion

The results obtained from the Vision Transformer models trained to detect lighting and weather conditions highlight several important aspects of their performance and generalization capabilities.

First, the model specialized in lighting conditions performs notably well on clearly defined classes like `daytime` and `night`, with near-perfect precision. However, there is notable confusion in the `dawn_dusk` class, which shares visual features with both daytime and nighttime scenes, making discrimination more difficult. This suggests that the class is inherently ambiguous.

On the other hand, the weather classification model performs well on categories like `snowy` and `rainy`. However, there is some overlap between classes like `overcast` and `partly cloudy`, likely due to visual similarities such as cloud cover, diffuse lighting, and low contrast.

It is also evident that classes with fewer training samples [9], such as `foggy` or `sand`, tend to exhibit higher error rates. This behavior aligns with the general requirement of deep learning models for large and diverse datasets to build robust representations of each category.

Finally, the modular use of independent models for lighting and weather conditions enables a multimodal assessment of each scene. However, it also introduces challenges when combining predictions that may not be entirely orthogonal. A promising future improvement involves exploring multitask learning approaches [10] with shared architectures to exploit correlations between the two dimensions.

5 Conclusions

Based on the development and evaluation of the proposed system, the following conclusions can be drawn:

- Pretrained Vision Transformer models can be effectively adapted for specialized classification tasks involving environmental and lighting conditions, achieving high accuracy with well-curated and balanced datasets.
- Knowledge transfer from an initial dataset (DAWN) to a more diverse one (BDD100K) improves generalization and facilitates adaptation to varied visual contexts.
- Visually distinctive classes (e.g., `daytime`, `snowy`) are recognized with high accuracy, while overlapping ones (`dawn_dusk`, `overcast`, `partly cloudy`) remain challenging.
- The modular design, with models specialized by dimension, enables independent updates or enhancements depending on new datasets, needs, or operational contexts.
- Future work will explore multitask classification and data augmentation to improve class separation and enhance performance in real-world scenarios.

5.1 Limitations, Challenges, and Future Work

The system presents several limitations. Its reliance on medium-scale datasets (DAWN and BDD100K) restricts generalization to conditions not represented in training. The exclusive use of Vision Transformers limits comparison with other model families and may reduce applicability in resource-constrained environments.

Environmental variability remains a challenge, affecting robustness in real-world scenarios. Misclassification between visually similar categories (e.g., overcast vs. partly cloudy, dawn/dusk vs. daytime) highlights the need for more balanced datasets. Real-time optimization is also pending, as latency is critical in practical deployments.

Future directions include integrating adaptive image enhancement techniques (dehazing, rain removal, low-light enhancement) as preprocessing, exploring multitask architectures for joint weather and illumination classification, and adopting multimodal approaches that combine complementary information (e.g., text or sensor data) to strengthen contextual understanding. Extending evaluation to real-world and real-time settings is essential to validate applicability in autonomous and surveillance contexts.

Acknowledgments. We thank the Integral Naval Education Command of the Colombian Navy for providing the resources required for this study and the Naval Technological Development Center for offering a suitable environment to carry it out. We also acknowledge the School of Digital Transformation, Universidad Tecnológica de Bolívar, for its academic support, as well as the Artificial Intelligence Laboratory VerbaNex (https://github.com/VerbaNexAI) for their contributions to this project.

References

1. Ibrahim, M.R., Haworth, J., Cheng, T.: Weathernet: recognising weather and visual conditions from street-level images using deep residual learning (2019). https://arxiv.org/abs/1910.09910
2. Dosovitskiy, A., et al.: An image is worth 16x16 words: transformers for image recognition at scale (2021). https://arxiv.org/abs/2010.11929
3. Raghu, M., Unterthiner, T., Kornblith, S., Zhang, C., Dosovitskiy, A.: Do vision transformers see like convolutional neural networks? (2022). https://arxiv.org/abs/2108.08810
4. Vaswani, A., et al.: Attention is all you need (2023). https://arxiv.org/abs/1706.03762
5. Najjar, E., Breesam, A.: Supervised machine learning a brief survey of approaches, (12), 71 – 82 (2023)
6. Zhou, H.-Y., Lu, C., Yang, S., Yu, Y.: Convnets vs. transformers: whose visual representations are more transferable? (2021). https://arxiv.org/abs/2108.05305
7. Raw, N.: Fine-tune ViT for image classification with transformers (2022). https://huggingface.co/blog/fine-tune-vit
8. Phi, H., Tran, N.: Evaluation of vision transformer on weather image recognittion. Tra Vinh Univ. J. Sci. (09) (2023). ISSN: 2815-6072; E-ISSN: 2815-6099
9. Manivannan, P., Palaniyappan, S., Jayakumar, V., Chandrasekaran, J., Srinivasan Ananthanarayanan, B., Sayeed, M.: Weather classification for autonomous vehicles under adverse conditions using multi-level knowledge distillation. Comput. Mater. Continua **81**(12), 4327–4347 (2024)
10. Saleem, H., Salim, F., Purcell, C.: STC-VIT: Spatio temporal continuous vision transformer for weather forecasting (2024). https://arxiv.org/abs/2402.17966
11. Mohammed, S., et al.: The effects of data quality on machine learning performance on tabular data. Inf. Syst. **132**, 102549 (2025). https://doi.org/10.1016/j.is.2025.102549
12. KENK, M.: DAWN. https://data.mendeley.com/datasets/766ygrbt8y/3
13. Yu, F., et al.: Bdd100k: a diverse driving dataset for heterogeneous multitask learning (2020). https://arxiv.org/abs/1805.04687
14. Wang, D., et al.: Advancing plain vision transformer towards remote sensing foundation model (2022). https://arxiv.org/abs/2208.03987
15. He, K., Chen, X., Xie, S., Li, Y., Dollár, P., Girshick, R.: Masked autoencoders are scalable vision learners (2021). https://arxiv.org/abs/2111.06377
16. Mao, A., Mohri, M., Zhong, Y.: Cross-entropy loss functions: theoretical analysis and applications (2023). https://arxiv.org/abs/2304.07288
17. SabrePC: Epochs, batch size, and iterations in machine learning (2021). https://www.sabrepc.com/blog/Deep-Learning-and-AI/Epochs-Batch-Size-Iterations. Accessed 01 Sept 2025

18. Liu, Z., et al.: Swin transformer: hierarchical vision transformer using shifted windows (2021). https://arxiv.org/abs/2103.14030
19. Assion, F., et al.: A-BDD: leveraging data augmentations for safe autonomous driving in adverse weather and lighting (2024). https://arxiv.org/abs/2408.06071
20. Tharwat, A.: Classification assessment methods: a detailed tutorial. Appl. Comput. Inf. **17**(1), 168–192 (2021)

F. Human-Computer Interaction and Simulation

Characterizing the Use of Generative Artificial Intelligence in Software Usability Testing

Juan Diego Eraso Muñoz[1], Johan Manuel Alvarez Pinta[1], Marta Cecilia Camacho Ojeda[1(✉)], María Clara Goméz Álvarez[2], and Juan Pablo Machado Vélez[1]

[1] Institucion Universitaria Colegio Mayor del Cauca, Carrera 7 # 2–41 Edificio Bicentenario, Popayán, Colombia
{juaneraso,cecamacho}@unimayor.edu.co
[2] Facultad de Minas, Universidad Nacional de Colombia Sede Medellín, Carrera 80 No 65–223 - Campus Robledo, Medellín, Colombia
https://unimayor.edu.co/web/ , https://minas.medellin.unal.edu.co/

Abstract. This study presents a systematic review on the use of Generative Artificial Intelligence (GenAI) in software usability testing. Combined methodologies of systematic mapping and PRISMA were applied, analyzing articles selected between 2021 and 2025 through inclusion and exclusion criteria, searches in academic databases (Google Scholar, IEEE, EBSCOhost) and the Snowballing method. A total of 18 relevant articles were identified, from which the type of GenAI used, the tasks of the testing process in which it was used, the types of software used, and the effects observed were characterized. The most common model was ChatGPT 3.5. The most frequent task was user simulation, especially during test design. The most analyzed types of applications were mobile and web. Among the most reported positive effects are the expansion of test coverage, reduction of time and effort, improvement in efficiency, analysis of results, and reduction of cognitive load. Limitations were also identified, such as loss of context, need for professional supervision, and incomprehensive responses. The findings show that although GenAI does not completely replace testing with real users, it represents a valuable tool for optimizing usability processes. As a future work, it is proposed to apply these technologies in the development of emergency reporting software in a university context.

Keywords: Generative Artificial Intelligence · Usability testing · UX evaluation · software development

1 Introduction

Nowadays, the design of Web/Mobile applications has gained great relevance in the digital environment, Web/Mobile applications are those that integrate

W. J. Sarmiento and E. L. Sierra-Ballén (Eds.): CCC 2025, CCIS 2692, pp. 63–79, 2027.
https://doi.org/10.1007/978-3-032-28447-1_5

a web application for information management and a mobile application for end users [12]. The usability of these applications is a determining factor for their success, as it directly influences the user experience, their adoption and retention. Usability testing represents a critical stage in software development, as it allows the identification of navigation barriers, possible failures and areas for improvement in the user-interface interaction. Usability testing allows to identify how users interact with interfaces, detecting potential glitches, confusion, or navigation barriers that could affect user satisfaction and overall application performance. As these applications become more complex, usability testing needs to be comprehensive enough to cover multiple devices, operating systems, and user flows.

Traditionally, the usability test design process includes creating test cases that are designed based on the expected user experience. These test cases are used to simulate how users interact with the interface, evaluating accessibility, navigation, and ease of use [6]. However, this test case design process can be laborious, subjective, and limited by the creativity of the test team and the availability of data on actual user behavior [3].

In this context, GenAI emerges as a promising tool to optimize the design of usability tests [17]. Capable of generating content from the processing of large volumes of data, GenAI offers the possibility of automating the creation of realistic and diversified test cases, overcoming the limitations inherent in traditional manual design, such as subjectivity, the effort required, and the constraints derived from human creativity [2].

This article presents a systematic review on the use of GenAI in software usability testing, which presents GenAI tools used within different usability testing tasks and the positive and negative effects reported. The structure of the article is as follows: Sect. 2 presents the methodology used; Sect. 3, the review protocol; Sect. 4, the results; and Sect. 5, their discussion. Ultimately, Sect. 6 is where the conclusions of the study are presented.

2 Methodology

The methodology followed to carry out this systematic review on the use of GenAI in software usability testing two approaches were followed, the first is "Systematic Mapping Studies in Software Engineering", which defines guidelines for systematic mapping [22] and the second is "PRISMA" which aims to "help authors improve the writing of the publication of systematic reviews and meta-analyses" [18]. This methodology is composed of 6 stages, the definition of research questions, execution of the search, selection of relevant articles, search for key concepts, mapping process with data extraction and PRISMA. The Fig. 1 shows the phases in the upper tables and in the lower table the result obtained after carrying it out.

Initially, the research questions were established, and the scope of the study was delimited. The objective of this review is to contextualize where GenAI is used in software usability testing. Subsequently, the search for articles in the databases described in the point "2.2 Search String and Databases", using the search strings and applying the inclusion and exclusion criteria specified in Table 2 in addition to the Snowballing method [24] the relevant articles were selected. After compiling the pertinent studies, the articles were read more thoroughly to determine which ones answered the research questions, thus obtaining the key concepts. A detailed analysis of the responses for mapping and data extraction to generate categories continued, thus obtaining qualitative data depending on how many articles coincided with the answers of others and to see if they were related to each other and to know the results given from the use of the GenAI. Finally, PRISMA was applied in the distribution and writing of the report of the results obtained and for the writing of this article.

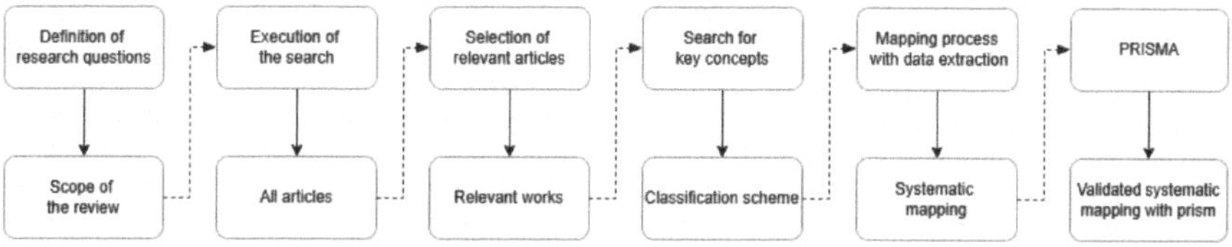

Fig. 1. Methodology.

3 Review Protocol

The present systematic review was conducted in accordance with the guidelines for conducting literature reviews in software engineering, as presented in the previous session. The following section outlines the steps that have been taken..

3.1 Research Questions:

Table 1. Research Questions

Research Questions	Motivation
What types of GenAI are being used in usability testing?	One would expect to know the types and tools that have been used in usability testing to facilitate the selection of the same to be used in usability testing.
What usability testing tasks are they being used for?	Classify the tools that were identified for usability testing in the tasks that can be used
In what types of applications Is GenAI used for usability testing?	Relate the identified tools to the type of software in which they can be used for usability testing
What effects are achieved with the use of GenAI in usability testing?	Know what effects were achieved by recognizing the benefits and limitations of the results obtained when using the tools of GenAI

The objective of this review is to contextualize works where GenAI are used in software usability tests, to know how they are used and what the results were obtained. Identifying which GenAI tools can be performed on which tasks in the testing process and the expected results would make it easier for the development team to make decisions about whether or not to use these tools in this process.

For the study and selection of material, four research questions were used, which are in Table 1

3.2 Search Strings and Databases:

To define the search string, the keywords referring to the objective of the review and the questions asked were considered, the terms were organized in 3 sections separated by the logical operator AND where each one has 2 terms separated by OR. The first part is focused on generative artificial intelligence and LLM (Large Language Models) searches, the second section focuses on usability testing and UX (User Experience) evaluation and finally a section for the engineering and software approach. We also used exclusion terms to avoid results that were not really related to the review.

The search strings were adapted according to the database consulted, without affecting the terms, but rather the syntax and logical operators used by the search engine of each database.

Google Scholar: It was selected as a database option or search engine because of its wide number of varied results from different databases, giving us results from ACM, IEEE, Arxiv and others, which allows us to have better general coverage of the expected results. The search string used was:

"Generative Artificial Intelligence" OR "LLM" AND "usability testing" OR "UX evaluation" AND "software development" OR "software engineering" - education -students -curriculum -teaching

EBSCOhost: It is a platform recognized for offering access to a wide variety of databases. It was chosen for its ability to provide software engineering and user experience articles. The search chain used was:

("Generative Artificial Intelligence" OR "LLM" AND "usability testing" OR "UX evaluation" AND "software development" OR "software engineering") NOT education NOT students NOT curriculum NOT teaching

IEEE: It is a very popular source for systematic reviews, especially around technology, so it was considered a reliable and necessary source for this review. The search string used was as follows:

("All Metadata":"Generative Artificial Intelligence" OR "All Metadata": "LLM") AND ("All Metadata":"usability testing" OR "All Metadata":"UX evaluation") AND ("All Metadata":"software development" OR "All Metadata":"software engineering")

3.3 Inclusion and Exclusion Criteria:

The selection of articles in this systematic review was determined according to the inclusion and exclusion criteria described in Table 2.

Initially, it had been considered to use the number of citations as a quality criterion for the selection of articles. However, since the Generative Artificial Intelligence such as ChatGPT, Gemini, among others, is a relatively recent technology, many of the relevant articles still have few or no citations from other sources. For this reason, this criterion was not applied in this review.

Table 2. Inclusion and exclusion criteria

Inclusion Criteria	Exclusion Criteria
- **Language:** English only. - **Availability:** Full item available. - **Type:** Journal articles, event reports, doctoral and master's theses. - **Filter:** The article must answer at least three of the research questions - **Years:** 2021–2025. - **Title and abstract:** Both the title and the abstract of the article are in accordance with the as expected according to the criteria of the researchers.	- **Language:** Any language other than English. - **Availability:** The article was not found available in open access or with access to the universities of the group - **Type:** Undergraduate monographs, books or book chapters - **Filter:** The article does not answer at least three of the research questions. - **Years:** Articles prior to 2021 - **Title and abstract:** Both the title and the abstract of the article are not in accordance with what is expected according to the criteria of the researchers.

4 Review Results

Making use of the criteria such as the search strings and database defined, the review was carried out through 4 phases, applied to each database as follows.

Review of Google Scholar:

Phase 1: In this first phase, only the number of results were reviewed using the appropriate search string for the database, where 56 were obtained by applying the acceptance criterion of year from 2021.

Phase 2: Of the 56 results were filtered through the abstract and the title, in addition to applying the language and type criteria that provide us with enough content to know the focus of the article, in this process 11 results were obtained that went to the next phase.

Phase 3: In this phase, the availability acceptance criterion was applied so that they could be reviewed in the next phase, from this stage 7 results that were available or accessible came out.

Phase 4: In this phase, the available articles from phase 3 are already submitted to the research questions where they must fully answer at least 3 of the 4 established. In this last stage, the results for the database are already defined where 6 were achieved in this case.

IEEE Review:

Phase 1: In this first phase they only obtained 2 results with the search string established using the filter or criterion of the year from 2021 onwards.

Phase 2: The 2 results reviewed by abstract and title that were in accordance with what is to be obtained, which went to the next phase. In this phase, the criteria of language, type and revision of abstract and title were applied.

Phase 3: In this phase, the availability acceptance criterion was applied so that they could be reviewed in the next phase, from this stage only 1 available could be obtained.

Phase 4: In phase 4, the available article from phase 3 is submitted to the research questions where they must fully answer at least 3 of the the 4 established, of which the only result to be reviewed correctly solved the research questions. So finally, a result was obtained from this database.

Revision de EBSCOhost

Phase 1: The search string was adapted to the characteristics of the database for a better search, complemented by the use of the filters provided by the advanced search option (Language: English; Subject: generative artificial intelligence, artificial intelligence, language models, machine learning, natural language process, inguser interfaces, agile software development, knowledge management, computer science, information science, Academic Journal; Publication Date: Last 5 years.). As a result, 293 matches were obtained.

Phase 2: Of the 293 results, we were manually reviewed through the abstract and the title, in addition to applying the language and type criteria that provide us with enough content to know the focus of the article, in this process we managed to obtain 8 results that continue to the next phase.

Phase 3: The criterion of availability of the article was reviewed, in which of the 8 results, they were reduced to 6 articles.

Phase 4: Finally, the 6 articles were subjected to the filter of answering at least 3 of the 4 research questions previously proposed, in this case, they ended up passing only 2 articles.

After carrying out all this process with each database, the result was 9 articles, which passed the filters of each phase. Due to the low number of approved articles, it was decided to use an alternative that would allow this result to be expanded.

Snowballing Method: Considering the above, it was decided to apply the Snowballing Method. This method helps to broaden the search for relevant studies from the articles already selected, with the review of the citations they contain (Backward Snowballing) and the review of other articles or works that cite the articles selected in the review (Forward Snowballing) [24]. Therefore, for this procedure, 3 articles that had other works that cited them were selected, to review both Backward Snowballing and Forward Snowballing. With this procedure, 9 more articles were achieved that passed the acceptance criteria, expanding our number of results to 18.

4.1 Results

This section presents the results obtained from the systematic review process. As illustrated in Table 3, a comprehensive overview of the numerical outcomes of the article selection process across various databases is provided. This includes the number of records that were screened at each stage of the selection process. Table 4 provides a overview of the 18 studies that met the established inclusion criteria and were ultimately selected.

Table 3. Total results obtained per phase, together with the snowball method

Data Bases	Ph 1	Ph 2	Ph 3	Ph 4	Snowballing	Total
Google Scholar	56	11	7	6	9	18
EBSCOhost Research Database	293	8	6	2		
IEEE Computer Society Digital Library	2	2	1	1		

This section presents a synthesis of the answers obtained for the research questions of the 18 final articles obtained, in this way, it is possible to correctly classify the answers to visualize them in a better way.

Question 1: What types of GenAI are being used in usability trials?
For this research question, a review of the tools or types of GenAI used by the authors in their articles was carried out. A graph was generated from the information from the 18 studies analyzed.

As can be seen in Fig. 2 there are the models and the number of articles that used them, there, there is also a special category that we call "Own" which refers to the number of articles that developed their own model and, finally, it can be obtained that the model most used by the studies was ChatGPT 3.5 (GPT-3.5) with 5 appearances.

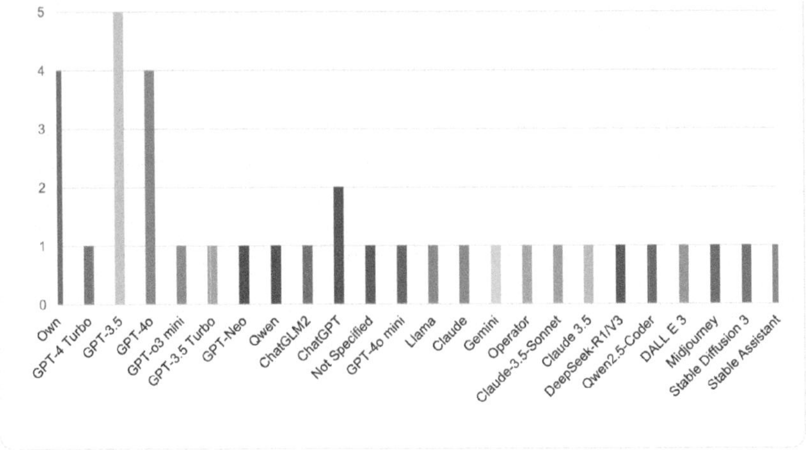

Fig. 2. All specific GenAI models used by the articles.

Question 2: What usability testing tasks are they being used for?

In this question, the following classifications or labels were used for the usability testing tasks, according to the answers obtained in the results that were analyzed. The results can be seen in the Fig. 3.

- **Interface evaluation:** Describes the tasks of usability testing that are carried out at the time of evaluating the interface, an example of these tasks are the correct use of colors or metaphors, they are carried out by UX experts. [17,19,20]
- **Usability testing design:** Describes the phase in which a usability testing process is designed or prepared, where it is established which tools or parameters are going to be used, whether they are surveys, interviews, peer review, heuristics or others. [5,8,13,14,16,21,23,26]
- **Generation of usable interfaces:** This task in usability tests is carried out based on the knowledge of the feedback or the results given by the test user, so new interfaces are generated or modified. [1,22]
- **User simulation:** This classification was used since there are works where GenAI is used to simulate users who evaluate software to anticipate possible unforeseen situations and reduce the time required. [4,5,8,9,11,13,14,21,23,25,26]
- **Evaluation of results:** This task was included since there are situations in which the GenAI is used to evaluate the results of surveys or tests already carried out to determine an action to be taken or suggestions that can be given. [5,10,13,14,23,26]

Figure 3 shows the usability tasks in which the GenAI has been used, according to the articles analyzed. It is observed that the task with the highest recurrence is "User simulation" with 11 articles, although 8 of this report the use of

Table 4. Title and year of the 18 articles obtained

Tittle	Year	Ref
UXAgent: An LLM-Agent-Based Usability Testing Framework for Web Design	2025	[13]
Does GenAI Make Usability Testing Obsolete?	2025	[17]
Capabilities of Large Language Models in Web Accessibility Evaluation: A Design Science Approach	2024	[19]
Genie: Reimagining Home-screen Interfaces using Multi-Agent AI Systems	2025	[22]
Combating Missed Recalls in E-commerce Search: A CoT-Prompting Testing Approach	2024	[25]
Exploring consumer behaviors in e-commerce using machine learning	2024	[9]
LLM-Based Labelling of Recorded Automated GUI-Based Test Cases	2025	[5]
AI-MDD-UX: Revolutionizing E-Commerce User Experience with Generative AI and Model-Driven Development	2025	[1]
Towards Designing a Set of Usability and Accessibility Heuristics Focused on Cognitive Diversity: An Exploratory Case Study with Generative Artificial Intelligence	2024	[16]
Make LLM a Testing Expert: Bringing Human-like Interaction to Mobile GUI Testing via Functionality-aware Decisions	2024	[11]
Agents for Automated User Experience Testing	2021	[4]
AXNav: Replaying Accessibility Tests from Natural Language	2024	[21]
Large Language Model-Brained GUI Agents: A Survey	2025	[26]
WebGen-Bench: Evaluating LLMs on Generating Interactive and Functional Websites from Scratch	2025	[14]
AgentA/B: Automated and Scalable Web A/B Testing with Interac- tive LLM Agents	2025	[23]
Enhancing UX Evaluation Through Collaboration with Conversa- tional AI Assistants: Efects of Proactive Dialogue and Timing	2024	[10]
onversational AI-Powered Design: ChatGPT as Designer, User, and Product	2023	[8]
How well can generative AI design and evaluate user interfaces?	2025	[20]

this in the "Design of usability tests" phase, a category that is especially relevant for the present study.

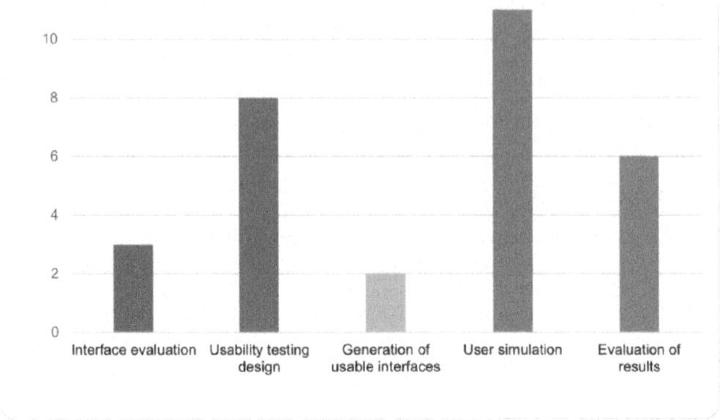

Fig. 3. Usability tasks.

Question 3: In what types of applications are GenAI used for usability testing?

In this question, types of software were found where these software tests accompanied by GenAI were used, such as: Web applications, Mobile applications, VR, Desktop, Video Games or Unspecified. In this way, the following graph was obtained. It can be seen in Fig. 4 that the tests carried out on the articles focus mainly on 2, which are the "Web Applications" and "Mobile Applications".

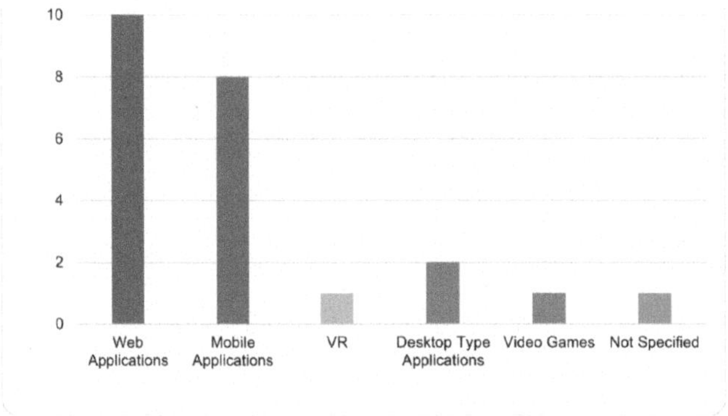

Fig. 4. Types of Software Used.

Question 4: What effects are achieved with the use of the GenAI in usability tests?

The rankings for question 4 on the effects achieved with the use of GenAI in usability tests were extracted from the answers given, and from these the key points of their results were taken and represented in Fig. 5. Each classification is numbered in the text to be correctly Fig. 5. Therefore, when looking at the figure and understanding it, you must check which of the effects are being referred to.

- **Greater user coverage (1):** Increase in the number of test users or greater diversity of users who are usually difficult to recruit for usability tests, simulated with the help of GenAI. [4,17,19,23]
- **Facilitates the analysis of test results (2):** GenAI helps in the processing of usability test results, such as analyzing surveys or video descriptions of tests. Giving different perspectives that some professionals overlook or omit. [21,26]
- **Decreases time (3):** This refers to the fact that by using GenAI, it allows for a significant reduction in the time to identify usability problems in the early stages of development. [5,17,19,22]
- **Facilitates data collection at early stages (4):** The use of GenAI in usability testing allowed to accelerate iterative design cycles, facilitating data collection at early stages of the study. [13]
- **Identify design improvements (5):** Using GenAI to perform interfaces with specific parameters helps to improve interface design. [1,8,11,13,14,16,20,23]
- **Facilitates the analysis of behavioural patterns (6):** Access to a considerable amount of simulated data facilitated the analysis of behavioural patterns and supported informed decision-making before conducting studies with humans. [1,4,13]
- **Greater coverage of tests (7):** The use of GenAI shows an increase in the amount of evidence collected, some of which were not seen by trained personnel, allowing other points of view to be realized. [4,5,11,13,16,17,19,21,23,25,26]
- **Improves efficiency (8):** GenAI improves efficiency as it allows better results to be obtained with less effort and in shorter times and is very feasible to use in small teams. [5,10,17,19,22,26]
- **Useful for small teams (9):** As mentioned above, using GenAI is feasible for small teams because they streamline some processes and have a good percentage of successes, making them able to use fewer resources in these areas. [17]
- **Decreases cognitive load (10):** Objective assessments showed that the reports generated with GenAI were consistent and detailed, thus reducing the cognitive load on the assessors. [5,16,19,22,26]
- **UX improvement (11):** When using GenAI in the generation of usability tests, an improvement in the UX is observed, since they were detected and improved in an optimal way with GenAI, which helped users to experiment better efficiency, effectiveness and satisfaction. [1,11] [23]
- **Improves the identification of errors (12):** The GenAI manages to identify errors, which a small percentage are false positives and helps in the debugging of this, which improves the identification of errors. [5,11,25]
- **Increased customer satisfaction (13):** Shows results that end users can detect within the software, especially on e-commerce platforms where customer satisfaction is crucial. [1,9,10,21]

- **Improved accessibility (14):** The GenAI shows the ability to identify accessibility problems that were overlooked and not considered, which helped in their correction. [16,19,21,22,26]
- **Increased engagement (15):** Within articles that involved e-commerce, there was talk about "engagement", which was increased with more engaged users on the platforms. [1,9]
- **Decreased manual effort (16):** The use of GenAI represented less manual effort on the part of the professionals in charge of usability testing, in tasks such as test design or execution. [16,26]
- **Reduces costs (17):** By being able to generate usability tests with GenAI and obtain good results, small companies can use this technology and reduce expenses. [23]

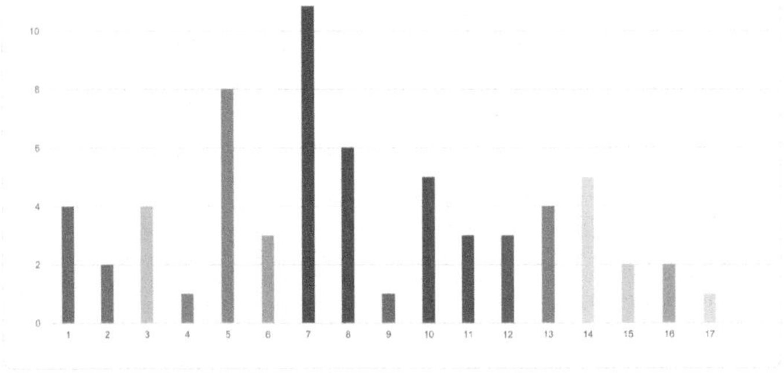

Fig. 5. Classification of the results in the articles analyzed.

When studying Fig. 5 obtained from the responses of the articles to question 4, we can notice that the studies that used GenAI concluded mostly with results of "Greater coverage of tests".

A study was also carried out of those answers to question 4 involving conclusions or results that can be considered negative or regular, from which the following classifications were obtained. In the same way, enumerations of the classifications were made to make a better Fig. 7.

- **Loss of context (1):** In very long conversations, the GenAI lost the initial context of what a request was about. [8]
- **Difficulty with the processing of some requests (2):** There were requests that were made by the GenAI that could be non-specific or very complex, which in some results represented problems at the time of making a request. [8]
- **Needs someone professional to trust the answers (3):** The need for professional staff or knowledge of the subject to be able to fully trust the GenAI responses was highlighted. [1,8,16,20]

- **Not very exhaustive answers (4):** There were answers given by the GenAI in very general or shallow that resulted in distrust of them. [8]
- **Non-Real User Testing (5):** No testing with real users, only testing with simulated users. [8,13]

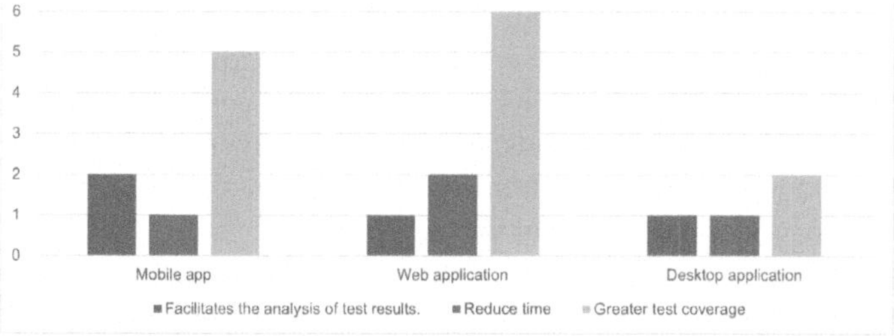

Fig. 6. Comparison of reported effects in types of applications.

Effects were selected, and the articles were classified according to the application or applications in which IAGen was used. The greatest coverage of tests is superior to the other effects. Figure 6

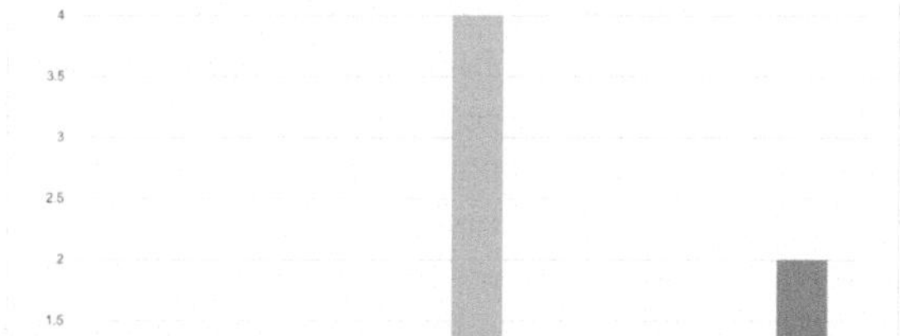

Fig. 7. Classification of results in the articles analyzed.

There is evidence of a greater number of cases where GenAI presents an inconvenience of "Needs someone professional to trust the answers", which means that when operating with GenAI it is still necessary to have the supervision of an expert in the field.

5 Discussion

GenAI has become a tool that optimizes usability testing by reducing time, effort, and cognitive load on the evaluator, while increasing the efficiency and coverage of the process. Its application allows for the generation of dynamic test scenarios, the simulation of user profiles, and the automated analysis of interactions, offering recommendations aligned with frameworks and theories such as Nielsen's heuristics [15]. For example, when evaluating the visibility of the system status, IAGen can simulate interactions to detect the absence of feedback or confirmation of actions. Similarly, it aligns with the ISO 9241-11:2018 [7] standard by analyzing efficiency through interaction times or number of steps. This makes it possible to detect system visibility issues, explore design alternatives more quickly, and enrich the feedback process in user-centered software development. However, it is a tool that must be used in a supervised manner and with the understanding that there are limitations that must be considered and that it does not replace testing with real users.

This review provides an analysis of the use of the GenAI in usability tests, providing a contribution on the tools used, in which tasks of the testing process and in what types of applications it has been used, and especially presents the results reported, classifying them in such a way that the reader can know what benefits to expect, but also the limitations to be faced. However, the reader should consider the scope and restrictions of this work, in which articles in languages other than English were not considered, although initially it had been proposed to consider a criterion of quality of the articles considering the citations of these, however it cannot be done due to the recentness of the works found.

Based on the findings of this review, the authors believe that the integration of IAGen into the design and execution of usability tests will allow testers to focus on important aspects. The time saved in test design and behavior prediction can be used to develop more customized tests with greater follow-up. Greater test coverage will allow developers to focus on creating apps that are more accessible to different types of users, customizable to users' preferences or interests, and even adaptive apps that modify themselves by learning from users' behavior. Testing with automatic interaction analysis will make it easier to detect errors or potential improvements. These aspects will improve the user experience and functionality of future software applications.

6 Conclusions

The GenAI is a tool that optimizes the usability testing process. The benefits reported in its use include the reduction of time, effort and cognitive load of the testers, while increasing the efficiency and coverage of the tests. The potential advantages of using the GenAI in the testing process are obvious, but to overcome the reported limitations it is necessary for the usability tester to have previous knowledge and experience in usability testing, to know the objective of the application to be tested, its users and its characterization, in addition to using strategies for the generation of proms, seeking to avoid the loss of context and the difficulty of processing in the interaction with AI, to achieve responses of a higher level of depth and relationship.

At the user level, using GenAI represents benefits, because of the use for the predictability of behaviors and therefore covers aspects of usability that may not be considered. In addition, it decreases the number of tests that must be carried out with real users, which is a benefit for testers since it is not easy to recruit them for testing and for users not having to perform an exhaustive number of tests decreases the cognitive load and therefore prioritizes the tests they perform.

As a future work, it is proposed to use GenAI for the usability testing process for an emergency reporting software for the health area of Institutional Welfare of a University, software in which usability is a crucial feature to achieve its acceptance, commitment and effectiveness. During this application, it is expected to report on the effects achieved specifically in the design of usability tests.

Acknowledgments. This work has been funded by the project ID IUCMC 2024.GII.010, within the framework of the 2024 call for research projects of the Colegio Mayor del Cauca University Institution.

References

1. Alti, A., Lakehal, A.: AI-MDD-UX: revolutionizing E-commerce user experience with generative AI and model-driven development. Future Internet **17**(4), 180 (2025). https://doi.org/10.3390/fi17040180. https://www.mdpi.com/1999-5903/17/4/180
2. Bajaj, Y., Samal, M.: Accelerating software quality: unleashing the power of generative ai for automated test-case generation and bug identification. Int. J. Res. Appl. Sci. Eng. Technol. (2023). https://api.semanticscholar.org/CorpusID:259584359
3. Dunn, P.G., Hayes, A.: The problems with usability testing. In: Kurosu, M. (ed.) Human-Computer Interaction. Design and User Experience, pp. 420–430. Springer International Publishing, Cham (2020)
4. Fernandes, P.M., Lopes, M., Prada, R.: Agents for automated user experience testing. In: 2021 IEEE International Conference on Software Testing, Verification and Validation Workshops (ICSTW), pp. 247–253. IEEE, Porto de Galinhas, Brazil (2021).https://doi.org/10.1109/ICSTW52544.2021.00049. https://ieeexplore.ieee.org/document/9440143/
5. Franzosi, D.B., Alégroth, E., Isaac, M.: LLM-based labelling of recorded automated GUI-based test cases. In: 2025 IEEE Conference on Software Testing, Verification and Validation (ICST), pp. 453–463. IEEE, Napoli, Italy (2025).https://doi.org/10.1109/ICST62969.2025.10988984. https://ieeexplore.ieee.org/document/10988984/
6. Hass, C.: A practical guide to usability testing, pp. 107–124. Springer International Publishing, Cham (2019). https://doi.org/10.1007/978-3-319-96906-0_6
7. International organization for standardization: Iso 9241-11:2018 ergonomics of human-system interaction — part 11: usability: definitions and concepts. international organization for standardization (2018), disponible. https://www.iso.org/standard/63500.html
8. Kocaballi, A.B.: Conversational AI-powered design: ChatGPT as designer, user, and product
9. Kolluru, V., Chintakunta, A.N., Nuthakki, Y., Koganti, S., Mungara, S.: Exploring consumer behaviors in E-commerce using machine learning. SSRN Electr. J. (2024). https://doi.org/10.2139/ssrn.4988848. https://www.ssrn.com/abstract=4988848

10. Kuang, E., Li, M., Fan, M., Shinohara, K.: Enhancing UX evaluation through collaboration with conversational AI assistants: effects of proactive dialogue and timing. In: Proceedings of the CHI Conference on Human Factors in Computing Systems, pp. 1–16. ACM, Honolulu HI USA (2024). https://doi.org/10.1145/3613904.3642168
11. Liu, Z., et al.: Make LLM a testing expert: bringing human-like interaction to mobile GUI testing via functionality-aware decisions. In: Proceedings of the IEEE/ACM 46th International Conference on Software Engineering, pp. 1–13. ACM, Lisbon Portugal (2024). https://doi.org/10.1145/3597503.3639180
12. Loayza Zambrano, J.R., Loor Campues, F.D.: Panel de control para la gestión de aplicaciones móviles de la espam mfl (2022). https://repositorio.espam.edu.ec/handle/42000/1860
13. Lu, Y., et al.: UXAgent: an LLM agent-based usability testing framework for web design. In: Proceedings of the Extended Abstracts of the CHI Conference on Human Factors in Computing Systems, pp. 1–12 (2025). arXiv:2502.12561 .https://doi.org/10.1145/3706599.3719729
14. Lu, Z., et al.: WebGen-bench: evaluating LLMs on generating interactive and functional websites from scratch (2025). arXiv:2505.03733 .https://doi.org/10.48550/arXiv.2505.03733
15. Nielsen, J.: Usability Engineering. Morgan Kaufmann, San Francisco, CA (1994)
16. Pascual Almenara, A., Sayago Barrantes, S.: Towards Designing a Set of Usability and Accessibility Heuristics Focused on Cognitive Diversity: An Exploratory Case Study with Generative Artificial Intelligence. Information **15**(12), 769 (2024). https://doi.org/10.3390/info15120769. https://www.mdpi.com/2078-2489/15/12/769
17. Pourasad, A.E., Maalej, W.: Does GenAI make usability testing obsolete? (2025). arXiv:2411.00634. https://doi.org/10.48550/arXiv.2411.00634
18. Royo, M.: BiblioGuías: revisiones sistemáticas: PRISMA 2020: guías oficiales para informar (redactar) una revisión sistemática. https://biblioguias.unav.edu/revisionessistematicas/guias_oficiales
19. Sergelius, S.: Capabilities of large language models in web accessibility evaluation: a design science approach
20. Sun, Z., Baber, C.: How well can generative AI design and evaluate user interfaces? contemporary ergonomics and human factors
21. Taeb, M., Swearngin, A., Schoop, E., Cheng, R., Jiang, Y., Nichols, J.: AXNav: replaying accessibility tests from natural language. In: Proceedings of the CHI Conference on Human Factors in Computing Systems. pp. 1–16. ACM, Honolulu HI USA (2024). https://doi.org/10.1145/3613904.3642777, https://dl.acm.org/doi/10.1145/3613904.3642777
22. Thomsy, N.: Genie: reimagining home-screen interfaces using multi-agent AI systems (2025). https://summit.sfu.ca/item/39371, publisher: Simon Fraser University
23. Wang, D., et al.: AgentA/B: automated and scalable web A/B testing with interactive LLM agents (2025). arXiv:2504.09723. https://doi.org/10.48550/arXiv.2504.09723
24. Wohlin, C.: Guidelines for snowballing in systematic literature studies and a replication in software engineering. In: Proceedings of the 18th International Conference on Evaluation and Assessment in Software Engineering, pp. 1–10. ACM, London England United Kingdom (2014).https://doi.org/10.1145/2601248.2601268. https://dl.acm.org/doi/10.1145/2601248.2601268

25. Wu, S., et al.: Combating missed recalls in e-commerce search: a CoT-prompting testing approach (2024). arXiv:2406.19633. https://doi.org/10.48550/arXiv.2406.19633
26. Zhang, C., et al.: Large language model-brained gui agents: a survey (2025).https://doi.org/10.48550/arXiv.2411.18279

Design of Digital Twins for Behavior Analysis in Naval Radar Simulations Using Multi-agent Systems

Oscar Eduardo Olaya Vera(✉), J. L. Villa, Juan Carlos Martinez-Santos, and Jairo E. Serrano

Universidad Tecnologica de Bolivar, Cartagena, Colombia
{oolaya,jvilla,jcmartinezs,jserrano}@utb.edu.co

Abstract. This article presented a radar simulation system design based on a multi-agent architecture for analyzing plane behavior in naval environments. The system modeled the dynamic movement of a plane and its detection by radar-type sensors, represented by the Navigation Radar and the Fire Control Radar. The simulation outputs were processed to analyze movement patterns, idealizing the physical variables that influenced the detection and tracking of targets. The proposed design was a foundation for developing a scenario emulator to create digital twins of sensors employed on military vessels, including navigation radars, surveillance radars, fire-control radars, and electro-optical systems. The generated data would contribute to structuring a scenario simulator to instruct and train officers in the Colombian National Navy. In addition, the integration of data fusion and correlation techniques was outlined, along with a prediction module and a central arbiter, to enhance the detection and tracking of contacts in threat situations. The results demonstrated that this multi-agent approach effectively emulated realistic plane movements and enabled efficient agent communication. This framework not only improved radar detection and tracking capabilities but also provided a robust platform for future enhancements, such as incorporating additional sensors and machine learning techniques to refine the accuracy and efficiency of naval radar simulations.

Keywords: Multi-agent systems · radar digital twins · plane dynamic systems

1 Introduction

Radars play a significant role in modern surveillance, defense, and navigation by enabling the detection, tracking, and identification of objects in complex environments. These systems facilitate monitoring air, maritime, and terrestrial domains, providing real-time information for decision-making in applications such as air defense, air traffic control, and naval operations [5]. In response to the growing demand for precise and reliable data in dynamic, highly uncertain

W. J. Sarmiento and E. L. Sierra-Balln (Eds.): CCC 2025, CCIS 2692, pp. 80–93, 2027.
https://doi.org/10.1007/978-3-032-28447-1_6

scenarios, the correlation and analysis of information from multiple sensors have become notable challenges [11].

To address these challenges, digital twins have emerged as a suitable tool for evaluating radar systems. A digital twin is a virtual model that mirrors the behavior of a physical system, enabling the simulation of various operational scenarios and predicting future behaviors. In the context of radar sensors, these models make it possible to simulate and analyze the performance of individual or combined radars [1], which enhances data prediction and reduces the need for costly physical testing. Moreover, digital twins allow experimentation with advanced data fusion methods, enhancing information integration in critical scenarios.

This article proposes a simulation framework based on multi-agent systems (MAS) to address data fusion challenges from different types of radars. In this approach, we modeled a Fire Control Radar Agent (FCRA) and a Navigation Radar Agent (NRA) in a controlled maritime environment, together with the dynamic movement of an aircraft. Each radar is an autonomous agent within the system, providing data on position and speed [7]. Coordination among these agents seeks to improve detection and tracking accuracy, optimize information processing, and provide a more comprehensive view of the operational scenario.

Potential applications of this approach include naval personnel instruction and training, improvements in maritime surveillance and defense, and the optimization of air traffic control. The following sections discuss key concepts regarding digital twins in radar systems, their implementation through multi-agent systems, and the strategies used to improve data fusion and motion analysis within these environments.

2 Key Concepts and Technologies in Radar Digital Twins with Multi-agent Systems

Radar systems employed as digital twins have transformed instruction and training in civil and military applications, enabling real-time simulation and analysis without physical implementations. Combining digital twins, data fusion, and multi-agent systems has improved precision, efficiency, and safety in detecting and tracking targets.

2.1 Radar Systems

Radar systems are electronic devices that emit electromagnetic waves to detect, locate, and track objects. Depending on their application, we find different types of radars in maritime environments, such as:

- **Navigation Radar (NR)**: employed to prevent collisions and assist in safe navigation.
- **Surveillance Radar (SVR)**: used for monitoring and surveillance over wide areas.

- **Fire Control Radar (FCR)**: designed for target detection and high-precision tracking in defense systems.

Radar's operating principle emits electromagnetic waves and analyzes the echoes reflected by objects. We can determine the target's distance, speed, and direction from these echoes. In addition, we can evaluate the speed at which signals are transmitted and received to optimize detection and real-time tracking.

2.2 Digital Twins in Radar Systems

Digital twins enable the simulation of sensor operation under different operational scenarios. In the context of radar systems, they provide a platform for evaluating data fusion and detection strategies without physical testing [4].

Their integration with multi-agent systems improves the ability to manage large volumes of data. It optimizes detection in dynamic and complex situations. It facilitates experimentation with new sensor configurations and processing algorithms before implementing them in real systems [10].

2.3 Data Fusion

Data fusion combines information from multiple sources to generate a more accurate and coherent representation of the environment. In radar systems, this technique is essential for:

- Integrating data from various sensors, improving detection, and reducing uncertainty in target tracking [5,7].
- Optimizing signal processing through advanced algorithms that filter out interference and improve accuracy.
- Applying machine learning (ML) techniques to enhance data correlation and address high-dimensional problems, reducing the computational load of traditional simulations [6,13].

2.4 Multi-agent Systems (MAS)

Multi-agent systems (MAS) provide an efficient solution for coordinating multiple sensors and performing data fusion in radar systems. Their distributed approach allows autonomous agents to collaborate and provide an integrated view of the environment, enhancing information correlation and optimizing target detection and tracking [12].

MAS facilitates the modeling of complex systems through the interaction of independent agents, enabling the solution of distributed problems in dynamic environments [15]. Their application has been key in:

- **Simulation and analysis of naval radars**, improving threat detection and real-time response.

- **Optimization of air traffic control**, permitting the efficient management of multiple planes.
- **Automation of uncrewed units**, increasing accuracy and safety in automatic landings, even in adverse conditions [1,16].

Advances in digital twins, data fusion, and multi-agent systems have transformed radar systems, improving the detection and tracking of targets in complex operational scenarios [1,5]. These technologies have been fundamental in critical applications such as maritime navigation, fire control, and the management of uncrewed planes [7,12].

The following section presents the methodology used to design and validate each agent's roles, goals, and functions within the proposed system. This explanation provides a detailed view of the implementation process and its relevance in developing the simulation framework.

3 Design and Evaluation Strategies for Multi-agent Systems

The multi-agent system developed in this study simulates a plane's dynamic movement and detection by an NR and an FCR. The primary objective is to analyze radar systems' behavior in detecting and tracking planes and assess the impact of various factors on sensor accuracy.

3.1 Agent Identification and Definition

As the first stage of the design, the agents interacting within the system were identified (see Table 1), defining their skills, resources, interactions, and roles. This characterization is essential for structuring the tasks of plane detection and tracking.

3.2 Multi-agent System Goals

The main goals of the three agents (PA, NRA, and FCRA) are summarized in Table 2:

The following sections describe the agents created and their respective abilities, clearly understanding their roles and functionalities.

3.3 Skills and Capabilities of the Multi-Agent System

This section lists and describes the agents created and their respective abilities (see Table 3). The objective is to provide a clear and detailed view of each agent's roles and functionalities within the system, enabling an integrated understanding of their purpose and operations. Each agent plays a specific role in the digital radar ecosystem, contributing to the system's accuracy and effectiveness in detecting and tracking targets.

Next, we examine the data generation process based on dynamic systems and the interaction of each multi-agent element, highlighting how these components are relevant to the NR and FCR emulators.

Table 1. Summary of agents and their descriptions

Agent	Description
Plane Agent (PA)	Represents the plane in motion. It generates position and velocity data based on a flight model derived from dynamic equations. Its behavior depends on predefined parameters, including evasive maneuvers and programmed trajectories.
Navigation Radar Agent (NRA)	Simulates a navigation radar capable of long-range detection. It receives and processes the plane's reflected signals to determine its position. Its accuracy depends on distance and simulated environmental conditions.
Fire Control Radar Agent (FCRA)	Designed for high-precision target detection and tracking. It provides detailed data on the plane's speed and trajectory. It integrates motion prediction algorithms to enhance data fusion.

3.4 Dynamic Modeling and Detection of the Plane Agent

Dynamic systems enable the simulation of radar movement in space, the detection of objects, their tracking, the sending of position, velocity, and altitude data, and the inspection of contacts through the multiplication of the three rotation matrices, yielding the total rotation of the NR detection trace [8]. The system design considers two key aspects of the simulation:

- **Dynamic Movement of the Plane**: Modeled using differential equations representing flight dynamics, considering physical variables in stable environments [9]. The PA generates position and velocity data based on these equations, simulating changes in trajectory and speed. These data feed the radar systems, enabling precise tracking of the plane's movement.
- **Rotation Matrix for Detection Radar**: Applied to simulate radar sweeps and update the target's position within the detection trace. This method allows for analyzing localization accuracy and sensor responses to different motion patterns. The NRA processes reflected signals to determine the plane agent's position, adjusting for distance and environmental conditions. The FCRA integrates these data with motion predictions to track the plane, providing detailed information on its speed and trajectory.

Integration of Radars as Digital Twins: We integrated the PA dynamic data and the radar systems' detection capabilities as digital twins. This approach facilitates a comprehensive analysis of the plane movement and the radar's tracking accuracy. The NRA and the FCRA operate together to ensure continuous, accurate monitoring of the PA, thereby enhancing overall system performance and reliability.

Table 2. Summary of Tasks and Goals

Task	Description and Goal
Simulating Plane Movements	**Description:** The Plane Agent (PA) is responsible for modeling the aircraft's dynamic behavior in a virtual environment. It incorporates aerodynamic principles, environmental factors (e.g., wind resistance), and aircraft-specific parameters to generate realistic movement patterns. **Goal:** Simulate the following movements: forward, starboard (right), port (left), upward (increase in altitude), and downward (decrease in altitude).
Detecting the Plane's Movement and Parameters	**Description:**The Navigation and Radar Agent (NRA) performs detection of the aircraft's spatial and kinematic parameters. It uses simulated radar sweeps and sensor fusion to estimate position, velocity, and altitude with high fidelity. **Goal:** Determine the next movement position, the object's speed, and altitude.
Transmitting the Plane's Data	**Description:** The Flight Communication and Relay Agent (FCRA) ensures secure and structured transmission of flight data across the simulation network. It formats and relays telemetry, including position (XYZ), velocity (knots), and altitude (feet) to other agents and external systems. **Goal:** Transmit position (XYZ), speed (knots), and altitude (feet) data.
Contact Inspection and Tracking	**Description:** The NRA also performs continuous 360° surveillance to detect and classify external contacts. It applies tracking algorithms to monitor trajectories and assess potential threats or anomalies. **Goal:** Inspect and track contacts.

We carried out the PA motion simulation using dynamic systems, where the following Equations were employed to produce more realistic movements in the Plane Agent:

Table 3. Agents and Skills Overviews

Agent	Skills
Plane Agent	Move throughout any area in a continuous, smooth manner, adjusting its trajectory based on the environment's dynamic conditions. Simulate changes in speed and altitude in response to predefined evasion scenarios and maneuvers.
Fire Control Radar Agent	Detect the position of contact with high precision in combat or tactical surveillance environments. Track the previously detected contact, adjusting trajectory prediction based on detected changes in the object's speed and direction. Send information on the position, ensuring interoperability with other radar systems and data-analysis platforms. Apply filtering algorithms to enhance detection accuracy and reduce noise in the captured data.
Navigation Radar Agent	Detect the position of a contact under maritime navigation conditions, with the ability to adapt to changes in visibility and atmospheric disturbances. Track the previously detected contact, providing continuous data for information fusion with other sensors on the vessel. Implement interference suppression techniques to minimize false alarms and improve system reliability in adverse operational environments.

$$\dot{u} = rv - qw - g\sin(\theta) + \frac{1}{m}(T_x - D) \tag{1}$$

$$\dot{w} = qu - pv + g\cos(\theta) + \frac{1}{m}L \tag{2}$$

$$\dot{q} = \frac{1}{I_y}(M - (I_z - I_x)pr - I_{xz}(p^2 - r^2)) \tag{3}$$

$$\dot{\theta} = q \tag{4}$$

Where:

- u: Eq. 1 represents the forward velocity.
- w: Eq. 2 represents the vertical velocity.
- q: Eq. 3 is the rate of change of the pitch angle.
- θ: Eq. 4 is the angle of attack (related to the pitch angle).
- r: Angular velocity around the vertical axis.
- p: Angular velocity around the longitudinal axis.
- g: Acceleration due to gravity.

- m: plane mass.
- T_x: Thrust in the x direction.
- D: Aerodynamic drag.
- L: Lift force.
- I_y: Moment of inertia around the lateral axis.
- M: Moment applied around the lateral axis.
- I_z: Moment of inertia around the vertical axis.
- I_x: Moment of inertia around the longitudinal axis.
- I_{xz}: Product of inertia.

We obtained more accurate movements in the given environment using the Runge-Kutta 4 (RK4) numerical method to solve ordinary differential equations (ODEs) and discretize these equations.

In the NRA, we based the movement system on rotation matrices. We used the following rotation matrices to simulate the radar's movement in three-dimensional space. These matrices allow the rotation of a vector in space around the X, Y, and Z axes.

Rotation around the X axis (Rx):

$$R_x(\theta_x) = \begin{bmatrix} 1 & 0 & 0 \\ 0 & \cos(\theta_x) & -\sin(\theta_x) \\ 0 & \sin(\theta_x) & \cos(\theta_x) \end{bmatrix}$$

Rotation around the Y axis (Ry):

$$R_y(\theta_y) = \begin{bmatrix} \cos(\theta_y) & 0 & \sin(\theta_y) \\ 0 & 1 & 0 \\ -\sin(\theta_y) & 0 & \cos(\theta_y) \end{bmatrix}$$

Rotation around the Z axis (Rz):

$$R_z(\theta_z) = \begin{bmatrix} \cos(\theta_z) & -\sin(\theta_z) & 0 \\ \sin(\theta_z) & \cos(\theta_z) & 0 \\ 0 & 0 & 1 \end{bmatrix}$$

Total rotation, obtained by multiplying the three rotation matrices:

$$R = R_z(\theta_z) \cdot R_y(\theta_y) \cdot R_x(\theta_x)$$

These methods make it possible to simulate the radar's movement in space, detect and track objects, send position, velocity, and altitude data, and inspect contacts by multiplying the three rotation matrices, obtaining the total rotation of the NR detection trace.

3.5 Dynamic Data Generation

The general procedure and the interaction among the agents were shown to generate the data required for the NR and FCR emulators. These components were implemented using the GAMA Platform [12], an integrated development environment (IDE) framework designed for creating and simulating multi-agent systems. This platform was particularly effective for modeling scenarios involving both cooperation and competition among agents, offering robust tools for spatial reasoning, behavioral scripting, and real-time data visualization.

The dynamic data generation process was driven by continuous feedback among agents: The Plane Agent (PA) initiated movement patterns based on predefined aerodynamic models, grounded in classical principles of flight dynamics and control [2], the Navigation and Radar Agent (NRA) detected and tracked positional data in real time, and the Flight Communication and Relay Agent (FCRA) transmitted telemetry across the network. This architecture supports synchronized simulation and aligns with recent efforts to virtually validate GNSS signal reception under jamming, highlighting the value of agent-based environments for testing sensor resilience [14].

Figure 1 depicts the multi-agent process diagram describing the interaction and coordination between the two radars in the system. This diagram presents each radar as an autonomous agent performing specific tasks, such as object detection, trajectory tracking, and information exchange with other agents, to enhance the overall accuracy and efficiency of the system. The modularity of this approach allows for scalability and adaptability, making it suitable for future integration with real-world naval platforms and sensor fusion architectures.

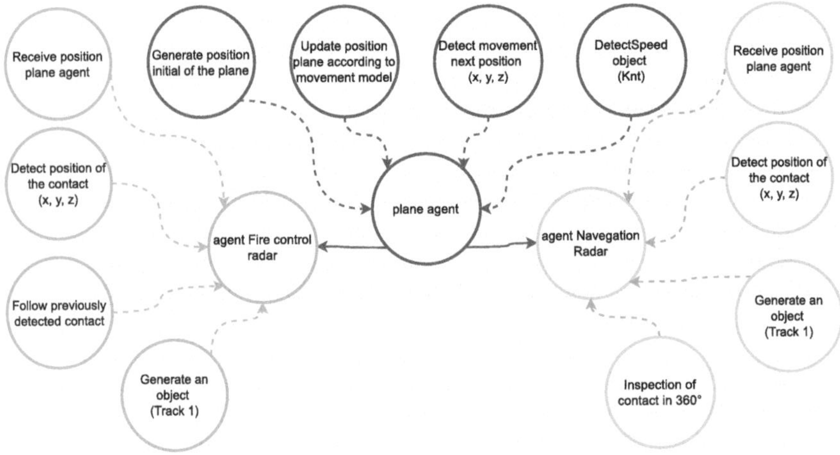

Fig. 1. General Multi-Agent Skills and Capability Model.

The multi-agent system allowed each radar to act as an independent agent while collaborating with the other to ensure continuous detection and tracking of objects within its operational area.

The following section presents the results obtained from implementing these methodologies and techniques, highlighting the use of multi-agent systems for evaluating radar dynamics on the Colombian National Navy's military vessels.

4 Results

The results obtained in this application stemmed from the programming of the Plane's dynamic motion equations and their integration with the developed radar agents. Through various figures and analyses, the simulated Plane's behavior was illustrated, along with the effectiveness of the radar agents in detecting, tracking, and communicating data. These outcomes demonstrated the system's ability to emulate realistic plane movements and the efficient interaction among the different agents, thus providing a solid foundation for future research and improvements in Naval Radar environments.

By implementing the dynamic motion equations of a plane, stated in Eqs. 1, 2, 3, and 4, we achieved a simulation of movements in a configured environment. This movement is shown in Fig. 2 b, where the red point indicates the contact's motion and the blue point denotes the system's center (the "own ship") that detected said movement. We can also observe the trace of the previous ten positions.

Upon obtaining the plane position and transferring that data to the Navigation Radar Agent and the Fire Control Radar Agent, distinct behaviors under different detection capabilities were observed [5].

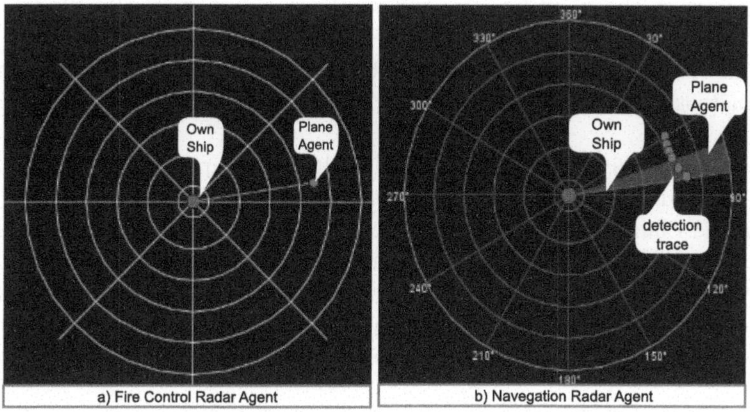

Fig. 2. Result NR and FCR.

In Fig. 2 b, we observed the detection of a contact when the Navigation Radar agent completed a 360° rotation in four seconds. It made it possible to reveal both the behavior and detection of contacts and the inefficiency in plotting a contact exhibiting rapid movements.

Using the Fire Control Radar enabled constant tracking, as illustrated in Fig. 2 a. The contact was locked onto irrespective of its maneuvers since the radar did not rotate continuously but instead tracked the mentioned target [3].

Once the agents were developed and initialized, the next step was establishing efficient communication between the Plane Agent and the Radar Agent, as shown in Fig. 3.

Table 4. Data frame of Plane Agent (X, Y, Z)

Description	X	Y	Z
New position:	46.06302689372297	43.64658275891241	10.015000000000002
New position:	45.64513746370388	42.56893365730817	10.020000000000003
New position:	45.352130094215546	41.553265804612785	10.025000000000004
New position:	45.17397823259887	40.59466571996685	10.030000000000005
New position:	45.1014579425137	39.68861461468774	10.035000000000005
New position:	45.12608369497559	38.83095681450772	10.040000000000006
New position:	45.2400492960492	38.017870708053	10.045000000000007
New position:	45.43617354026581	37.24584201946435	10.050000000000008

The primary objective of this interaction was to ensure that the Plane Agent continuously transmitted the detected contact's position data to the Radar Agents (see Table 4). It allowed the radars to conduct contact detection in parallel, optimizing the identification and tracking processes of the Plane or any given target.

In this manner, we transmitted contact location details and synchronized detection tasks, allowing a faster and more precise response to any required situation.

These results highlighted the use of multi-agent approaches in simulating movements and detecting contacts within Naval Radar environments. The emulation of the Plane's movements and their transfer to the NR and FCR agents allowed precise observation of the differences in detection and tracking capabilities between the two systems [14]. As shown in the figures, integrating communication among the agents underscored the importance of continuous synchronization and cooperation for optimizing the identification and tracking of moving targets.

These findings lead to the following section, which discusses the study's conclusions and emphasizes potential contributions and future work based on the obtained results.

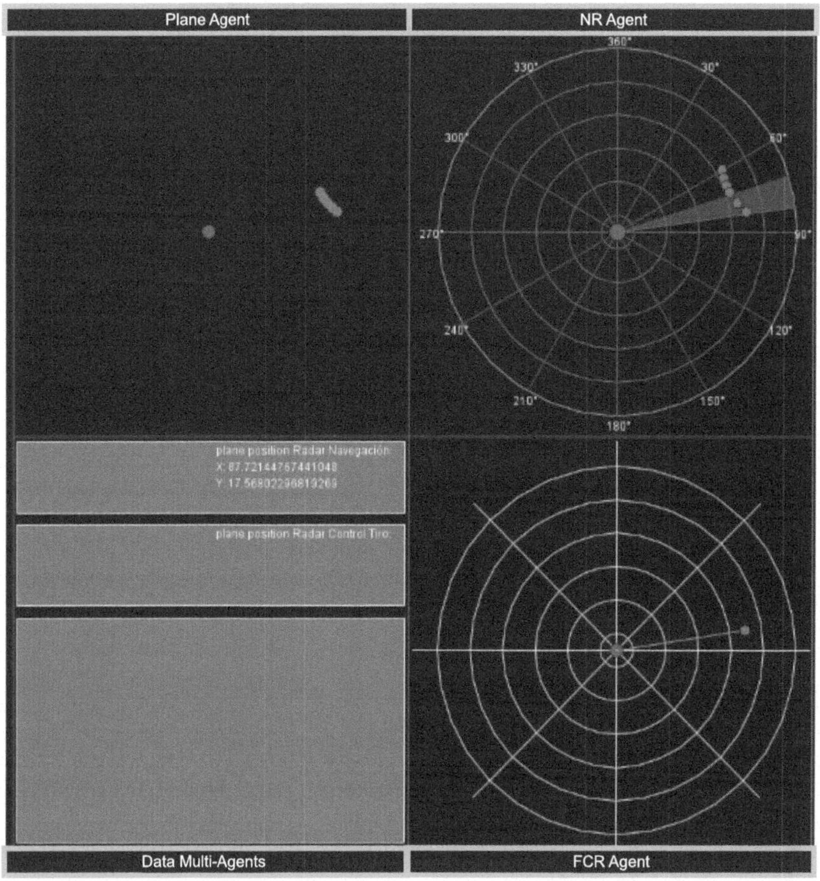

Fig. 3. Communication between Agents.

5 Conclusions

Using multi-agent systems proved an effective strategy for enhancing data compensation and coordination among radars. The collaboration among autonomous agents led to more precise detection and tracking of dynamic targets, optimizing the system's ability to operate in complex and varied environments. This flexible approach facilitated the adaptation and exploration of diverse scenarios and conditions.

Integrating digital twins and dynamic models was key to achieving high-fidelity simulation of radars and contacts. Nevertheless, although many users adopted the GAMA Platform internationally, its primary challenge was not local adoption but rather the lack of accessible information online. Like other multi-agent simulation tools, such as NetLogo or AnyLogic, this limits the availability of documentation, tutorials, and support forums. Although references existed, finding and understanding them was a challenge, which complicated the implementation and optimization of the models. Despite these limitations, the simulations highlighted the value of these approaches in improving the efficiency and reliability of the developed systems.

This work opened up new avenues for future research, suggesting the incorporation of advanced techniques, such as machine learning, to optimize movement prediction and target discrimination. Furthermore, we recommended broadening experimental validation in real-world environments and exploring the integration of additional sensors, including optical cameras or surveillance radars, which could enrich the data fusion process and enhance the system's accuracy and robustness.

Acknowledgment. This research was supported through a strategic collaboration between the Colombian National Navy and the Universidad Tecnológica de Bolívar. The validation and technical review of the simulation framework were carried out with the support of the Centro de Desarrollo Tecnológico Naval (CEDNAV), whose contributions were essential in aligning the system's capabilities with operational requirements. We also acknowledge the academic leadership and guidance provided by UTB professors, whose expertise in modeling, simulation, and systems integration played a key role in the development of this work.

References

1. Alatorre, A., Carino, J., Castillo, P., Lozano, R.: Dynamic trajectory for landing an aerial vehicle on a mobile. IFAC PapersOnline (2022). https://doi.org/10.1016/j.ifacol.2022.07.597
2. Etkin, B., Reid, L.D.: Dynamics of flight - stability and control. John Wiley & Sons Inc, New York (1996)
3. Green, W., Oh, P.: A mav that flies like an airplane and hovers like a helicopter. In: Advanced Intelligent Mechatronics Proceedings, 2005 IEEE/ASME International Conference on, pp. 693–698 (2005)
4. Kumar, A., Singh, S.: Design and simulation of autopilot control system for stratospheric airship. J. Aerosp. Eng. (2023). https://doi.org/10.1016/j.jaeroeng.2023.123456
5. Kumar, M., Mondal, S.: Recent developments on target tracking problems: A review. Ocean Eng. **236**, 109558 (2021). https://doi.org/10.1016/j.oceaneng.2021.109558
6. Li, J., Du, X., Martins, J.R.R.A.: Machine learning in aerodynamic shape optimization. Prog. Aerosp. Sci. (2022). https://doi.org/10.1016/j.paerosci.2022.100849
7. Lim, Y., Gardi, A., Sabatini, R., et al.: Avionics human-machine interfaces and interactions for manned and unmanned planes. Prog. Aerosp. Sci. (2018). https://doi.org/10.1016/j.paerosci.2018.05.002
8. Liu, L., Wang, Y., Zhang, Z.: A review of control strategies used for morphing plane applications. Aerosp. Sci. Technol. **112**, 106625 (2021). https://doi.org/10.1016/j.ast.2021.106625
9. Milne, B., Samson, S., Carrese, R., Gardi, A., Sabatini, R.: High-fidelity dynamics modelling for the design of a high-altitude supersonic sounding rocket. Designs **7**(1), 32 (2023). https://doi.org/10.3390/designs7010032

10. Partovi, A.R., Wang, X., Lum, K., Lin, H.: Modeling and control of a small-scale hybrid plane. In: Proceedings of the 18th World Congress The International Federation of Automatic Control,Milano, Italy (2011)
11. Piacenza, I.A., Giulietti, F., Avanzini, G.: Inverse simulation of unconventional maneuvers for a quadcopter with tilting rotors. IFAC PapersOnline (2013). https://doi.org/10.3182/20131120-3-FR-4045.00035
12. Salazar-Cruz, S., Lozano, R., Escarco, J.: Stabilization and nonlinear control for a novel tritrotor mini-plane. Control. Eng. Pract. **17**(8), 886–894 (2009). https://doi.org/10.1016/j.conengprac.2009.02.013
13. Scardaoni, M.P., Magnacca, F., Massai, A., Cipolla, V.: Plane turnaround time estimation in early design phases: simulation tools development and application to the case of box-wing architecture. J. Air Transp. Manag. (2021). https://doi.org/10.1016/j.jairtraman.2021.102122
14. Tripathi, V., Caizzone, S.: Virtual validation of in-flight GNSS signal reception during jamming for aeronautics applications. Aerospace **11**(3), 204 (2024). https://doi.org/10.3390/aerospace11030204
15. Weiss, G.: multiagent systems: a modern approach to distributed artificial intelligence, MIT Press (2000)
16. Zhang, J., Hu, H., Zhang, G., Zhang, Z., Wang, Z.: Deep reinforcement learning-based ditching vertical velocity control for a multi-fault wing-in-ground craft. Appl. Ocean Res. **136**, 103585 (2023). https://doi.org/10.1016/j.apor.2023.103585

A4SGD: Accessibility for Serious Game Design – A Conceptual Framework

Ismar Frango Silveira(✉) and Valéria Farinazzo Martins

Mackenzie Presbyterian University, São Paulo, Brazil
{ismar.silveira,valeria.martins}@mackenzie.br

Abstract. Serious games are games developed with purposes beyond entertainment, including applications in education, health, professional training, rehabilitation, and social awareness. The design of such games presents unique challenges that go beyond those of commercial entertainment games. Designers must balance ludic appeal with instructional goals, simplify complex topics without distorting their meaning, and often work under limited resources and tight development schedules. Moreover, accessibility remains an often-overlooked dimension, despite its critical importance in ensuring that users of all abilities can benefit from serious games, particularly in educational and therapeutic contexts. In this context, the present paper discusses the need for accessible serious game design and introduces a conceptual framework, A4SGD (Accessibility for Serious Games Design), that outlines key considerations and strategies for integrating accessibility into the design process from the outset. The framework is then applied to a set of case studies to evaluate its practical relevance and potential for guiding inclusive, effective, and engaging serious game development.

Keywords: Accessibility · Serious games · Game Design

1 Introduction

Serious games are games designed for purposes beyond entertainment. Examples of their applications include education, health, training, social awareness, and rehabilitation, among others. They use the motivational, ludic and engaging power of games to support learning and promote behavior change. This definition comes from the early work of Abt [1], as mentioned by Wilkinson [2], and such terminology has grown among game developers, teachers and, therapeutics and policy makers, to cite a few.

The design process of serious games encounters the same obstacles as that of entertainment games, along with some additional challenges. One of them is how to balance between fun and serious goals: it's well-known how challenging is to make a game that is both ludic, enjoyable and educational [26]. Besides, it is not straightforward that having fun while playing a serious game ensures learning [25]. Too much focus on educational content can make the game feel boring and uninspiring, while focusing too much on fun can water down the learning goals. In this sense, designers must be able to carefully integrate educational content without compromising the game's enjoyment or flow.

W. J. Sarmiento and E. L. Sierra-Ballén (Eds.): CCC 2025, CCIS 2692, pp. 94–111, 2027.
https://doi.org/10.1007/978-3-032-28447-1_7

Another key point relies on the complexity of the serious context: some subject matter is difficult to translate into engaging gameplay. Simplifying complex topics without distorting their meaning is a fine and dangerous line. Also, if compared to commercial entertainment games, serious games usually suffer with limited budgets and shorter development cycles, especially in educational or research-driven projects. These factors make the design of serious games a uniquely complex and multidisciplinary task.

Besides, like all digital (and non-digital) products, serious games can unintentionally (or no) exclude players with disabilities if accessibility is not a point of fully consideration from the scratch [23, 24]. Making serious games accessible ensures that all users, regardless of their physical, sensory, or cognitive abilities, can benefit from them. This is especially important when these games target learning, professional development, or therapy, among others.

Considering all these aspects, one important question that must be posed is: how can serious games become accessible? Without claiming to provide a comprehensive answer to this complex question, the following section presents a few key points to aid the understanding, presenting a conceptual framework named A4SGD (Accessibility for Serious Games Design), which underpin key aspects of the game design process in the context of serious games, having accessibility issues in mind. Afterwards, some case studies are presented and their adherence to this framework is studied.

2 Accessibility for serious games design: The A4SGD Framework

Taking into account the importance of accessibility in the process of designing digital games in general, and particularly serious games, there is a need for conceptual frameworks that can guide designers, developers, and researchers in embedding accessibility from the very beginning of the design process. Such frameworks help structure decision-making, encourage inclusive design practices, and support the evaluation of accessibility features in a systematic and scalable way. By framing accessibility as a core principle rather than a simple add-on or a reactive fix, conceptual models can foster more equitable, adaptable, and effective serious games which can truly fulfill their mission to educate, treat, heal, train, or inform, without exclusion.

Related works can be seen in [19], which brought a literature review on accessibility in web-based serious games; or [13], which proposes two frameworks to guide teams in reviewing the issues involved in accessibility and related aspects. There are more specific approaches, focusing on one kind of device, purpose or impairment. Some examples include the work [20], which proposed a framework for improving accessibility of serious games using augmented reality in handheld devices; some approaches are more specific, or [21], which focuses on physical rehabilitation, and the work of [22], which deals with serious games for low-vision children.

A4SGD is a conceptual framework that delineates the key aspects of the design process for serious games having Accessibility on mind. It provides structured guidance for integrating accessibility aspects into the creation of educational, therapeutic, or training-oriented games, emphasizing adaptability, accessibility, and pedagogical relevance throughout all stages of design and development. Based on [3–6], Fig. 1 depicts the seven key points of A4SGD framework, which are explained afterwards:

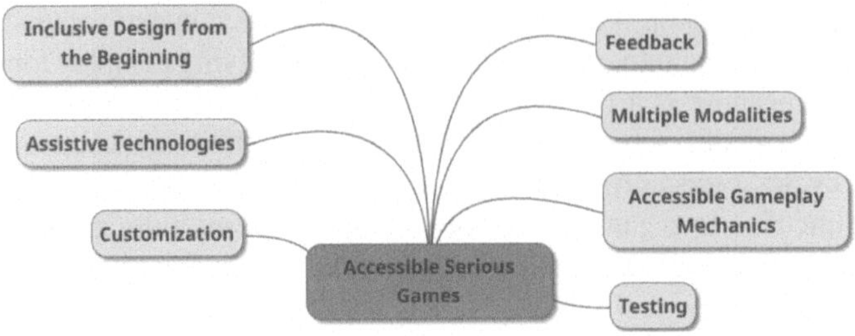

Fig. 1. The A4SGD framework

Whilst these considerations are also present in commercial entertainment games, they take on particular significance, special colors and unique nuances when applied to serious games. In this context, the need to balance ludic experiences and engagement with educational or behavioral objectives adds significant layers of complexity to design choices. Designers must ensure that accessibility, inclusivity, and user experience not only support enjoyment but also effectively promote learning outcomes or desired behavioral changes for a diverse audience, making these elements even more critical. The points depicted in Fig. 1 are expanded in the following subsections.

2.1 Inclusive Design from the Beginning

Accessibility should be embedded in the design phase, as an integral part of it, never added later as a plus or a patch [13, 14]. Early inclusion means working with users with disabilities, involving accessibility experts (meaning that co-design strategies are desirable), and following Universal Design (UD) principles. Early integration of accessibility ensures that the entire game experience (involving mechanics, interface, narrative, and feedback systems, among others) is conceived with the diverse needs of all players in mind. Achieving this requires actively engaging users with disabilities throughout the design and testing stages, as well as collaborating with accessibility specialists to identify potential barriers and opportunities. This leads to codesign approaches, in which users and designers work together iteratively, are particularly effective in this regard. Furthermore, adherence to UD principles helps to create solutions that are flexible and usable by the widest possible audience, promoting both equity and quality in serious games.

Some practical key elements for this point can be seen at Fig. 2, with things to consider and implement during the accessible serious game design process:

The following items detail the points of Fig. 2.

a) **Involve Users with Disabilities:** this includes to conduct interviews, workshops, requirements elicitation and codesign sessions with this audience, which should be wide in terms of the impairments to be supported. Besides, interactive tests must be performed with a diverse audience.
b) **Consult Accessibility Experts:** specialists should be engaged during concept and prototype phases. Accessibility tests must be also performed throughout development.

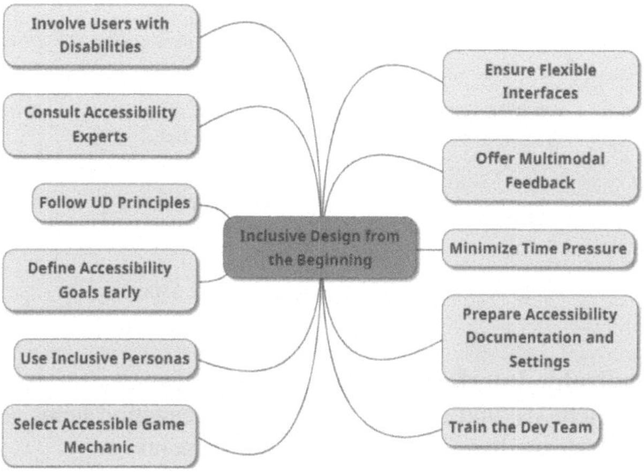

Fig. 2. Key aspects of Inclusive Design from the Beginning

c) **Follow UD Principles**: avoid the one-size-fits-all approach, designing for flexibility, forecasting multiple ways of interaction.
d) **Define Accessibility Goals Early:** set clear objectives for accessibility from project kickoff, and document all accessibility requirements in the Game Design Document (GDD)
e) **Use Inclusive Personas:** propose player personas that reflect a wide range of abilities, including sensory, motor, and cognitive diversity.
f) **Select Accessible Game Mechanics:** prioritize those mechanics that can be adapted to different abilities. Also provide different options for each game mechanics (e.g., turn-based play, adjustable difficulty, input remapping).
g) **Ensure Flexible Interfaces:** give support to alternative input devices (switches, eye-tracking, adaptive controllers). Also provide scalable UI elements (adjustable text size and contrast, zoom able menus, colorblind-friendly palettes, resizable icons, dyslexia-friendly fonts, among others).
h) **Offer Multimodal Feedback:** provide feedback via text, audio, and visual cues. Give support to subtitles, captions, and audio descriptions
i) **Minimize Time Pressure:** allow players to pause, extend, or slow down gameplay. Also avoid mandatory fast reflex actions unless alternatives exist to it.
j) **Prepare Accessibility Documentation and Settings:** provide an Accessibility Options menu and clearly communicate accessibility features in game descriptions and tutorials
k) **Train the Dev Team:** indoctrinate the team about creating awareness about accessibility standards (e.g., WCAG, Game Accessibility Guidelines). Encourage an inclusive mindset in all design decisions.

2.2 Assistive Technologies

Serious games should be compatible with screen readers, alternative input devices, and voice control, so players with limited mobility or vision can interact with them [11, 12]. Figure 3 synthesizes these aspects in a mindmap.

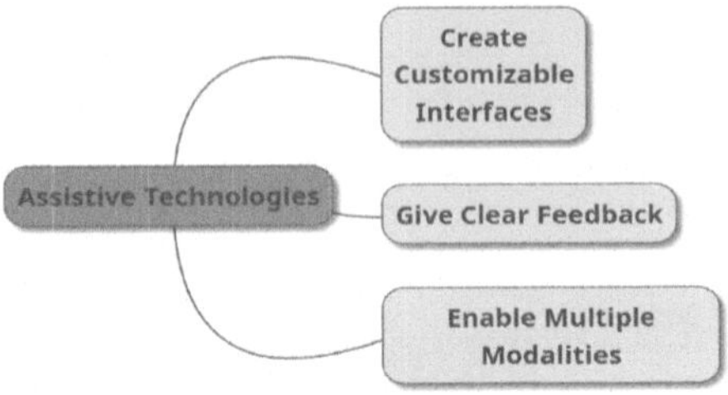

Fig. 3. Assistive Technologies - key aspects

a) **Create customizable Interfaces:** Allow players to adjust text size, contrast, audio settings, input methods, and control schemes.tre Adaptation helps address diverse needs, such as dyslexia, low vision, or motor impairments.
b) **Give Clear Feedback**: Accessible games should minimize the mental effort required to interpret what's happening. Clear feedback—such as explicit indicators of success, failure, or required actions—helps players stay oriented without confusion, which is particularly helpful for players with learning disabilities or neurodivergence, for instance.
c) **Enable Multiple Modalities**: Key information should be presented in more than one way—visual, auditory, and haptic—to ensure it's perceivable by everyone

2.3 Customization and Adaptation

Digital games can be made more accessible, inclusive, and engaging by implementing mechanisms that either allow player-driven configuration (customization) or enable the system to intelligently respond to player needs (adaptation). When dealing with serious games, they present distinct goals and user profiles that make customization and adaptation especially critical. These approaches serve different functions, but both aim to optimize the player's experience across ability levels and contexts of use.

At this point, a discussion about the differences between Customization and Adaptation is required. Game customization is different from Adaptation: both are strategies to enhance player experience and accessibility in serious games, but they differ fundamentally in terms of control, implementation, and responsiveness. Customization supports preference, while adaptation responds to need. Whilst customization is always player-initiated, adaptation involves the system automatically adjusting game mechanics, interface elements, or difficulty based on the player's behavior, performance, or detected needs, often using AI-based strategies.

Customization empowers player autonomy and supports a wide range of needs without requiring system intelligence. However, it requires players to be aware of their needs, besides being capable of navigating through settings, which may not be feasible for all users, especially those with cognitive impairments or first-time players. On the other hand, Adaptation can proactively support players without requiring their intervention; it is especially useful for players who may not know what settings they need. Nonetheless, it brings an important risk of misinterpretation, reduced transparency, and loss of player agency if poorly implemented. The key elements for Customization and Adaptation are:

a) **Control Remapping:** players should be able to reassign input commands to accommodate preferred layouts or support assistive technologies, according to physical and cognitive characteristics.
b) **Adjustable Difficulty Levels:** customizable challenge settings help accommodate cognitive diversity and allow players to progress at their own pace without frustration or disengagement. Also, this adjustability could be achieved by means of Adapatation, through the use of DDA (Dynamic Difficulty Adjustement) [16].
c) **Scalable User Interface:** the option to resize, reposition, or simplify UI elements (menus, HUDs - Heads-Up Displays, text boxes, among others) is essential for players with visual impairments or attention-related conditions.
d) **Text and Color Customization:** options for font size, style, spacing, and contrast improve readability for users with dyslexia or low vision. The same way, providing color palette adjustments or filters ensures that players with various forms of color vision deficiency can distinguish key game elements.
e) **Audio Settings:** independent volume controls for music, effects, and voice, as well as the option to enable subtitles or visual sound indicators, benefit players who are deaf, hard of hearing, or neurodiverse. Adaptive audio techniques can also be used to improve immersion and adjust mood [17, 18].
f) **Gameplay Speed Control:** Allowing players to slow down or pause certain game elements can help individuals who need more time for input, processing, or decision-making. This can be achieved by customizing controls or in automatic, adaptive ways.
g) **Alternative Input Methods:** Support for different devices (e.g., touch, switch controls, eye tracking) ensures the game can be accessed through various means, depending on the player's motor skills and assistive technologies.
h) **Customizable Game Avatars and Representations:** Enabling players to create avatars that reflect their identity, including visible disabilities (e.g., prosthetics, mobility aids), promotes inclusion and self-expression, which is especially important for serious games, mainly rehab and educational games.
i) **Checkpoint Flexibility:** Players should be able to save progress frequently and return to the game without penalty, supporting those who may need breaks due to fatigue or external interruptions.

Figure 4 depicts these aspects related to Customization and Adaptability:

In a nutshell, customization respects user autonomy and self-knowledge, while adaptation ensures continuous support and challenge. Customization guarantees the game is usable from the start, and adaptation guarantees continued accessibility as user needs change over time or during gameplay. In serious games, Customization helps learners feel in control and reduces cognitive load. Adaptation ensures the learning curve is balanced, offering the right challenge level to maintain engagement and prevent dropout. Thus, together they can foster engagement, equity, and meaningful outcomes, which are pillars for the success of serious games.

2.4 Feedback

Feedback is a key component of games in general, as well as in education and rehabilitation. With serious games, whose purposes vary from learning and skills development

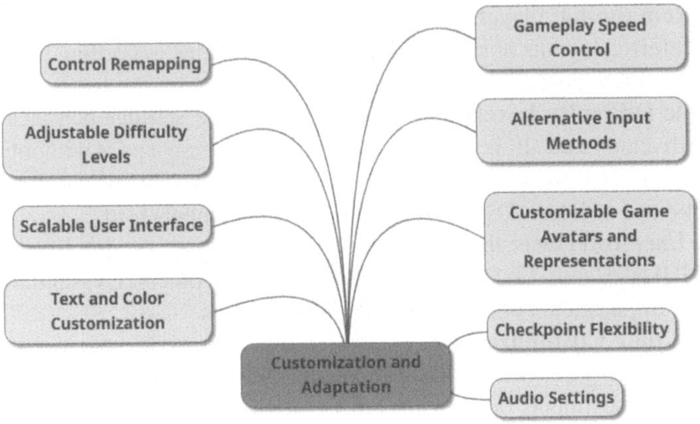

Fig. 4. Key aspects for Customization and Adaptation in Serious Games

to rehabilitation and behavior change, its importance is even greater. In every game, having feedback is essential to the player's experience, but in serious games, it becomes a pedagogically and functionally central element. In serious games, feedback is not only an element of gameplay mechanics but also a critical learning and motivational tool. Well-designed feedback bridges the gap between game mechanics and learning (or other serious) objectives, guiding users toward desired outcomes while maintaining engagement, accessibility, and efficacy.

Here are the key aspects for Feedback regarding accessibility in Serious Games Design (Fig. 5):

a) **Multimodal channels**: feedback must be tailored to individual needs, which ensures that players with sensory or cognitive impairments can perceive and respond to in-game events.
b) **Context-sensitiveness:** Feedback should be relevant to the player's context and skill level. It must adapt to their performance without overwhelming or distracting them.
c) **Clarity and Simplicity**: clear, unambiguous feedback is always required, besides being easy to interpret, avoiding complex visual or verbal content that may overwhelm users with cognitive or learning disabilities.
d) **Well-timed responses**: real-time feedback is essential for flow keeping in games. Especially in serious games, instant reinforcement helps players understand cause-effect relationships to make decisions and perform rapid corrections.
e) **Consistency and Predictability:** feedback patterns should be internally consistent, helping players learn through repetition and predictability. On the other hand, consistency supports neurodivergent players who rely on stable, structured environments.
f) **Constructiveness**: instead of giving feedback that some task has failed, it must be pointed out how to proceed and what to improve. Feedback should always focus on encouragement, suggestions, and growth, rather than punishment or frustration. This is a key point for players with low self-confidence or learning impairments.

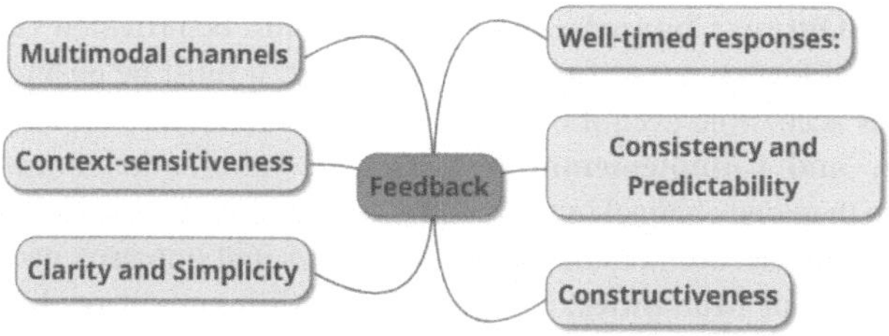

Fig. 5. Key aspects for Feedback in Designing Accessible Serious Games

Effective feedback in accessible serious games must address diverse players' needs, align with learning or therapeutic goals, and adapt to varying interaction styles. When well-implemented, feedback becomes a powerful enabler of inclusive, meaningful, and successful gameplay experiences.

2.5 Multiple Modalities

In the context of accessible serious game design, the use of multiple modalities (which refers to the integration of diverse input and output channels) plays an important role in ensuring that games are inclusive and usable by all individuals, with varied abilities. Such a multimodal approach allows players to interact with and receive feedback from the game through different sensory and interaction pathways, including visual, auditory, haptic, speech-based or brain-computer interfaces (BCIs), among others.

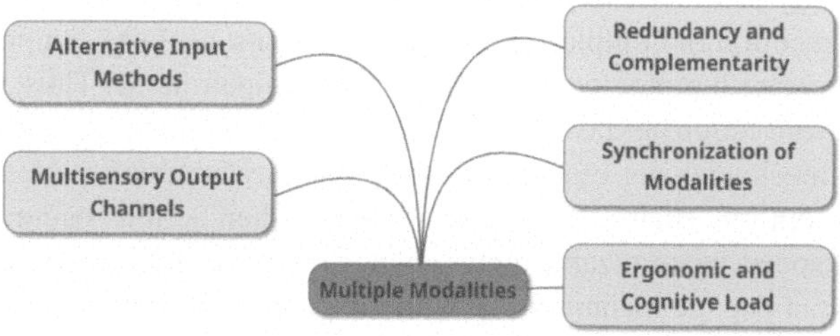

Fig. 6. Key elements of Multiple Modalities

By providing redundant, complementary interaction channels, multimodal systems increase the robustness and accessibility of the game experience. In serious games, multiple modalities also support better learning outcomes, as they align with diverse cognitive realities and reduce barriers to engagement, immersion, agency, and effectiveness, which are key elements of successful serious games. These are the key aspects for this, depicted at Fig. 6.

a) **Alternative Input Methods:** serious games must consider give support for voice commands, gesture recognition, eye-tracking, switch devices, and BCIs. This would enable users with motor impairments or limited dexterity to interact with the game.

b) **Multisensory Output Channels:** game feedback must be provided via visual, auditory, haptic, and textual cues, if possible. Moreover, it must be ensured that critical information is accessible even if one modality is unavailable.
c) **Redundancy and Complementarity:** The same information must be delivered through more than one channel (e.g., sound + subtitles + vibration). This guarantees accessibility while accommodating user preferences and impairments.
d) **Synchronization of Modalities:** feedback across different channels must be well-timed and coordinated, to avoid confusion. This is important for maintaining immersion and consistency.
e) **Ergonomic and Cognitive Load:** While multiple modalities can enhance accessibility and inclusiveness, they also introduce the risk of cognitive and sensory overload if not designed carefully. In serious games, where players may be engaged in complex tasks for learning, training or rehabilitation, for instance, the simultaneous use of visual, auditory, and haptic channels must be carefully orchestrated to avoid overwhelming the user.

When dealing with multiple modalities for serious games, the key is to balance variety with simplicity: deliver essential information using diverse channels, but avoid excesses. Multiple modalities should complement each other and align with the user's context, task, and preferences. Usability testing with diverse users is vital to fine-tune this balance.

2.6 Accessible Gameplay Mechanics

Gameplay mechanics, whose definition apparently has no accepted standard within the broader community [10], should be designed to avoid requiring rapid reflexes, precise timing, or the execution of complex simultaneous actions, as these can pose significant barriers for players with motor, cognitive, or sensory impairments. Thus, Game Design is supposed to avoid requiring fast reflexes or complex simultaneous actions.

When such mechanics are essential to the gameplay experience, designers should offer alternative options such as turn-based modes (when actions occur once a time), adjustable game speed, customizable input schemes, or extended time windows, to ensure that all players can engage meaningfully with the game, which gains more importance when dealing with serious games.

Turn-based modes in games refer to a gameplay style where players take turns performing actions, instead of acting in real time. In this mode, the game pauses between turns, allowing each player (or the system or some AI-based character) to think, plan, and execute actions at their own pace.

Incorporating these options supports broader inclusivity and aligns with UD principles, enhancing both usability and enjoyment for diverse audiences. The key points for Accessible Gameplay Mechanics are thus depicted in Fig. 7:

For a better understanding, the three points presented in Fig. 7 are better explained as follows:

a) **Sequential Play**: One player (or AI) acts at a time, followed by the next.
b) **No Time Pressure**: Players can take as long as needed to decide on their next move or action.

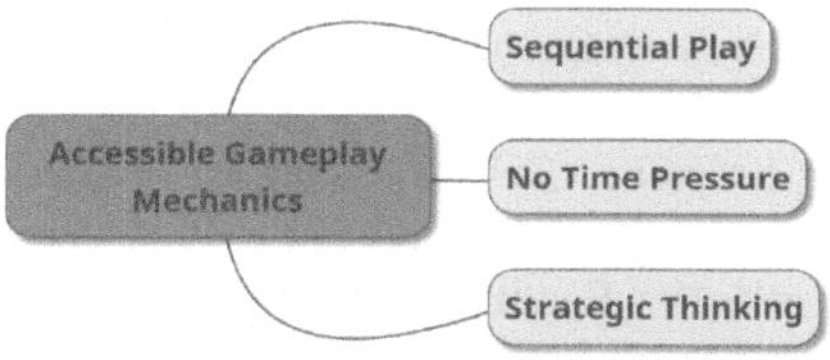

Fig. 7. Key points of Accessible Gameplay Mechanics

c) **Strategic Thinking**: Emphasis is on planning and decision-making rather than speed or reflexes.

All of these features would reduce pressure from players with impairments, allowing the use of specific assistive technologies without rush. These aspects also allow the use of asynchronous play, where users engage at their own pace and schedule.

2.7 Testing with Diverse Players

Iterative testing with users having various disabilities ensures true, real-world accessibility—not just theoretical compliance which is obtained relying solely on standards or guidelines. Only direct engagement with users can reveal practical barriers and usability issues [8, 9]. Involving a wide range of players leads to a deeper understanding of different needs and contributes to a more inclusive and equitable game design.

These are some key aspects to be considered (Fig. 8):

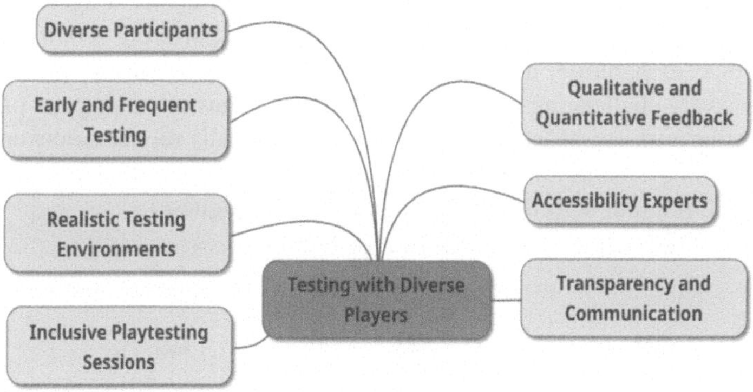

Fig. 8. Aspects involving testing with distinct participants

a) **Diverse Participants**: include in the test's users with a variety of disabilities (visual, hearing, motor, cognitive, neurodivergent). According to the expected audience, also consider to engage users from different age groups, cultural backgrounds, and levels of gaming experience.
b) **Early and Frequent Testing**: begin testing accessibility features during early prototypes, conducting multiple rounds of testing throughout development (iterative testing).

c) **Realistic Testing Environments:** test on the devices and assistive technologies players actually use.
d) **Inclusive Playtesting Sessions:** Design sessions so testers can comfortably give honest feedback, with the appropriate support for the testers.
e) **Qualitative and Quantitative Feedback:** Use surveys, interviews, and observation to gather insights; record success and fail rates, time to complete tasks, cognitive gains and player satisfaction. In the Iterative process, systematically track issues and resolutions at each iteration. Re-test after changes to verify improvements.
f) **Accessibility Experts**: make a partnership with specialists in accessibility to guide test design and interpret results. Also validate findings against established accessibility standards (such as WCAG or the Game Accessibility Guidelines [7]).
g) **Transparency and Communication**: Clearly communicate accessibility features and limitations to players

3 Case Studies

Here are presented three case studies of serious games focused on specific audiences. Afterwards, the A4SGD framework will be applied over them to verify the adherence to the frameworks' aspects.

3.1 Serious Game for the Literacy of Children with Dyslexia

The game [27] is a digital application developed for mobile devices with internet access, designed to support the literacy process of children with dyslexia aged 4 to 8. The game combines playful elements with accessible features, offering activities based on rhyming, voice recognition, and both visual and auditory feedback. It is organized into progressively challenging levels in which children must identify pairs of rhyming words—starting with visual and textual support and gradually relying solely on auditory cues.

Figure 9 shows four screenshots of the game (in Portuguese).

Fig. 9. (a) Level 1; (b) Level 2; (c) Level 3; (d) Review screen using voice recognition

The gameplay revolves around colorful virtual cards that represent words with similar endings. These colors act as visual reinforcement to help establish correct associations, providing extra support for children with phonological processing difficulties. As players progress, the challenges increase in complexity, culminating in a review phase where they are required to recognize and pronounce the learned words using the device's microphone. Immediate feedback is provided through narration, images, and animations that indicate correct or incorrect responses.

The interface design follows the HEDEG guidelines (Heuristics for Evaluation of Digital Educational Games) [Valle et al., 2013], with a focus on legibility, clarity of instructions, reduced visual distractions, and the use of the Open Dyslexic font. Navigation is optimized for touchscreen devices, and interactive elements—such as buttons, texts, and images—are scaled for easy interaction by young children. The implementation of a Voice User Interface (VUI) further enhances accessibility, enabling use by children with motor or typing difficulties and fostering an inclusive environment.

As a serious game, this application exhibits several core characteristics: a clearly defined pedagogical objective (improvement of reading skills), the incorporation of digital accessibility guidelines, expert validation by professionals in Speech Therapy and Computer Science, and ongoing refinement based on user testing with the target audience. Through the integration of responsive web technologies, gamification strategies, and inclusive design, the game aims to increase engagement and support the learning process of children with reading disorders, reinforcing the role of technology as an inclusive tool in educational contexts.

3.2 Serious Game for Children with Autism

The game proposed by [28] was developed as a complementary tool for speech therapy sessions targeting children diagnosed with Autism Spectrum Disorder (ASD). Its primary goal is to support auditory training in a context shaped by the Covid-19 pandemic, particularly addressing the communication difficulties caused by the widespread use of face masks.

The game is structured as a 2D platformer set across three familiar everyday environments: a bakery, a supermarket, and a school. In each setting, the child controls a masked child avatar who must navigate the environment and collect images associated with words from phonological minimal pairs (e.g., *pão* [bread] and *mão* [hand], or *uva* [grape] and *luva* [glove]) in a specific sequence given via audio instructions.

These instructions are recorded with a real speaker wearing a face mask, simulating the muffled speech conditions common during the pandemic. This setup encourages the child to practice sound discrimination without relying on visual cues such as lip reading or facial expressions.

Each scenario contains four levels with increasing phonological complexity. Players lose points for collecting items out of order or colliding with obstacles, but earn stars when they succeed. The controls are simple—arrow keys for movement and the spacebar to jump—and the inclusion of enemies requiring precise jumps helps maintain engagement.

The game was designed with accessibility in mind for children with ASD, emphasizing:

- **Clear visual design**, free of distracting stimuli, aligned with the sensory preferences of children on the spectrum;
- **Familiar environments**, supporting predictability and anticipation;
- **Realistic auditory stimuli**, to strengthen listening skills in pandemic-altered communication contexts;
- **Intuitive interface**, offering both visual and auditory feedback and simple navigation;
- **Validation by professionals and users**, including testing with nine speech therapists and fourteen autistic children during therapy sessions, with positive results in motivation, comprehension, and clinical applicability.

Due to its educational intent, direct alignment with therapeutic goals, and focus on developing specific skills for a population with special needs, the game qualifies as a serious game with a strong emphasis on accessibility. Its practical application in clinical contexts during the uniquely challenging period of the Covid-19 pandemic further underscores its potential as an inclusive and supportive therapeutic tool. Figure 10 shows some screenshots of this game (in Portuguese).

Fig. 10. (a, upper-left) Control panel; (b, upper-right) Splash screen; (c, lower-left) Word search at the bakery; (d, lower-right) Final screen of bakery phase (in Portuguese).

3.3 Game for Children with Intellectual Impairment

The game [29] is a digital puzzle game developed as a playful and pedagogical resource for children with intellectual impairments, aged between 4 and 7 years. The game's primary objectives are to stimulate cognitive development, promote the recognition of shapes and figures, and provide an accessible, enjoyable, and meaningful learning experience for this audience.

Game mechanics are based on assembling images from pieces that must be dragged to their corresponding locations. Each image is divided into three parts, and at the end of each level, the child is rewarded with visual and auditory animations, providing positive reinforcement for their performance. The images used reflect elements from the child's everyday life—such as animals, fruits, vehicles, human figures, and common objects—helping to strengthen the connection between the game content and real-world experiences. Figure 11 shows some game's screenshots.

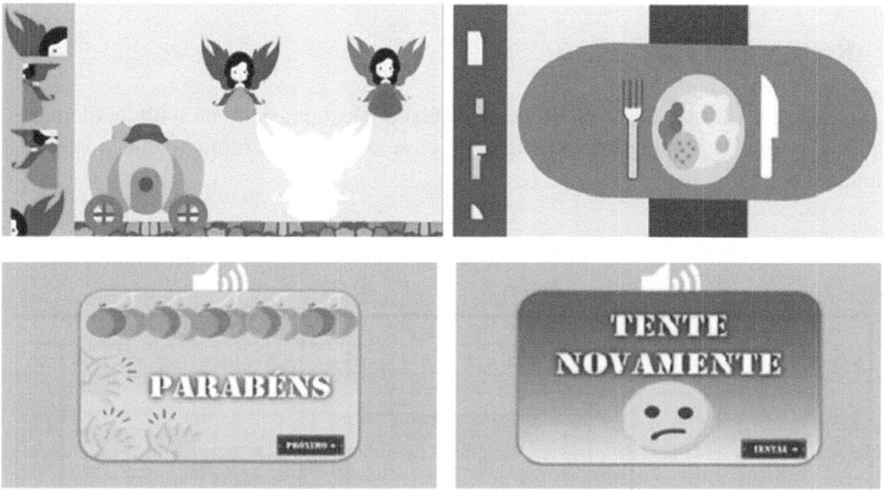

Fig. 11. (a, upper-left and b, upper-right) Puzzle assembly for "Fairy" and "Knife" ("Fada" and "Faca" in Portuguese); (c, lower-left and d, lower-right) success and failure screens, in Portuguese.

Accessible via web browsers, the game is designed to run on computers, tablets, and smartphones. It was developed using the Unity framework, and its interface was carefully designed to be clear and intuitive, featuring large buttons and clean visual elements to support interaction by children with cognitive or motor difficulties. No reading skills are required to play, which increases accessibility for users with limited literacy.

The development process followed accessibility guidelines and universal design principles, with the following key features:

- **Simplified and visually organized interface**, minimizing simultaneous stimuli to meet the concentration and sensory processing needs of children with ID;
- **Immediate audiovisual feedback**, to support action comprehension and reinforce learning;
- **Touch- or click-based interaction**, with intuitive drag-and-drop mechanics and no complex commands;
- **Colorful and animated characters**, to stimulate engagement and sustain user interest;
- **Gradual difficulty progression**, encouraging the development of cognitive and fine motor skills;
- **Responsive design**, ensuring compatibility with various devices and screen sizes.

In addition, the game was validated with input from specialists in Special Education and Assistive Technology and tested with children with intellectual disabilities in a supervised setting. The results indicated high levels of acceptance, ease of use, and user motivation.

With its explicit pedagogical approach, attention to the specific needs of the target audience, and commitment to digital accessibility principles, the game qualifies as an accessible serious game. It aligns with inclusive education guidelines and the ethical use of technology for educational purposes.

4 Results and Discussion

Regarding the seven aspects for Accessible Game Design, each one with its elements, Table 1 brings an evaluation of the three case studies according to the theoretical foundation.

Table 1. Evaluation of accessibility game design aspects according to A4SGD Framework.

Game 1	a	b	c	d	e	f	g	h	i	j	k	sc
1												0,64
2												0,67
3												0,33
4												0,67
5												0,80
6												1,00
7												0,43
												0,65

Game 2	a	b	c	d	e	f	g	h	i	j	k	sc
1												0,82
2												0,67
3												0,44
4												0,67
5												0,80
6												1,00
7												0,71
												0,73

Game 3	a	b	c	d	e	f	g	h	i	j	k	sc
1												0,73
2												0,67
3												0,22
4												0,67
5												0,80
6												1,00
7												0,43
												0,64

Table 1 must be read as follows: lines, numbered from 1 to 7, represent the seven key aspects of A4SGD (explained in sections from 2.1 to 2.7). Columns represent the specific elements of each aspect. For instance, line 1, column a represents Inclusive Design from the beginning (aspect 1) and Involve Users with Disabilities (element a). The green color indicates if this element was covered during game design process (receiving score 1); red color indicates the absence of this element during the game design (score 0); white color (no score) indicates the absence of labeled elements per aspect (for instance, the aspect 2, Assistive Technologies, has only 3 elements, from a to c).

The columns labeled as *sc* represent scores for each element. They are marked with the yellow color if the average count of all elements for each aspect is in the range $0.25 \leq sc < 0.75$, and green if the average is above; just one average was less than 0.25 in the three games studied, marked in red – it refers to the Customization and Adaptation dimension for Game 3. It must be noted that only the sixth A4SGD dimension was fully covered in all games (Accessible Gameplay Mechanics), showing that gameplay mechanics played a central role in the accessible game design for all developers.

At the bottom of the table lie the final average scores for each game, showing that Game 2 had better evaluation regarding the aspects covered by A4SGD. However, the results indicate that future game design initiatives must invest greater effort to more effectively address the A4SGD dimensions.

5 Conclusions and Further Work

Discussing accessibility in serious games design is essential because these games are often used in contexts where inclusion is not optional, but foundational. Whether applied in education, rehabilitation, training, or public awareness, serious games aim to generate real-world impact. If accessibility is not considered from the outset, large groups of potential users (particularly those with physical, sensory, or cognitive impairments) may be unintentionally excluded. This undermines the game's effectiveness, equity, and reach. Thus, it can be affirmed that accessibility is not an add-on: it is a core design responsibility in the development of serious games. Ultimately, making serious games accessible not only benefits people with physical or cognitive disabilities. It often improves usability for all players and expands the game's reach and impact.

Accessibility features such as customizable UI, multiple input methods, adjustable difficulty levels, and multimodal feedback mechanisms are not solely useful for users with impairments; they also are helpful to diverse learning preferences and realities, age groups, cultural contexts, and play environments. For instance, voice commands or larger UI elements may assist a user with visual or motor impairments but also benefit a learner/player using a mobile device in a noisy environment. In this way, accessible design practices foster universal usability, contributing to a smoother and more flexible user experience for all players. Moreover, by lowering entry barriers and accommodating a wider range of user needs, accessible serious games tend to achieve greater reach, sustained engagement, and improved learning outcomes, which are important metrics in educational, therapeutic, and training contexts. Thus, accessibility is not merely a compliance requirement or a specialized design concern: it is a strategic component of inclusive game development that amplifies the game's impact, scalability, and social value.

In this sense, this paper presented A4SGD, a conceptual framework that is meant to be a guideline for the process of serious games design with accessibility aspects being considered from the scratch. Some games were used as a sandbox for testing the framework's reach and completeness, bringing important practical results that arose from the analysis of these games under the lights of A4SGD. Further work involves the application of this frameworks in other serious games and also its usage as a cornerstone for developing new games.

References

1. Abt, C.C.: Serious Games. University Press of America (1987)
2. Wilkinson, P.: A brief history of serious games. In: Dörner, R., Göbel, S., Kickmeier-Rust, M., Masuch, M., Zweig, K. (eds.) Entertainment Computing and Serious Games. Lecture Notes in Computer Science, vol. 9970, pp. 17–41. Springer, Cham (2016). https://doi.org/10.1007/978-3-319-46152-6_2
3. Ira, V.: The game accessibility special interest group (GA-SIG) of the international game developer's association (IGDA). Except. Parent **45**(6), 30–32 (2015)
4. Microsoft: Inclusive Design toolkit (2025). https://inclusive.microsoft.design/
5. Horowitz, A.: Accessibility by numbers: a critical review of game accessibility guidelines. In: Disability and Video Games: Practices of En-/Disabling Modes of Digital Gaming, pp. 259–294 (2023)

6. Brown, M., Anderson, S.L.: Designing for disability: evaluating the state of accessibility design in video games. Games Cult. **16**(6), 702–718 (2021)
7. Westin, T., Ku, J.J., Dupire, J., Hamilton, I.: Game accessibility guidelines and WCAG 2.0 – a gap analysis. In: Miesenberger, K., Kouroupetroglou, G. (eds.) Computers Helping People with Special Needs. ICCHP 2018. Lecture Notes in Computer Science, vol. 10896, pp. 270–279. Springer, Cham (2018). https://doi.org/10.1007/978-3-319-94277-3_43
8. Viudes-Carbonell, S.J., Gallego-Durán, F.J., Llorens-Largo, F., Molina-Carmona, R.: Towards an iterative design for serious games. Sustainability **13**(6), 3290 (2021)
9. Ben Amara, B., Mhiri Sellami, H., Ben Said, L.: An approach for serious game design and development based on iterative evaluation. J. Softw. Evol. Process **36**(10), e2680 (2024)
10. Lo, P., Thue, D., Carstensdottir, E.: What is a game mechanic?. In: Baalsrud Hauge, J., C. S. Cardoso, J., Roque, L., Gonzalez-Calero, P.A. (eds.) Entertainment Computing – ICEC 2021. ICEC 2021. Lecture Notes in Computer Science, vol. 13056, pp. 336–347. Springer, Cham (2021). https://doi.org/10.1007/978-3-030-89394-1_25
11. Krasniqi, V., Zdravkova, K., Dalipi, F.: Impact of assistive technologies to inclusive education and independent life of down syndrome persons: a systematic literature review and research agenda. Sustainability **14**(8), 4630 (2022)
12. Cañete, R., Peralta, E.: Assistive technology to improve collaboration in children with ASD: state-of-the-art and future challenges in the smart products sector. Sensors **22**(21), 8321 (2022)
13. Cezarotto, M., Martinez, P., Chamberlin, B.: Developing inclusive games: design frameworks for accessibility and diversity. In: Game Theory-from Idea to Practice. IntechOpen (2022)
14. Sin, J., L. Franz, R., Munteanu, C., Barbosa Neves, B.: Digital design marginalization: new perspectives on designing inclusive interfaces. In: Proceedings of the 2021 CHI Conference on Human Factors in Computing Systems, pp. 1–11, May 2021
15. Valle, P.H.D., Vilela, R.F., Parreira Junior, P.A., Inocencio, C.G.: Hedeg-heurísticas para avaliação de jogos educacionais digitais. Nuevas Ideas en Informática Educativa TISE **12**, 2013 (2013)
16. Zohaib, M.: Dynamic difficulty adjustment (DDA) in computer games: a review. Adv. Hum. Comput. Interact. **2018**(1), 5681652 (2018)
17. Moustakas, N., Rovithis, E., Vogklis, K., Floros, A.: Adaptive audio mixing for enhancing immersion in augmented reality audio games. In: Companion Publication of the 2020 International Conference on Multimodal Interaction, pp. 220–227, October 2020
18. Stensgaard, P., Alléus, A., Palm, J.: Adaptive Mood Audio: Rethinking Audio for Games (2013)
19. Salvador-Ullauri, L., Acosta-Vargas, P., Luján-Mora, S.: Web-based serious games and accessibility: a systematic literature review. Appl. Sci. **10**(21), 7859 (2020)
20. Žilak, M., Car, Ž: A framework for improving accessibility of serious games in handheld augmented reality based on user interaction data. Appl. Sci. **15**(4), 2161 (2025)
21. Beristain-Colorado, M.D.P., et al.: Standardizing the development of serious games for physical rehabilitation: conceptual framework proposal. JMIR Serious Games **9**(2), e25854 (2021)
22. Othman, N.I., Mohamed, H., Mat Zin, N.A.: Serious games accessibility design model for low-vision children. Adv. Hum. Comput. Interact. **2023**(1), 9528294 (2023)
23. Shrestha, A., Karki, A.: Accessibility in Video Games: A Review (2022)
24. Zheng, J.: Human factors in game design: the importance of accessibility. Appl. Comput. Eng. **42**, 102–110 (2024)
25. Iten, N., Petko, D.: Learning with serious games: is fun playing the game a predictor of learning success? Br. J. Edu. Technol. **47**(1), 151–163 (2016)
26. Middlebrooks, A.E.: The paradox of serious fun. In: Beghetto, R., Sriraman, B. (eds.) Creative Contradictions in Education. Creativity Theory and Action in Education, vol. 1, pp. 265–280. Springer, Cham (2017). https://doi.org/10.1007/978-3-319-21924-0_15

27. Figueiredo, K.O., Cardenuto, R.R., Amato, C.A., Martins, V.F.: Kera Puzzle: jogo digital educacional para apoio à intervenção fonoaudiológica. Revista Ibérica de Sistemas e Tecnologias de Informação **E41**, 503–515 (2021)
28. Oikawa, V., Amato, C.A.H., Martins, V.F.: A serious game for the daily care of autistic children in times of Covid-19. Educ. Knowl. Soc. **26**, 1 (2025)
29. Oliveira, A.A., Amato, C.A.H., Martins, V.F.: DysLEK: um jogo digital para treino de leitura em crianças com dislexia. In: Proceedings of the 2024 19th Iberian Conference on Information Systems and Technologies (CISTI 2024), Salamanca (2025, in press)

Impact of Gamification and Immersive Technologies on Programming Education: A Systematic Review

Jheimer Julián Sepúlveda-López(✉), Luz Angela Aristizábal-Quintero, Tatiana Katerine Guzmán-Cruz, and Julián López-Marulanda

Universidad Nacional de Colombia – Sede Manizales, Manizales, Colombia
{jjsepulvedal,laaristizabalq,tguzmanc,jullopezm}@unal.edu.co

Abstract. Programming teaching faces multiple challenges related to the abstraction of concepts, the lack of individualization, and limited student motivation. In this context, gamification and immersive technologies, such as augmented reality and virtual reality, have emerged as alternatives to improve the learning experience. This paper presents a systematic literature review aimed at identifying the impact of these technologies on programming teaching, analyzing both their benefits and limitations. The literature review was developed through a structured methodology that included the definition of keywords, the construction of search equations, and the application of inclusion criteria in specialized databases. The results show that gamification and immersive technologies contribute significantly to increasing motivation, improving academic performance, developing soft skills, and providing immediate feedback. However, limitations such as high costs, technological dependence, potential distractions, and technical barriers are also evident. This analysis leads us to conclude that, while these technologies offer great potential to transform programming instruction, strategies must be implemented to mitigate their disadvantages and ensure their balanced and pedagogically effective use.

Keywords: Gamification · Immersive technologies · Virtual reality · Programming teaching · Systematic review

1 Introduction

Programming is a fundamental skill in the 21st century, but traditional teaching can be ineffective. Immersive technologies offer an engaging alternative to enhance the learning experience. These technologies seek to motivate and engage students in programming, as well as improve learning and understanding of complex concepts and develop critical thinking and problem-solving skills. Based on the above, there is a need to review how gamification techniques and emerging technologies are impacting university students' programming learning.

Programming education faces several significant challenges, such as the required abstraction and lack of individualization in large classes. Traditional approaches do not always consider students' learning styles or take advantage of the necessary dynamics.

W. J. Sarmiento and E. L. Sierra-Ballén (Eds.): CCC 2025, CCIS 2692, pp. 112–123, 2027.
https://doi.org/10.1007/978-3-032-28447-1_8

It has been identified that traditional programming teaching faces several challenges, among which the following stand out:

Abstraction: Programming concepts can be abstract and difficult for students to grasp, especially those without prior experience in areas such as logic and mathematics.

Lack of individualization: Large classes make it difficult to individualize learning, which can lead to frustration for students who progress at different rates or have different learning styles.

Traditional approaches: Traditional approaches to teaching programming often rely on passive methods such as theoretical exposition and exercise solving, which can be slow and unmotivating for students.

However, immersive technologies such as virtual and augmented reality offer a possible solution; these dynamic and visually stimulating tools can improve understanding by allowing practical interactions in addition to comprehension and engagement, the use of video games as an educational tool promotes skills such as reflection, problem solving and self-esteem, while encouraging imagination and social interaction, as evidenced by various research. Some research highlights the positive impact that gamification processes and technologies have on enhancing programming learning in high school and university students. However, it is necessary to analyze the true impacts of these technologies on these teaching-learning processes and based on this, define platforms that improve processes and have a positive impact on students (Resninck et al., 2009; Klopfer et al., 2009; Gee, 2003; Wouters et al., 2013).

Finally, it is worth concluding that immersive technologies and video games represent promising tools for improving programming instruction. However, further research and evaluation of their real impact on teaching and learning processes is needed to develop educational platforms that fully leverage the potential of these technologies and generate a positive impact on students.

2 Background and Related Work

Programming education presents several challenges due to its abstract, logical, and sequential nature. Traditionally, this approach has focused on theoretical instruction and technical exercises, which has proven insufficient to motivate and retain students, especially in the first semesters of university. In this context, innovative approaches have emerged that integrate emerging technologies with active methodologies to enhance the learning experience and respond to the learning styles of so-called "digital natives" (Prensky, 2001).

One of the most prominent approaches is gamification, understood as the use of game design elements in non-game contexts (Deterding et al., 2011). This strategy seeks to increase students' intrinsic motivation through the use of rewards, missions, levels, immediate feedback, and immersive narratives. Gamification, applied in education, has been shown to improve not only student engagement and participation, but also their academic performance and satisfaction with learning (Hamari et al., 2014; Seaborn & Fels, 2015). Particularly in programming, it allows students to approach complex problems gradually, encouraging perseverance and autonomous challenge resolution.

On the other hand, immersive technologies such as augmented reality (AR) and virtual reality (VR) have gained prominence as tools that enable the simulation of abstract or complex environments, offering highly visual and interactive learning experiences. According to Wouters et al. (2013), these technologies generate greater knowledge retention by actively engaging students in meaningful learning scenarios. In programming education, these tools can facilitate the visualization of logical structures, algorithms, and control flows, allowing students to "see" and "manipulate" concepts that are traditionally only taught textually.

Furthermore, the use of serious games, i.e., those developed for educational or training purposes, complements this approach by allowing the student to learn through experimentation in a controlled environment.

Together, these approaches respond to the demands of the current educational paradigm, which recognizes that learning is most effective when it is contextualized, experienced, and emotionally connected to the student. In this sense, constructivism (Piaget, 1975; Vygotsky, 1978) and meaningful learning (Ausubel, 1963) provide a solid theoretical foundation, suggesting that knowledge is actively constructed when students relate new concepts to prior experiences and actively participate in their learning process.

Finally, recent studies in the context of higher education in computer science (Papastergiou, 2009; Howard & Davis, 2023; Swacha & Szydlowska, 2023) have shown that the integration of gamification and immersive technologies in programming courses can contribute to reducing dropout rates, improve conceptual understanding and encourage the development of transversal skills, such as creativity and collaboration.

3 Research Methodology

To carry out the systematic literature search process, the following steps were taken, which were in line with the objectives set out in this document:

- Step 1: The problem to be solved was formulated through research. To do this, a research question was defined on the topic of interest to the authors. Following this, the components associated with that question were defined.
- Step 2: The search strategy is created and established. In this process, the associated keywords were defined for each component. Likewise, the criteria for selecting or rejecting documents were defined, the search equation was established, and the information source where this process would be carried out was defined.
- Step 3: The search equation was applied to the defined database, and the quantity and specificity of the results were reviewed. At this point, it should be noted that if these two aspects are not balanced, it would be necessary to return to step 2.
- Step 4: Having found the balance between quantity and specificity of the results found, the classification, critical analysis, bibliometric analysis and presentation of the results were carried out.

This bibliographic review methodology allowed for ordering the review process and provides elements that justify the process. Below, each step of the review methodology is outlined, along with a detailed application to the specific case of a literature review on the use of gamification and immersive technologies in programming instruction.

Problem Definition
The first step of the methodology is to define the problem or questions to be answered

based on the review of the material found; define what answers would be expected from conducting the search; and review the literature found. For the specific case of this document, the following question is posed:

> What has been the impact of gamification and immersive technologies on programming teaching?

Definition of Components and Keywords

Based on the statement of the question/problem, the components to be used in the query and the keywords related to each component are established.

Each component is grouped into a category of keywords that are of interest to the particular study being conducted.

Definition of the Search Equation

After establishing the components and keywords to be used in the query, you must construct the search equation to be applied to the different information sources. Each component is linked to the other by an AND, and the keywords within each component are linked by an OR:

> Gamification OR Serious Games OR " immersive technology" OR "augmented reality" OR "virtual reality" (Title) AND Instruction OR Education OR Training OR Pedagogy OR Tutoring OR Coaching OR Learning (Title) AND coding OR "Software Development" OR "Computer Coding" OR "Algorithm Development" OR "Software engineering" OR "Code development" OR "Computer programming" OR Programming (Title) and Education Educational Research or Computer Science Interdisciplinary Applications or Education Scientific Disciplines or Computer Science Information Systems or Computer Science Software Engineering or Engineering Electrical Electronic or Computer Science Theory Methods or Engineering Multidisciplinary or Computer Science Cybernetics or Multidisciplinary Sciences or Communication or Robotics (Web of Science Categories).

Definition of Selection Criteria and Sources of Information

As a fundamental part of the systematic information search process, establishing article selection criteria must justify why articles are included or excluded from the search. These criteria are established by the researcher based on their objectives and prior experience:

- Articles in English.
- Articles that contain the words indicated in the search equation.

For this specific case the selected information sources where Web of Science is a tool that allows researchers to search current or retrospective information related to science, social sciences, arts, and humanities from approximately 12,300 of the world's most prestigious and high-impact research journals.

4 Results

4.1 Bibliometric Analysis on Gamification and Immersive Technologies for Teaching Programming

To perform the bibliometric analysis, we used the Biblioshiny platform, a graphical interface that allows you to use the Bibliometrix software to perform bibliometric analyses. This means you can import bibliographic data, perform statistical analyses, and view the results in the form of graphs and maps. The main results provided are listed below:

As can be seen in Fig. 1, a compilation of literature is made from a period between 2009 and 2024. There are 64 journals in which 89 documents referring to the subject have

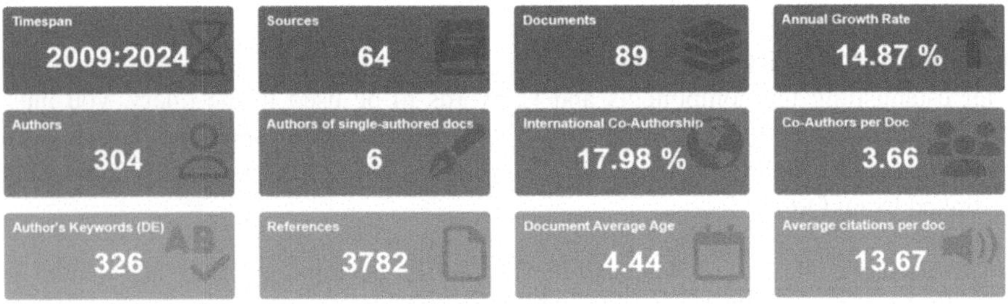

Fig. 1. Overview of the identified documents

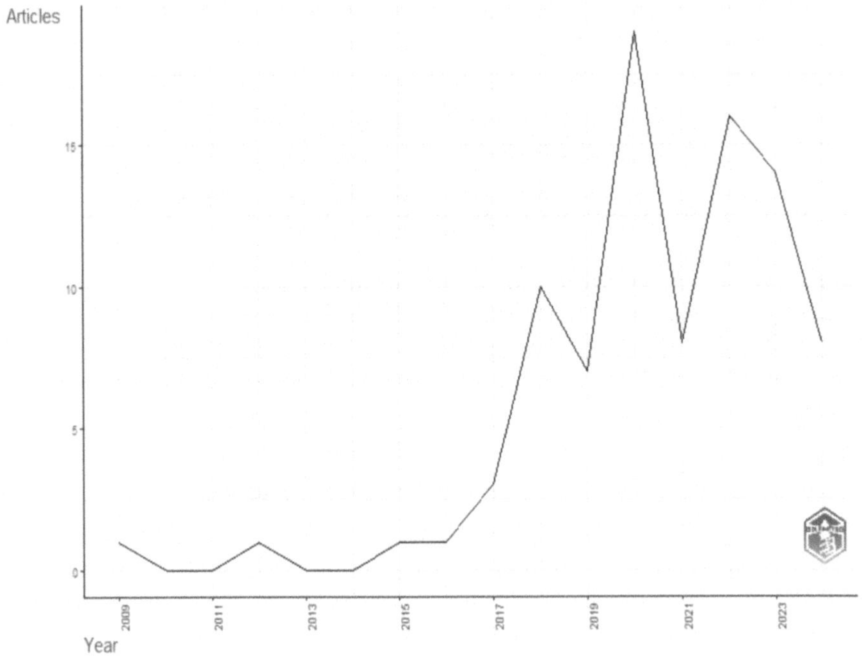

Fig. 2. Annual scientific production

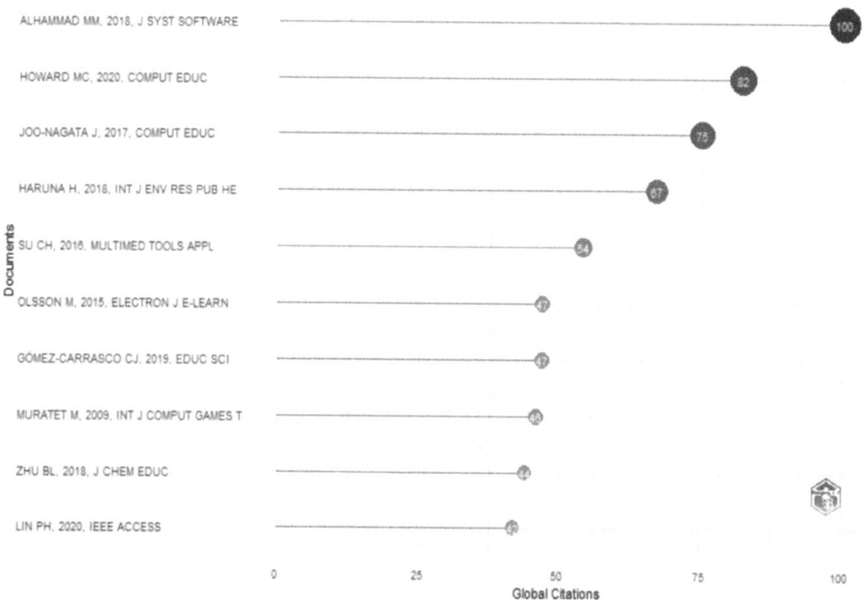

Fig. 3. Most cited documents

been published, something that is striking is the average annual growth of publications which was 14.87%, indicating a sustained increase in interest in the subject. In addition, it is reflected that 17.98% of the documents are with international co-authorship.

Figure 2 shows the evolution of annual scientific production between 2009 and 2023. A significant increase in the number of articles published is observed starting in 2017, with a notable peak in 2020, the year in which the highest number of publications was reached. Subsequently, although production remains relatively high, it fluctuates, with a new upswing in 2022 before declining again in 2023.

Figure 3 shows the documents with the highest number of global citations within the analyzed corpus. At the top of the list is the article by Alhamad MM (2018), published in the Journal of Systems and Software, with a total of 100 citations, indicating its high impact on the scientific community. This is followed by the works of Howard MC (2020) and Joo-Nagata A (2017), both published in Computers & Education, with 82 and 72 citations respectively.

This type of analysis allows us to identify the most influential studies in the field, revealing both the topics of greatest interest and the authors and journals with the greatest academic relevance.

4.2 Positive and Negative Aspects of Using Gamification and Immersive Technologies for Teaching Programming

Several studies have explored the positive and negative impacts of using gamified tools on the learning process, considering aspects such as motivation, skill development, academic performance, and associated technical and economic challenges. Table 1 summarizes the main dimensions and aspects identified in the literature, providing an overview of the benefits and limitations of these innovative methodologies, as well as the authors who have contributed to their analysis.

The evidence gathered by various authors highlights the need for a balanced pedagogical design that maximizes the benefits of these tools while mitigating their potential drawbacks.

Table 1. Dimensions identified in the literature

Dimensions / Aspects	Guy	Definition	Authors
Increased motivation	Positive	Developing serious games can increase student motivation and engagement by allowing them to apply their knowledge in a practical and fun context	(Muratet et al., 2009) (Howard & Davis, 2023) (Rodrigues et al., 2022) (Lin & Chen, 2020) (Xinogalos & Tryfou, 2021) (Tasadduq et al., 2021) (Swacha & Szydlowska, 2023) (Cabada et al., 2020) (Sideris & Xinogalos, 2019) (Tsai & Lai, 2022) (Abbasi et al., 2021) (Schez-Sobrino et al., 2020)
Limitations techniques	Negative	Limitations in current hardware and software, such as imperfect spatial mapping and narrow fields of view, were identified that can affect the effectiveness of AR programs	(Howard & Davis, 2023) (Segura et al., 2020) (Sunday et al., 2023) (Kazimoglu, 2020) (Akbulut et al., 2018)
Skill development soft	Positive	Serious games encourage critical thinking and problem-solving, essential skills in programming and in professional life	(Muratet et al., 2009) (Rodrigues et al., 2022) (Xinogalos & Tryfou, 2021) (Tasadduq et al., 2021)

(*continued*)

Table 1. (*continued*)

Dimensions / Aspects	Guy	Definition	Authors
Performance improvement academic	Positive	It has been observed that students who use gamified platforms tend to perform better on their assessments and final exams compared to those who do not	(Cabada et al., 2020) (Abbasi et al., 2021) (Toukiloglou & Xinogalos, 2022) (Huesca et al., 2023) (Akbulut et al., 2018) (Teng et al., 2018) (Su, 2016)
Solutions very expensive	Negative	It is mentioned that gamification can entail additional costs in terms of time and resources, which can be a barrier to its adoption in some educational institutions	(Alhammad & Moreno, 2018) (Wee et al., 2022)
Dependence technological	Negative	Some students may become dependent on technology to solve problems, which could limit their ability to address challenges without the support of digital tools	(Rodrigues et al., 2022) (Lin & Chen, 2020) (Tasadduq et al., 2021) (Tsai & Lai, 2022) (Toukiloglou & Xinogalos, 2022) (Akbulut et al., 2018)
Feedback immediate	Positive	The implementation of immediate feedback allows students to know their performance in real time, which can improve their learning and motivation	(Tasadduq et al., 2021) (Toukiloglou & Xinogalos, 2022) (Yang et al., 2023)
Possible distraction	Negative	Some students may get distracted by the game elements and not focus on learning programming concepts	(Tasadduq et al., 2021) (Huesca et al., 2023) (Soboleva V et al., 2021) (Su, 2016)
Learning experiential	Positive	VR allows students to learn through direct experience, which can improve information retention and understanding of complex concepts. Interacting with a 3D environment provides a practical context that facilitates learning	(Segura et al., 2020) (Teng et al., 2018)

(*continued*)

Table 1. *(continued)*

Dimensions / Aspects	Guy	Definition	Authors
Adapting to different learning styles	Positive	VR can cater to diverse learning styles, allowing students who learn best through hands-on experience to benefit from this methodology	(Alhammad & Moreno, 2018) (Revelo Sanchez et al., 2018) (Olsson et al., 2015) (Lin & Chen, 2020) (Howard & Davis, 2023)
Learning environment attractive	Positive	Serious games provide an environment where students can learn and practice OOP concepts in a stimulating and immersive way, helping to reduce anxiety and fear of failure	(Abbasi et al., 2021) (Soboleva V et al., 2021)

5 Discussion and Conclusions

A detailed analysis of the literature on gamification and immersive technologies in programming education identifies several positive and negative impacts that enrich the discussion about their applicability in educational contexts. Among the most notable positive impacts is increased student motivation and engagement. The development of serious games and immersive environments creates engaging contexts where students feel more involved and stimulated to learn. This aspect is crucial, especially in programming education, where dropout rates are often high due to the perceived difficulty of the subject. Furthermore, gamification and immersive technologies contribute significantly to the development of soft skills, such as problem-solving, critical thinking, and teamwork. These competencies are essential not only for programming but for professional practice in general. Students exposed to these environments learn to tackle complex challenges more autonomously and reflectively.

Another widely reported benefit is improved academic performance; students who interact with gamified platforms tend to perform better on their assessments and final exams compared to those who use traditional methods. The presence of immediate feedback facilitates this process, allowing students to adjust and improve their learning strategies in real time, fostering active knowledge construction. Likewise, immersive technologies such as virtual reality enable experiential learning that increases knowledge retention and facilitates the understanding of complex concepts through the practical manipulation of virtual environments. This approach also adapts to different learning styles, offering greater flexibility and accessibility for diverse student profiles.

However, the review also exposes limitations and risks that must be carefully considered. These include technical limitations related to the availability of suitable hardware, spatial mapping issues, and field-of-view limitations, which can affect the effectiveness of immersive technologies. Furthermore, the high cost of implementing and maintaining these solutions can represent a significant barrier, especially for institutions with limited resources. Another critical aspect is technological dependence, where some students may develop an overreliance on digital tools to solve problems, which can limit their ability to confront challenges without the support of such technologies. It was also identified that if playful elements are not properly integrated into educational content, there is a risk that students will become distracted and lose focus on the learning objectives.

In summary, gamification and immersive technologies have significant potential to transform programming teaching, generating more motivating, dynamic, and adaptive environments. However, their implementation must be carefully planned, considering technical, financial, and pedagogical aspects to maximize their benefits and mitigate their drawbacks. Future work could focus on the development of sustainable implementation models, as well as on teacher training that ensures effective and balanced use of these technologies. This study contributes to a more comprehensive view of the impact of these tools on education, providing key elements for decision-making in educational innovation processes in programming.

References

Muratet, M., Torguet, P., Jessel, J.-P., Viallet, F.: Towards a serious game to help students learn computer programming. Int. J. Comput. Games Technol. **2009** (2009). https://doi.org/10.1155/2009/470590

Howard, M.C., Davis, M.M.: A Meta-analysis of augmented reality programs for education and training. Virtual Reality **27**(4), 2871–2894 (2023). https://doi.org/10.1007/s10055-023-00844-6

Wee, C., Yap, K.M., Lim, W.N.: IProgVR: design of a virtual reality environment to improve introductory programming learning. IEEE Access **10**, 100054–100078 (2022). https://doi.org/10.1109/ACCESS.2022.3204392

Rodrigues, L., et al.: Are they learning or playing? Moderator conditions of gamification's success in programming classrooms. ACM Trans. Comput. Educ. **22**(3) (2022). https://doi.org/10.1145/3485732

Lin, P.-H., Chen, S.-Y.: Design and evaluation of a deep learning recommendation based augmented reality system for teaching programming and computational thinking. IEEE Access **8**, 45689–45699 (2020). https://doi.org/10.1109/ACCESS.2020.2977679

Xinogalos, S., Tryfou, M.M.: Using greenfoot as a tool for serious games programming education and development. Int. J. Serious Games **8**(2), 67–86 (2021). https://doi.org/10.17083/ijsg.v8i2.425

Tasadduq, M., Khan, M.S., Nawab, R.M.A., Jamal, M.H., Chaudhry, M.T.: Exploring the effects of gamification on students with rotate learning background while learning computer programming. Comput. Appl. Eng. Educ. **29**(6), 1871–1891 (2021). https://doi.org/10.1002/cae.22428

Segura, R.J., del Pino, F.J., Ogayar, C.J., Rueda, A.J.: VR-OCKS: a virtual reality game for learning the basic concepts of programming. Comput. Appl. Eng. Educ. **28**(1), 31–41 (2020). https://doi.org/10.1002/cae.22172

Cabada, R.Z., Barron Estrada, M.L., Rios Felix, J.M., Alor Hernandez, G.: A virtual environment for learning computer coding using gamification and emotion recognition. Interact. Learn. Environ. **28**(8), 1048–1063 (2020). https://doi.org/10.1080/10494820.2018.1558256

Swacha, J., Szydlowska, J.: Does gamification make a difference in programming education? Evaluating FGPE-supported learning outcomes. Educ. Sci. **13**(10) (2023). https://doi.org/10.3390/educsci13100984

Sideris, G., Xinogalos, S.: PY-RATE ADVENTURES: A 2D platform serious game for learning the basic concepts of programming with Python. Simul. Gaming **50**(6), 754–770 (2019). https://doi.org/10.1177/1046878119872797

Tsai, C.-Y., Lai, Y.-C.: Design and validation of an augmented reality teaching system for primary logic programming education. Sensors **22**(1) (2022). https://doi.org/10.3390/s22010389

Sunday, K., Oyelere, S.S., Joseph, A.F., Aliyu, M.B., Balogun, O.S., Bouali, N.: Usability evaluation of imikode virtual reality game to facilitate learning of object-oriented programming. Technol. Knowl. Learn. **28**(4), 1871–1902 (2023). https://doi.org/10.1007/s10758-022-09634-6

Abbasi, S., Kazi, H., Kazi, A.W., Khowaja, K., Baloch, A.: Gauge object oriented programming in student's learning performance, normalized learning gains and perceived motivation with serious games. Information **12**(3) (2021). https://doi.org/10.3390/info12030101

Schez-Sobrino, S., Vallejo, D., Glez-Morcillo, C., Redondo, M.Á., Castro-Schez, J.J.: RoboTIC: a serious game based on augmented reality for learning programming. Multimedia Tools Appl. **79**(45–46), 34079–34099 (2020). https://doi.org/10.1007/s11042-020-09202-z

Toukiloglou, P., Xinogalos, S.: Ingame worked examples support as an alternative to textual instructions in serious games about programming. J. Educ. Comput. Res. **60**(7), 1615–1636 (2022). https://doi.org/10.1177/07356331211073655

Kazimoglu, C.: Enhancing confidence in using computational thinking skills via playing a serious game: a case study to increase motivation in learning computer programming. IEEE Access **8**, 221831–221851 (2020). https://doi.org/10.1109/ACCESS.2020.3043278

Alhammad, M.M., Moreno, A.M.: Gamification in software engineering education: a systematic mapping. J. Syst. Softw. **141**, 131–150 (2018). https://doi.org/10.1016/j.jss.2018.03.065

Yang, F.-C.O., Lai, H.-M., Wang, Y.-W.: Effect of augmented reality-based virtual educational robotics on programming students' enjoyment of learning, computational thinking skills, and academic achievement. Comput. Educ. **195** (2023). https://doi.org/10.1016/j.compedu.2022.104721

Huesca, G., Campos, G., Larre, M., Perez-Lezama, C.: Implementation of a mixed strategy of gamification and flipped learning in undergraduate basic programming courses. Educ. Sci. **13**(5) (2023). https://doi.org/10.3390/educsci13050474

Tsiotras, D., Xinogalos, S.: Investigating the perceived player experience and short-term learning of the text-based Java programming serious game "rise of the java emperor". Inf. Educ. **20**(1), 153–170 (2021). https://doi.org/10.15388/infedu.2021.08

Soboleva, E.V., Suvorova, T.N., Grinshkun, A.V., Bocharov, M.I.: Applying gamification in learning the basics of algorithmization and programming to improve the quality of students' educational results. Eur. J. Contemp. Educ. **10**(4), 987–1002 (2021). https://doi.org/10.13187/ejced.2021.4.987

Maskeliunas, R., Kulikajevas, A., Blazauskas, T., Damasevicius, R., Swacha, J.: An Interactive serious mobile game for supporting the learning of programming in javascript in the context of eco-friendly city management. Computers **9**(4) (2020). https://doi.org/10.3390/computers9040102

Rojas-Lopez, A., Rincon-Flores, EG, Mena, J., Jose Garcia-Penalvo, F., Soledad Ramirez-Montoya, M.: Engagement in the course of programming in higher education through the use of gamification. Univ. Access Inf. Soc. **18**(3, Si), 583–597 (2019). https://doi.org/10.1007/s10209-019-00680-z

Akbulut, A., Catal, C., Yildiz, B.: On the effectiveness of virtual reality in the education of software engineering. Comput. Appl. Eng. Educ. **26**(4), 918–927 (2018). https://doi.org/10.1002/cae.21935

Teng, C.-H., Chen, J.-Y., Chen, Z.-H.: Impact of augmented reality on programming language learning: efficiency and perception. J. Educ. Comput. Res. **56**(2), 254–271 (2018). https://doi.org/10.1177/0735633117706109

Ozer, H.H., Kanbul, S., Ozdamli, F.: Effects of the gamification supported flipped classroom model on the attitudes and opinions regarding game-coding education. Int. J. Emerg. Technol. Learn. **13**(1), 109–123 (2018). https://doi.org/10.3991/ijet.v13i01.7634

Revelo Sanchez, O., Collazos Ordonez, C.A., Jimenez Toledo, J.A.: Gamification as a didactic strategy for teaching/learning programming: a systematic mapping of the literature. Lampsakos Digit. J. **19**, 31–46 (2018). https://doi.org/10.21501/21454086.2347

Yassine, A., Chenouni, D., Berrada, M., Tahiri, A.: A serious game for learning c programming language concepts using solo taxonomy. Int. J. Emerg. Technol. Learn. **12**(3), 110–127 (2017). https://doi.org/10.3991/ijet.v12i03.6476

Su, C.-H.: The effects of students' motivation, cognitive load and learning anxiety in gamification software engineering education: a structural equation modeling study. Multimedia Tools Appl. **75**(16), 10013–10036 (2016). https://doi.org/10.1007/s11042-015-2799-7

Olsson, M., Mozelius, P., Collin, J.: Visualization and gamification of e-learning and programming education. Electr. J. e-Learn. **13**(6), 441–454 (2015)

Gee, J.P.: What video games have to teach us about learning and literacy. Comput. Entertainment (CIE) **1**(1), 20 (2003)

Wouters, P., Van Nimwegen, C., Van Oostendorp, H., Van Der Spek, E.D.: A meta-analysis of the cognitive and motivational effects of serious games. J. Educ. Psychol. **105**(2), 249–265 (2013)

Klopfer, E., Osterweil, S., Salen, K.: Moving learning games forward: obstacles, opportunities, and openness. Educ. Arcade MIT, 1–29 (2009)

Resnick, M., et al.: Scratch: programming for all. Commun. ACM **52**(11), 60–67 (2009)

Prensky, M.: Digital natives, digital immigrants part 1. On the Horizon **9**(5), 1–6 (2001)

Deterding, S., Dixon, D., Khaled, R., Nacke, L.: From game design elements to gamefulness: defining "gamification". In: Proceedings of the 15th International Academic MindTrek Conference, pp. 9–15. ACM (2011). https://doi.org/10.1145/2181037.2181040

Hamari, J., Koivisto, J., Rot, G., Saari, T.: Does gamification increase student engagement in a learning management system? An affordance perspective. Comput. Educ. **68**, 41–51 (2014)

Seaborn, K., Fels, D.I.: Gamification in theory and action: a survey. Int. J. Hum. Comput. Stud. **74**, 14–31 (2015). https://doi.org/10.1016/j.ijhcs.2014.09.006

Piaget, J.: The Equilibration of Cognitive Structures. Ariel (1975)

Vygotsky, L.S.: Mind in Society: The Development of Higher Psychological Processes. Harvard University Press (1978)

Ausubel, D.P.: The Psychology of Meaningful Verbal Learning. Grune & Stratton (1963)

Papastergiou, M.: Digital game-based learning in high school computer science education: Impact on educational effectiveness and student motivation. Comput. Educ. **52**(1), 1–12 (2009). https://doi.org/10.1016/j.compedu.2008.06.004

G. Biomedical Signal Processing

ADVISE-BD: Clinical Deterioration Prediction

Lida J. Méndez[1], María Del P. Villamil[1(✉)], Ledys M. Izquierdo[2], and Luis F. Niño[3]

[1] Department of Systems and Computing Engineering, School of Engineering, Universidad de los Andes, Bogotá, Colombia
{lj.mendezo,mavillam}@uniandes.edu.co
[2] Unidad de Cuidado Intensivo Pediátrico, Hospital Militar Central, Bogotá, Colombia
lizquierdo@unal.edu.co
[3] Laboratorio de Investigación en Sistemas Inteligentes, Universidad Nacional de Colombia, Bogotá, Colombia
lfninov@unal.edu.co

Abstract. Monitoring patients in pediatric intensive care units is particularly challenging due to their fragile health and the shortage of medical staff in countries like Colombia. Automating data analysis can save lives in critical situations by providing early warnings. Our system, ADVISE-BD, detects vital sign anomalies before clinical deterioration occurs within four-time windows. The model combines feature engineering with machine learning classification (using K-Nearest Neighbors and AutoEncoders) to distinguish patient conditions. Using the metric F1-Score, it obtained values of 95.3% and 97.5% in the detection of critical deterioration events in older pediatric groups, outperforming similar studies, especially for adolescents. Data is available from the Hospital Militar Central in Colombia.

Keywords: UCI vital signs prediction · anomaly detection · Autoencoders

1 Introduction

Intensive Care Units (ICU) are a fundamental piece in our healthcare pipelines, in which vast quantities of patients are admitted each year, reaching around 5 million people per year just in United States [1]. Patients admitted at such young ages typically present respiratory, neurological, or cardiologic pathologies, among others [2]. The median length of stay is about seven days in approximately 35% of cases.

Such kind of patients should be carefully and constantly monitored given their critical health state. However, reality is far from perfect especially in developing countries as the amount of staff available is not enough to collect, process and interpret these large amounts of data produced by patient's vital signals 24 h a day, 7 days a week [2] becoming an overwhelming, stressful and difficult task to tackle.

Patient monitoring is therefore crucial to successful treatment outcomes [3]; however, some patients deteriorate due to insufficient human resources to monitor those at risk of death. Our research interest focus on predicting clinical deterioration, defined as the

W. J. Sarmiento and E. L. Sierra-Ballén (Eds.): CCC 2025, CCIS 2692, pp. 127–142, 2027.
https://doi.org/10.1007/978-3-032-28447-1_9

condition in which more than four out of six monitored biosignals fall outside the normal ranges for a patient's age group [4]. The prediction relies on detecting clinical anomalies, which we define as abnormal variations in vital signs from patient data streams that precede a deterioration event.[5].

Most existing studies focus only on adult patients. Moreover, few consider factors such as variability within patient groups or age-related differences in vital signs. Although work [5] addresses pediatric patients, the proposal also did not include age specific vital signs differences nor patient history information, and its metrics could be improved.

This research proposes ADVISE-BD (*Anomaly Detection on Vital Signs—Before Clinical Deterioration*), a system that uses machine learning and deep learning techniques to predict vital signs anomalies and try to prevent clinical deterioration. ADVISE-DB also explores the impact between vital signs of patients of different ages and the performance of selected machine learning algorithms.

2 Related Work on Anomaly Detection Methods

Vital signs are fundamental for enhancing patient monitoring tasks in ICUs due to their interpretability and ease of measurement in critical care settings [6]. From an AI perspective, diverse approaches exist, including statistical analysis via Hidden Markov Models (HMMs) [5, 7] and algorithmic methods using Support Vector Machines (SVM) [8]. These leverage parameterization, vital sign generation/comparison, and patient history particularities to detect critical events.

Additional research involves spike signal observation and characterization through neural networks [9, 10], along with autoencoders (AE)-based anomaly detection [11, 12].

Notable techniques include SVMs [13], Recurrent Neural Networks (RNN) [14], Long-Short Term Memory (LSTM) [15] and AE [16] which focus on detecting pre-critical event conditions. Several incorporate dimensionality reductions, enabling models to operate solely on the most salient input features through feature engineering.

Patient age variation represents a critical factor for parameterization, detection, and threshold definition [17] as Seidel et al. demonstrate, age variability in PICU patients correlates with significant changes across all measured physiological variables, complicating the establishment of absolute thresholds for alert generation. Few studies address these peculiarities in patients under 18 years – the population most affected by this data heterogeneity.

Early prediction of patient deterioration before critical events is imperative [6], as it enables targeted allocation of attention and resources to critically deteriorating patients, leveraging the correlation between pre-event and intra-event signal values. Anomaly detection facilitates identification of future behavioral patterns and early alert generation. Literature highlights SVMs and AEs for feature engineering/signal reconstruction, and RNNs/LSTMs for temporal dependency modeling. We note particular interest in autoencoders for dual-purpose applications.

Previous research in this field has a major limitation: most studies focus on adult patient data and overlook the specific characteristics of pediatric populations, such as the significant physiological variability linked to age. Furthermore, many works rely on limited datasets, use only a subset of vital signs, or present theoretical models without validation on real-world, which limits their clinical applicability.

The goal of this work is to overcome these limitations. It seeks to develop a method for the early detection of clinical deterioration specifically for the Pediatric Intensive Care Unit (PICU). To achieve this, the project applies machine learning techniques, to detect anomalies in the vital signs of pediatric patients in real time, while considering the patient's age and using actual clinical monitoring data. This enables the generation of early time alerts, allowing medical staff to intervene effectively.

3 Methodology

The methodology focuses on the construction and validation of the ADVISE-BD MODEL, an artificial intelligence system designed for the early detection of anomalies in pediatric PICU. The methodology consists of: selection of dependent variables based on understanding the data set, feature engineering to enrich the data used to train the algorithms, selection of metrics to evaluate the results, definition of a base line to compare the model performance, selection of the algorithms that provide better results compared to a base line and finally ensemble the model.

The proposed system (ADVISE-BD) integrates a machine learning model trained to identify the relationships and dependencies between biosignals and clinical variables, measuring their influence on the occurrence of anomalies. It is capable of recognizing patterns in physiological data features and infer the probability of an abnormal clinical event based on real-time physiological data, considering different age groups for a more precise and personalized prediction.

3.1 Data Understanding

The used dataset was taken from the medical practice and work done by [5]. A cross-sectional cohort of pediatric patients (under 18 years old) hospitalized in the PICU of the Hospital Militar Central in Bogotá, Colombia between January 2018 to January 2020 was made. The data, which was anonymized and approved by the Ethics Committee (Act 013-164-19), was collected from monitoring machines, in which values were categorized between normal clinical events and abnormal clinical events. A public dataset available on Kaggle[1] was used, comprising over 90,000 records from 90 patients. Six critical biosignals that reflect the patient's hemodynamic and respiratory status were selected: PAS (Systolic Arterial Pressure), PAD (Diastolic Arterial Pressure), PAM (Mean Arterial Pressure), FC (Heart Rate), FR (Respiratory Rate), and SpO2 (Peripheral Oxygen Saturation). Samples were taken with a one-minute granularity. A filtering process was implemented to remove missing values in long periods (5 to 10 min) without a signal, ensuring the integrity of the time series.

[1] https://www.kaggle.com/ledysmai5/critically-illpedi-atric-patients-in-picucsv.

A description of the demographic and clinical variables was performed using the most appropriate measures of central tendency and location based on the nature and distribution of each variable. An analytical analysis was conducted using artificial intelligence-based techniques, specifically machine learning, to associate variables, quantify their degree of correlation, and make predictions.

In addition, there was also a demographic data categorization between the selected records, collecting additional information such as age in months, weight in kilograms, height in centimeters, initial diagnosis, and complications during the morbility and/or mortality time. This categorization relied on patient-specific normality and specificity, which vary according to factors such as age, gender, and medical conditions. Pediatric patients' vital signs not only differ from adult patients but also vary across 5 distinct age groups defined for patients under 18 years. These groups are defined in the 'Name and description' column in Table 1, the maximum and minimum threshold parameters by age groups that are provided in [4] are also shown in Table 1, this serves as a reference for identifying when one vital sign value is abnormal, i.e. falls outside the defined thresholds [18].

Dataset Demographics

The data stream includes information about 90 patients, 60% of whom are male (54) and 40% of whom are female (36). Although the majority of infants are less than 60 months old, it is important to note that 43 patients fall under the 0 to 12 months old group range as defined by [4] (Fig. 1), followed by smaller groups in next group ages, being the 48 months to 60 months old the smallest of all with 4 patients. It is important to note that this sample may not be significant for the smallest age groups, i.e. those aged 48–60 and 61–144 months. Furthermore, three out of four patients in the 48–60 months old group died. Data inconsistencies were found in five patients. They were left out of the dataset because the original data was incomplete, reducing the sample to 85.

Median and average values were analyzed and compared against the minimum and maximum thresholds defined by PALS *(Pediatric Advanced Life Support Study),* for each age group [4] as shown in Fig. 2. The green line represents the mean values by group; while the red line represents the median values by age group. Since the median is less affected by outliers, it provides a more reliable measure for comparison.

These findings do not confirm large differences between groups, but they do not rule out behavioral differences. We are working under the theory that models can be made using population group differences by age and learning to detect normality and abnormalities in each group.

3.2 Definition of Target Variables

The most useful data comes from minutes before clinical deterioration events. So, it is important to characterize and analyze such data.

The initial dataset has labels indicating 4 different health statuses (clinical states): Stable, Potentially Unstable, Unstable and Critical. These labels were created using ML techniques in [5]. The records labeled as Critical were used to determine four time windows as: *One hour, Half hour, Ten minutes, and Five minutes* before clinical deterioration event happens (see "Target Variables" in Fig. 3). All records within the

Table 1. Minimum and maximum thresholds in vital signs for pediatric patients by age group.

Group	Name and description	Age (in months)		Respiratory frequency		Cardiac frequency		Systolic blood pressure		Diastolic blood pressure		Spo2
		Min	Max	Min	Max	Min	Max	Min	Max	Min	Max	Min
G1	Infant. Younger than 12 months of age	0	12	30	60	100	160	60	90	30	62	88
G2	Older infant. Between 1 and 4 years old	13	47	24	40	95	150	78	112	48	78	88
G3	Pre-scholar. Between 4 and 5 years old	48	60	22	34	80	140	85	114	52	85	88
G4	Scholar. Between 6 and 12 years old	61	144	18	30	70	120	95	135	58	88	88
G5	Adolescent. Between 13 and 18 years old	145	216	12	16	60	100	100	120	60	80	88

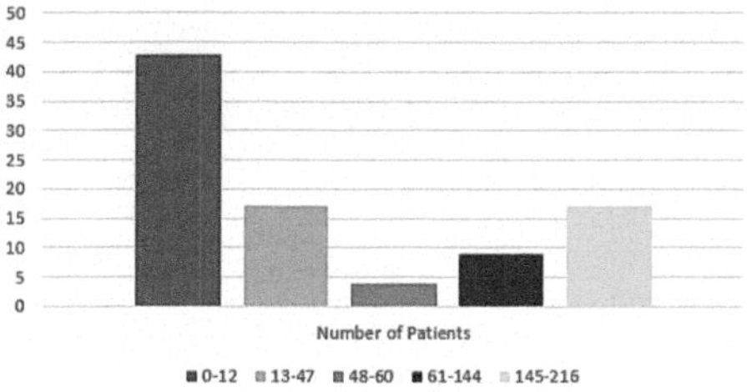

Fig. 1. Patients by age group

defined time windows preceding the moment of clinical deterioration (Critical status) were labeled, as they are important to identify patterns that, although not detectable by humans, an algorithm could recognize and characterize. This would enable patterns common to these records to be found and subsequently detected and classified. The above-mentioned time windows are then constituted as the dependent variables, these

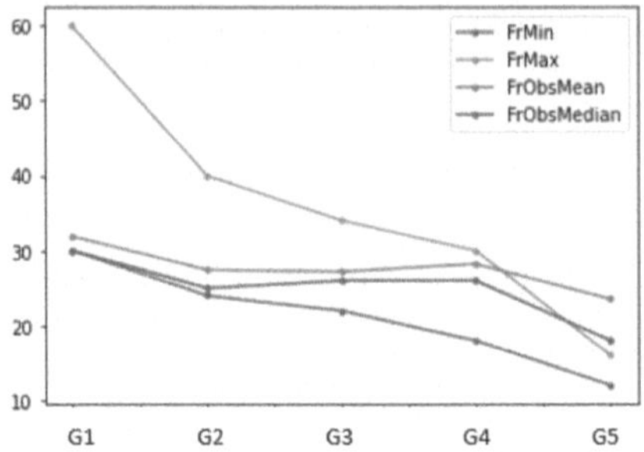

Fig. 2. Respiratory frequency values for different age groups.

being the ones that the algorithms want to learn to classify. Each window was modeled as a separate classification task, with independent hyperparameter searches and algorithm testing performed during model development.

3.3 Feature Engineering

The base dataset comprised demographic, vital signs, clinical label state and the target variables see Fig. 3(I). After reviewing the dataset, different strategies were applied to enrich the data used to train the algorithms. The first feature engineering process summarizes the patient's historical data on clinical states and preserves it in time windows from the moment they are connected to the monitor. For each time window delimited by critical events, a variable is created for each clinical state, showing how many times the patient has experienced that state. In the case of critical state labelling, the number of times these events occurred is recorded. For other states (i.e. stable, unstable and potentially unstable), this information is registered and reset until a critical state is detected (see "Feature Engineering 1" in Fig. 3(II)).

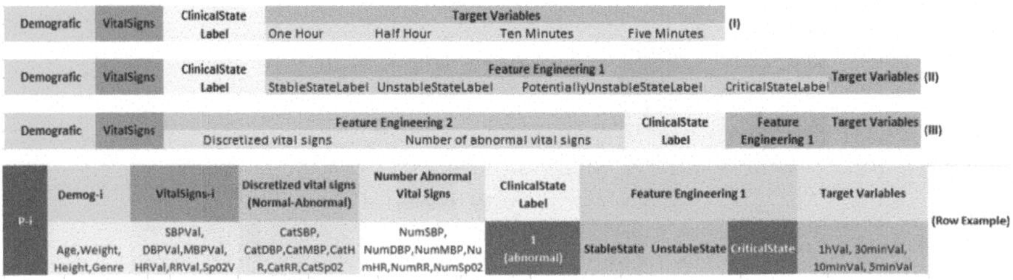

Fig. 3. Patient information for training processes.

A second feature engineering strategy adds seven new attributes to classify each vital sign per minute as normal or abnormal according to the thresholds in Table 1. (values 0 and 1, respectively). Then, for each minute, the number of abnormal labels was counted for each record (see "Feature Engineering 2" in Fig. 3(III)). At the end of these processes, we have a dataset with the information presented in the Fig. 3(Row example) to train the machine learning models. P-i in the figure shows the information collected from a patient before applying the machine learning process.

3.4 Evaluation Metrics

The metrics that are normally used to evaluate machine learning models are: Accuracy, F1 score (f1 score), precision and sensitivity (Recall) obtained from the confusion matrix. The positive class in the model to be tested are those records within the 4 defined time windows.

For this research, the F1-score, see Eq. 1, is used as a reference value to evaluate the results. This measure, which relates to precision and sensitivity, offers greater objectivity about the results when there is an imbalance between the classes that are to be predicted.

Equation 1: F1-Score

$$F1Score = \frac{2 * (Recall * Precisión)}{Recall + Precisión}(1) \quad (1)$$

Equation 1: F1-Score

3.5 Base Model Definition

KNN results are established as the base model as it is one of the most widely used machine learning models in anomaly detection problems. For this algorithm, the respective hyperparameter search is made using the Grid Search and cross validation methods [19] for each response variable (One Hr., Half an Hr., Ten Mins., Five Mins.).

KNN not provided good results with the original dataset (see Fig. 5), this because the health status labels used for creating the dependent variables were not consistent (for the same vital signs in the same age group or even the same patient, different health status labels were found). To solve this issue new health status labels, based on PALS, were re-calculated, using Table 1, that provides thresholds for every vital signs and age group, and then Table 2 which interprets biosignals states versus clinical events. A corresponding label for each record was created based on these tables (see "Clinical State Label" in Fig. 3), and with these new health status labels the dependent variables were created again and KNN algorithm was tested with this new dataset.

The health status labels were compared with the original label, revealing a significant difference (see Fig. 4), and as a greater number of critical events were identified resulting from the recalculation of labels, this also improved the results of the algorithm compared to using the original labels (see Fig. 5).

3.6 Machine Learning and Deep Learning Algorithm Selection

According to the literature review, the most used techniques for the type of problems addressed in this research work are of unsupervised nature, among which are found in reconstruction models using autoencoders (AE), followed by models under neural network architectures (ANN) and models performing supervised classification using

Table 2. Biosignals interpretation by clinical states given by the k-means algorithm by Izquierdo et.al. [5].

Bio-signal States	Clinical Event
All signals within normal ranges	Stable
1–2 abnormal bio-signals	Potentially Unstable
3–4 abnormal bio-signals	Unstable
>4 abnormal bio-signals	Critical

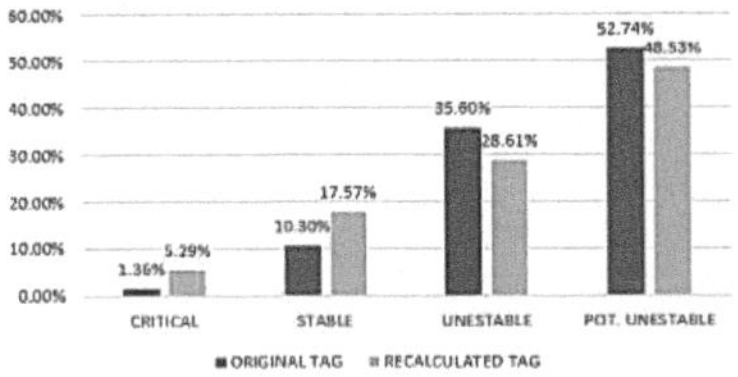

Fig. 4. Original vs recalculated labelling distribution by clinical state. Percentage of each health status tag

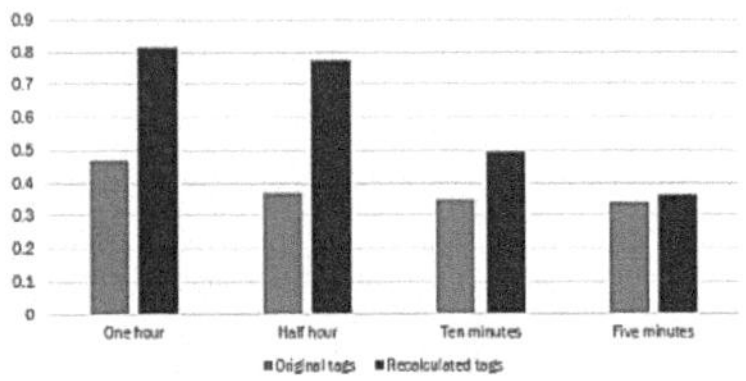

Fig. 5. Base model results (F1-score) with KNN original and recalculated labels

decision trees (DT). This motivates our decision to use a group of classification algorithms: K-Nearest Neighbors (KNN), Support Vector Machines (SVM), AE and Isolation Forest (IF) and, two versions of AE with 5 and 7 neural layers were tested, including data stream algorithms such as Hoeffding Adaptive Trees (HATC and HST) [20]. Long-Short Term Memory Networks (LSTM) are out of the scope of this research work as there is a separate group of research focused on working with this type of neural network architectures.

The next step is to make a limited search for hyperparameters that works best, with the available dataset and within the research time of this work, establishing whether an ensemble made up of some of these algorithms could give better results for the overall dataset or for subgroups of it.

3.7 Models Creation

The hyperparameter search and algorithm training use the dataset that have been created in feature engineering, using only the recalculated labels for the training phase.

Phase 1 involves testing the previously selected algorithms using the entire dataset for training and testing purposes.

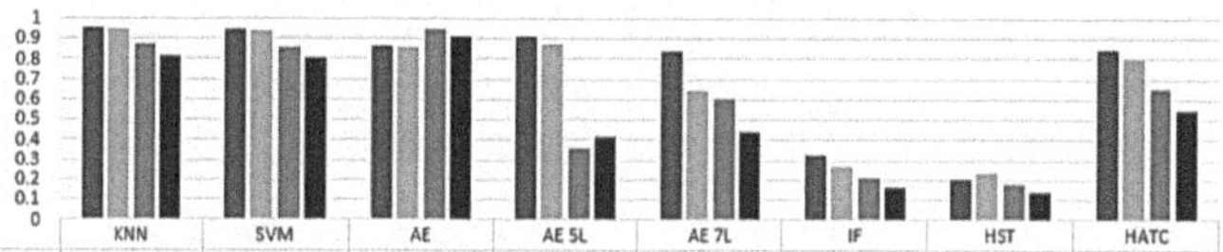

	KNN	SVM	AE	AE 5L	AE 7L	IF	HST	HATC
One hour	0.953	0.946	0.867	0.911	0.843	0.324	0.203	0.844
Half hour	0.944	0.935	0.857	0.872	0.645	0.271	0.232	0.796
Ten minutes	0.874	0.858	0.949	0.356	0.603	0.214	0.175	0.652
Five minutes	0.811	0.806	0.911	0.419	0.436	0.164	0.137	0.545

Fig. 6. F1-Score by trained algorithm in the train and testing dataset by time windows -Phase 1.

The best results are obtained with the KNN, SVM and AE Simple algorithms as seen in Fig. 6. KNN under one hour reaches 95.3% F1-Score, and 94.6% F1-Score in half-hour. The Autoencoders in predictions under 10 min reports 94.9% and 91.6% under 5 min, which is greatly improving compared to KNN's base model with 80% F1-Score. The worse results are reported by IF and HST models. Phase 1 demonstrates an evident improvement over the results (F1-Score) of the base model (see Table 3).

Phase 2 uses KNN, SVM and simple AE algorithms as they had the best results in Phase 1, and the dataset was separated into five groups according to age groups in PALS Table 1.

The results of the algorithms in phase 2 show in Table 3, in some cases, improve those of phase 1, green values in the table and in a few cases worsen, red values in the table. In the G5 group there are improvements in the four dependent variables. G5 group makes better discrimination of the time windows in comparison with the overall data (phase 1). The F1-score improved in the ten- and five-minute models of the G1 group. For the G2 group, the ten minutes model improved but the five minutes model worsen, as occurs in G3, where it is the only change. Finally, for the G4 group there is an improvement in the one-hour model and the rest remains very similar to the results of phase 1.

Results of group 5 (adolescents) are better in the single best model per group as compared with the base model, in which it is important to note a dramatic improvement over F1-Score of the Phase 2 best model (KNN) for the 5-min time windows (94.9%) vs. the baseline model (36.3%).

3.8 Ensemble of Classifiers

The proposed ensemble model allows the system to generate alerts that indicate whether the patient is presenting characteristics associated with clinical deterioration in their vital signs within a given time window. Two ensembles are proposed (Table 4), namely: (i) a general population ensemble that would be used for patients from age groups 1 to 4 and (ii) a specific ensemble for age group 5. The proposed ensemble architecture aims to achieve better results to reinforce the learning process according to the time-windows-based splitting. The idea then is to make a prediction in each model by using the incoming record. This allows the models to perform critical-condition predictions. Thus, each model will have a prediction in place to be voted in and weighted against the other models. The voting process generates a prediction for each record in which a

Table 3. Comparative performance evaluation of Phase 2 versus baseline (Phase 1) algorithms for predicting clinical deterioration, showing improvement (green) and deterioration (red) in F1 scores across demographic cohorts and prediction time windows.

KNN				SVM				AE			
OH	HH	TM	FM	OH	HH	TM	FM	OH	HH	TM	FM
Base model											
0.682	0.659	0.495	0.341								
Phase 1											
0.953	**0.944**	0.874	0.811	0.946	0.935	0.858	0.806	0.867	0.857	**0.949**	**0.911**
Phase 2											
G1 Age Group											
0.942	0.934	0.847	0.801	0.952	0.936	0.853	0.804	0.916	0.922	**0.972**	**0.933**
G2 Age Group											
0.904	0.778	0.61	0.497	0.896	0.742	0.58	0.287	0.688	0.615	**0.969**	**0.531**
G3 Age Group											
0.944	0.924	0.812	0.704	0.927	0.923	0.574	0.47	0.844	0.857	0.846	**0.763**
G4 Age Group											
0.963	0.905	0.808	0.787	0.948	0.888	0.765	0.609	0.875	0.882	0.7	0.818
G5 Age Group											
0.974	0.94	0.786	0.892	0.963	0.931	0.765	0.642	0.935	**0.967**	**0.975**	**0.949**

patient is categorized as close to clinical deterioration (1) or not (0) in each of the time windows along with each models' underlying probability. Then, the F1-Score is used as a weighting mechanism to check the prediction confidence of each model within a weighted majority voting scheme. Finally, the voting results are used to decide whether to generate or not the alert.

Table 4. Ensembles of algorithms proposed for age groups and response variables

Response variable	Algorithm ensemble G1–G4	Algorithm ensemble G5
OH	KNN	KNN
HH	KNN	AE
TM	AE	AE
FM	AE	AE

4 ADVISE-BD Model

Previous experiments paved the way to create the Anomaly Detection on Vital Signs – Before Clinical Deterioration (ADVISE-BD) system, an engineering pipeline consisting in inputting and pre-processing the monitoring vital signals and feed them into our trained machine learning model. The ADVISE-BD model comprises the following steps:

The vital signs monitor feeds a patient record into the pipeline, then carry out a feature engineering process for this record, label the incoming record by age group to validate the pertinence of applying an adolescents-based anomaly detection model (group 5) or applying the earlier-stages anomaly detection model (groups 1 to 4), run the models making part of the correspondent ensemble, perform the voting process in the ensemble model and define if the early warning should be generated, that is, if the record corresponds to the characteristics of the time windows prior to deterioration events.

5 Analysis of the Results

5.1 Machine Learning Model Construction

The primary objective of this study was to develop and validate a machine learning model capable of detecting clinical deterioration events in pediatric patients using real-time biosignal data. The results demonstrate the feasibility of this approach while highlighting key areas for future optimization.

The application of feature engineering significantly improved the performance of the classification algorithms, a finding consistent with the complex and high-dimensional nature of physiological data. Among the models tested, the KNN (K-Nearest Neighbors), SVM (Support Vector Machine), and simple autoencoder (AE) algorithms proved to be the most robust and suitable for this problem. The F1-Score was used as the main metric, achieving a critical deterioration detection with values up to 95.3% and 97.5% in the 30-min and one-hour time windows, respectively, especially in the older patient group (G5).

Surprisingly, the deep learning models, in their current configuration, did not outperform the traditional machine learning models. Contrary to initial expectations, no significant improvements in performance were observed. This limitation can be attributed to the intense computational load and time constraints that prevented an exhaustive hyperparameter search. We consider this a key opportunity for future research. A deeper exploration of the hyperparameter space and training with larger datasets could be the key to unlocking the true potential of these models.

A crucial finding was the impact of data cleaning and labeling. The quality of the initial labels, which sometimes showed inconsistencies for similar biosignals within the same age group, made it difficult to differentiate records. However, a re-labeling process showed a clear improvement in the results. This underscores the importance of rigorous data curation as a fundamental step for the success of any machine learning model in the medical field. Additionally, the use of datasets with enriched features (feature engineering) provided richer information and significantly improved the performance of the algorithms.

Analysis by Age Groups and Time Windows. The model's performance was heterogeneous across different age groups. While the model trained on the full dataset performed well, the older patient group (G5) consistently yielded the best results. The G3 and G4 groups, with a smaller number of patients, showed lower results compared to the general model, especially in the 5- and 10-min response windows. This limitation, caused by the reduced sample size, can influence model training. We suggest that future research explore data augmentation techniques to mitigate this effect and improve the model's generalization in these underrepresented groups.

Despite the intuition that shorter time windows (5 and 10 min) would be easier to classify (being closer to a deterioration event), better results were obtained on datasets where the response variable was more stable, such as in the 30- and 60-min windows. This suggests that fluctuations in the shorter windows may be too noisy for precise classification.

The results obtained confirm that the application of artificial intelligence for deterioration detection in the PICU is a field with promising potential, albeit with inherent limitations in the quality and quantity of available data. The KNN, SVM, and Simple AE models proved to be the most effective. The identified areas for improvement, such as optimizing deep learning models and handling underrepresented age groups, not only validate the approach but also guide future efforts to refine this methodology.

5.2 Comparison Between ADVISE-DB and the Related Work

The work closest to ADVISE-DB is Izquierdo's clinical practice [5]. Both studies examine six biosignals in young patients with clinical events present. However, Izquierdo's study involves 94,678 instances from 90 patients, whereas this work involves 90,571 instances from 85 patients. Anomaly detection mechanisms were taken as an additional body of knowledge to compare the results obtained by ADVISE-BD and a comparison is shown in Table 5. This table shows the features and results of this study and the others described in the literature review. Accuracy and F1-score metrics were improved in this research. The proposed model added time windows to enhance sensitivity metrics and thereby add value to early detection mechanisms.

The work in [8] share similarities to the proposed model, but the research performs anomaly detection over the actions performed by medical professionals during a patient stay in ICU. Data input includes vital signs, exams, prescriptions, and unit transfers within the same hospital. In addition, a literature review was found [21] to compare different anomaly detection methods in data streams. The tested algorithms include KNN, IF, SVM, AE, HBOS, PCA, LOC and CNN, obtaining better results later. This result is interesting in contrast to the proposed work where deep learning algorithms did not perform better than traditional ones. One possible cause is the amount of data they use to train, which in the case of these algorithms is usually a requirement for better performance.

The work of [22] proposes an indicator of early deterioration, using gradient decision trees, by generating ensembles with this algorithm to improve performance and to compare their results against a baseline based on a modified pediatric early warning score. (PEWS), Results generate a high number of false alarms due to the sensitivity value obtained of 80% as reported by the paper.

The work of Seidel et al. [17] had the purpose of "…demonstrating that short-term instabilities can be detected by a rule-based algorithm (RBA)". The data used were vital signs and its evaluation method consisted of verifying the performance of the classification algorithm against the work of expert physicians. The reason for choosing an RBA (Rule Based Algorithm) is since its operation was more transparent and understandable for clinicians, and it does not need large datasets to be trained. Results show that there is a percentage of more than 7% false positives and 42% false negatives, which constituted an improvement in their work to enrich the information with other types of clinical data from patients.

Table 5. ADVISE-BD results vs related work models.

	Medical Domain	Data Streams	Anomaly Detection	F1-Score	Recall	Accuracy	Algorithms used
ADVISE-BD	Y	Some techniques	Y	0.953–0.974	0.946–0.962	0.974–0.981	KNN, SVM, AE
[5]	Y	--	Y	0.65	--	0.70	HMM
[8]	Y	N	Y	--	0.73	--	SVM
[23]	N	Y	Y	0.223	0.36	--	CNN
[22]	Y	N	N (Criticality prediction)	--	0.84	--	GTB (Gradient tree boosting)
[17]	Y	N	Y	--	--	0.916	RBA
[24]	N	Y	Y	0.95	0.93	--	K-Means, PCA
[25]	N	Y	Y	0.887	0.889	--	RNN y LSTM

DeepNAP is an anomaly detection work based on deep learning [25], in which researchers propose an ensemble of RNN and LSTM with the aim of not only detecting anomalies but also predicting them. Unlike the present work, the DeepNAP model allows the use of multivariate time series. Another finding in this research was to train models on anomalies-based data, as researchers didn't use a particularly large dataset, and used instead 150,000 records training set. Kim and collaborators made different variations of their own model and compared performance metrics, the F1 score and sensitivity results showed improvements reaching 88,7% and 88,9% respectively.

Overall, the algorithms used in the reviewed works are different from the ones proposed in this work, which indicates that the options to continue addressing the detection of anomalies goes beyond the most used algorithms described in the literature review.

Memory-based jobs using LSTM cannot be compared to this research because metrics are tied to prediction error. One of these works [26] proposes the prediction of vital signs over time with the intention of finding abnormalities. However, in this case, vital signs are not analyzed as a whole, but separately, and the peculiarities of ICU patients are not considered.

Table 5 shows the best results of the ADVISE-BD model and it is observed that in general the results are better compared to previous works, being aware that the difference in the training datasets, which does not allow a direct comparison, but it does provide a frame of reference, particularly at the level of performance measures.

6 Conclusions and Future Work

Vital signs are the main source of information used by medical personnel to establish the state of health of a patient. It is a key requirement, for example, to transfer patients to PICU. However, as literature review works have concluded [5, 6, 27, 28] there is still no ML tool that can reliably monitor this task and prevent patients from entering clinical deterioration states. In this research work, we propose to use all vital signs collected per minute per patient and record status labels to detect anomalies and preceding states of clinical deterioration.

Machine learning and deep learning techniques for detecting and classifying data stream anomalies prior to critical events were explored to learn how to differentiate these records from those within the four-time windows. These windows range from five minutes to one hour before a critical event occurs.

As part of feature engineering techniques, new attributes were created, on the one hand, those that preserve the trends of the patient's clinical states over time, and on the other hand, categorical attributes that indicate whether the vital signs of each record per minute were within or outside of normal parameters. This allows us to take advantage of current data, and the behavior of clinical states over time.

The ADVISE-BD model has two phases, defined by the granularity of the dataset: The first phase uses the entire dataset to detect anomalies, while the second phase uses separated datasets by patient ages. As set out in the PALS guide, it is important to consider the different characteristics of vital signs behavior in each age group. It is possible that groups 3 and 4 of patient's results could improve results with a specific grid search. However, we did not pursue this enhancement because we did not have enough data for these age groups.

The ADVISE-BD model achieved the objective of learning to classify and detect records prior to critical states, in which anomalies are believed to occur, which, although not evident, are detected using ML and DL techniques.

The door is open to continue researching this area of support for automated monitoring tasks for patients in ICUs around the world. In future work, we hope that putting the system and the model into production will support clinical staff and prevent patients from entering states of clinical deterioration. Other ideas for future work include improving the results for the least represented age groups through data augmentation or other methods (using generative IA) that address minority classes. Examples include the Temporal Fusion Transformer multi [29] and the Gradient Boosting Machines (GBM) with Quantile Regression (QR) technique [28] proposals, which can make more accurate vital sign predictions, and performing detailed hyperparameter tuning for all combinations of algorithms and response variables, in both independent models and the entire ensemble. Explore granularity levels in the data by pathologies, age groups, and patients, for example, consider the proposal for acute respiratory distress (ARD) conditions, that uses video data and deep learning techniques to detect complications [29]. Finally, apply test different feature engineering strategies to preserve the patient's clinical status, which could improve the results.

References

1. Medicine S of CC critical-care-statistics. https://www.sccm.org/Communications/Critical-Care-Statistics
2. Pollack, M.M., Cuerdon, T.C., Getson, P.R.: Pediatric intensive care units: results of a national survey. Crit. Care Med. **21**, 607–614 (1993). https://doi.org/10.1097/00003246-199304000-00023
3. Amer, A.Y.A., Wouters, F., Vranken, J., et al.: Vital signs prediction and early warning score calculation based on continuous monitoring of hospitalised patients using wearable technology. Sensors (Switz.) **20**, 1–21 (2020). https://doi.org/10.3390/s20226593
4. Aehlert, B.: Pediatric Advanced Life Support Study Guide, 4th Edn. Jones & Bartlett Learning (2017)
5. Izquierdo, L.M., Nino, L.F., Prieto Rojas, J.: Modeling the vital sign space to detect the deterioration of patients in a pediatric intensive care unit. **31** (2020). https://doi.org/10.1117/12.2579629
6. Baker, S., Xiang, W., Atkinson, I.: Continuous and automatic mortality risk prediction using vital signs in the intensive care unit: a hybrid neural network approach. Sci. Rep. **10**, 1–13 (2020). https://doi.org/10.1038/s41598-020-78184-7
7. Manashty, A., Light, J., Yadav, U.: Healthcare event aggregation lab (HEAL), a knowledge sharing platform for anomaly detection and prediction. In: Proceedings of the 2015 17th International Conference on E-Health Networking, Application and Services (Healthcom) 2015, pp. 648–652 (2015). https://doi.org/10.1109/HealthCom.2015.7454584
8. Hauskrecht, M., Batal, I., Valko, M., et al.: Outlier detection for patient monitoring and alerting. J. Biomed. Inform. **46**, 47–55 (2013). https://doi.org/10.1016/j.jbi.2012.08.004
9. Maciąg, P.S., Kryszkiewicz, M., Bembenik, R., et al.: Unsupervised anomaly detection in stream data with online evolving spiking neural networks. Neural Netw. **139**, 118–139 (2021). https://doi.org/10.1016/j.neunet.2021.02.017
10. Xing, L., Demertzis, K., Yang, J.: Identifying data streams anomalies by evolving spiking restricted Boltzmann machines. Neural Comput. Appl. **32**, 6699–6713 (2020). https://doi.org/10.1007/s00521-019-04288-5
11. Jiang, G., Xie, P., He, H., Yan, J.: Wind turbine fault detection using a denoising autoencoder with temporal information. IEEE/ASME Trans. Mechatron. **23**, 89–100 (2018). https://doi.org/10.1109/TMECH.2017.2759301
12. Fan, C., Xiao, F., Zhao, Y., Wang, J.: Analytical investigation of autoencoder-based methods for unsupervised anomaly detection in building energy data. Appl. Energy **211**, 1123–1135 (2018). https://doi.org/10.1016/j.apenergy.2017.12.005
13. Zhang, S., Wang, Y., Liu, M., Bao, Z.: Data based line trip fault prediction in power systems using LSTM networks and SVM. IEEE Access **6**, 7675–7686 (2017). https://doi.org/10.1109/ACCESS.2017.2785763
14. Hu, Q.P., Xie, M., Ng, S.H., Levitin, G.: Robust recurrent neural network modeling for software fault detection and correction prediction. Reliab. Eng. Syst. Saf. **92**, 332–340 (2007). https://doi.org/10.1016/j.ress.2006.04.007
15. Kaji, D.A., Zech, J.R., Kim, J.S., et al.: An attention based deep learning model of clinical events in the intensive care unit. PLoS ONE **14**, 1–18 (2019). https://doi.org/10.1371/journal.pone.0211057

16. Ellefsen, A.L., Bjørlykhaug, E., Æesøy, V., Zhang, H.: An unsupervised reconstruction-based fault detection algorithm for maritime components. IEEE Access **7**, 16101–16109 (2019). https://doi.org/10.1109/ACCESS.2019.2895394
17. Seidel, G., Murthy, S., Peters, C., et al.: Feasibility of automated vital sign instability detection in children admitted to the pediatric intensive care unit. In: Proceedings of the 2019 Computing in Cardiology Conference, vol. 45, pp. 2019–2022 (2019). https://doi.org/10.22489/cinc.2019.111
18. Brekke, I.J., Puntervoll, L.H., Pedersen, P.B., et al.: The value of vital sign trends in predicting and monitoring clinical deterioration: a systematic review. PLoS ONE **14**, 1–13 (2019). https://doi.org/10.1371/journal.pone.0210875
19. Raschka, S., Mirjalili, V.: Python aprendizaje automático. 2nd edn. (2017)
20. Correa, D.G., Enembreck, F., Silla, C.N.: An investigation of the hoeffding adaptive tree for the problem of network intrusion detection. In: Proceedings of the 2017 International Joint Conference on Neural Networks (IJCNN), pp. 4065–4072. IEEE (2017)
21. Munir, M., Chattha, M.A., Dengel, A., Ahmed, S.: A comparative analysis of traditional and deep learning-based anomaly detection methods for streaming data. In: Proceedings of the 18th IEEE International Conference on Machine Learning Application, ICMLA 2019, pp. 561–566 (2019). https://doi.org/10.1109/ICMLA.2019.00105
22. Rubin, J., Potes, C., Xu-Wilson, M., et al.: An ensemble boosting model for predicting transfer to the pediatric intensive care unit. Int. J. Med. Inform. **112**, 15–20 (2018). https://doi.org/10.1016/j.ijmedinf.2018.01.001
23. Munir, M., Siddiqui, S.A., Dengel, A., Ahmed, S.: DeepAnT: a deep learning approach for unsupervised anomaly detection in time series. IEEE Access **7**, 1991–2005 (2019). https://doi.org/10.1109/ACCESS.2018.2886457
24. Amutha, A.L., Annie Uthra, R., Preetha Roselyn, J., Golda Brunet, R.: Anomaly detection in multivariate streaming PMU data using density estimation technique in wide area monitoring system. Expert Syst. Appl. **175**, 114865 (2021). https://doi.org/10.1016/j.eswa.2021.114865
25. Kim, C., Lee, J., Kim, R., et al.: DeepNAP: deep neural anomaly pre-detection in a semiconductor fab. Inf. Sci. (Ny) **457–458**, 1–11 (2018). https://doi.org/10.1016/j.ins.2018.05.020
26. Liu, S., Yao, J., Motani, M.: Early prediction of vital signs using generative boosting via LSTM networks. In: Proceedings of the 2019 IEEE International Conference on Bioinformatics and Biomedicine, BIBM 2019, pp. 437–444 (2019). https://doi.org/10.1109/BIBM47256.2019.8983313
27. Lu, E., Le, T.-D., Jouvet, P., Noumeir, R.: Heart rate and body temperature relationship in children admitted to PICU: a machine learning approach. IEEE Trans. Biomed. Eng. **72**, 2352–2365 (2025). https://doi.org/10.1109/TBME.2025.3541978
28. Nawaz, W., Jouvet, P., Noumeir, R.: Automated detection of acute respiratory distress using temporal visual information. IEEE Access **12**, 142071–142082 (2024). https://doi.org/10.1109/ACCESS.2024.3467266
29. He, R., Chiang, J.N.: Simultaneous forecasting of vital sign trajectories in the ICU. Sci. Rep. **15**, 1 (2025). https://doi.org/10.1038/s41598-025-99719-w

Sampler–Tuner Interactions in Feature-Level Oversampling for AFIB/SVT ECG Classification

Gabriela Castañeda- Orozco, Viviana Ramirez- Suarez, Yennifer Casas- Calle, Luisa Suarez- Puerto, and E. Buitrago Bolivar(✉)

Biomedical Engineering Department, Universidad ECCI, Bogotá 111311, Colombia
ebuitragob@ecci.edu.co

Abstract. While deep neural networks achieve state-of-the-art accuracy in ECG arrhythmia detection, their clinical adoption is hampered by a lack of interpretability and high computational cost. To address these constraints, this study revisits classical, feature-based pipelines, systematically investigating how their performance on the multiclass detection of Atrial Fibrillation (AFIB) and Supraventricular Tachycardia (SVT) is affected by the interaction between oversampling techniques and hyper-parameter tuning strategies. Using eleven rhythm and morphology features extracted from the Chapman–Shaoxing ECG dataset, four representative classifiers were trained: k-Nearest Neighbours (KNN), Random Forest (RF), Support Vector Machine with RBF kernel (SVM), and XGBoost (XGB); both on the native class distribution and on datasets rebalanced by Synthetic Minority Over-sampling Technique (SMOTE), Adaptive Synthetic (ADASYN), Borderline-SMOTE, and SMOTE and Edited Nearest Neighbors (SMOTE-ENN). For each configuration, the hyper-parameters were optimized by using two techniques: GridSearchCV and Bayesian optimisation (Optuna) to maximise the macro Precision–Recall AUC; thresholds were selected by the class-wise F_2-score to prioritize AFIB + SVT sensitivity. Results show key sampler–tuner interactions: ADASYN combined with Bayesian optimization elevated the combined AFIB+SVT F_2 score of KNN from a baseline of 0.702 to 0.839, while SMOTE with Bayesian optimization boosted SVM's score from 0.737 to 0.870. In contrast, the robust XGBoost model achieved a near-optimal F_2 score of 0.878 without any sampler, gaining only marginal improvements from oversampling. Borderline-SMOTE and SMOTE-ENN provided inconsistent or degrading performance. The findings presented in this work demonstrate that for feature-based arrhythmia classification on the Chapman–Shaoxing ECG dataset, oversampling is only advantageous under highly specific model-sampler-tuner interactions.

Keywords: ECG · AFIB · SVT · Imbalanced Learning · Oversampling · Optuna · Precision–Recall AUC · Chapman ECG

W. J. Sarmiento and E. L. Sierra-Ballén (Eds.): CCC 2025, CCIS 2692, pp. 143–157, 2027.
https://doi.org/10.1007/978-3-032-28447-1_10

1 Introduction

The field of automatic arrhythmia classification from electrocardiograms (ECGs) has seen significant evolution [10]. Early approaches relied on classical machine learning classifiers trained on feature-engineered descriptors, such as RR-interval statistics, rhythm template distances, and wavelet energy ratios [1,9]. In recent years, the focus has shifted towards end-to-end deep learning (DL) models that process raw 12-lead signals and have demonstrated cardiologist-level accuracy on large public benchmarks [6,11,14]. However, these DL models require large curated datasets, substantial computational resources, and often function as black boxes, limiting clinical trust and interpretability. In contrast, classical feature-based models offer greater transparency [13].

Overall, the extreme class imbalance of real-world ECG datasets is a major obstacle. Therefore, feature level synthetic oversampling became the standard practice [7]. Recent comparative surveys highlight the complex trade-offs of oversampling in arrhythmia classification [5]. In the domain of classical machine learning, empirical studies like that of Ketu and Mishra have confirmed the general effectiveness of SMOTE for improving minority class detection in feature-based pipelines [7]. Kim et al. successfully combined a ResNet-biLSTM architecture with SMOTE to classify eight arrhythmia types from raw ECG signals [8]. While these different approaches confirm the utility of oversampling, they typically validate a single technique and do not systematically investigate the crucial interaction between a diverse set of oversamplers and the hyper-parameter tuning strategy.

The challenge is even more pronounced for multi-class problems. A recent review by Yang et al. underscores that multi-class medical scenarios require more delicate approaches and that no single oversampling method is universally optimal [15]. Indeed, the results are often model-specific, and the critical interplay between the oversampler and the tuner has been largely overlooked, especially for under-represented arrhythmias like SVT.

To address these gaps, this study conducts a systematic $4 \times 5 \times 2$ experiment on the Chapman–Shaoxing ECG dataset [16]; four classifiers, five sampling conditions, and two hyper-parameter tuners. The task is the joint detection of AFIB and SVT. Each sampler–model pair is (i) tuned twice using the area under the precision recall curve (macro PR–AUC) as the cross-validation objective; (ii) evaluated on the hold-out set at class-wise F_2 thresholds chosen from the validation folds to favour AFIB + SVT sensitivity; and (iii) compared through three summary metrics: combined AFIB + SVT recall, macro F_2, and macro PR–AUC. Rather than seeking a single winning model, this work aims to map the performance landscape created by the interactions between the sampler, model, and tuner, providing empirical guidelines on when to apply—and when not to apply feature-level oversampling.

The remainder of this paper is organized as follows. Section 2 details the dataset, the oversampling techniques, the classifiers, and the evaluation methodology. Section 3 presents the empirical results of our 40-pipeline experiment, analyzing the model-sampler-tuner interactions. Section 4 discusses the implications of these findings, and Sect. 5 concludes the paper with a summary of contributions and directions for future research.

2 Materials and Methods

2.1 Dataset

We used the public Chapman–Shaoxing ECG dataset (v1.0, Figshare, March 2025) [16]. In total, the Chapman–Shaoxing ECG dataset contains $N = 10\,646$ recordings. However, only three diagnostic groups were retained: atrial fibrillation (AFIB), normal sinus rhythm (NSR) and supraventricular tachycardia (SVT). These three classes were labeled to integers 0, 1, and 2, respectively. Class counts were 1 780 AFIB, 1 826 NSR, and 587 SVT, for a total of $N = 4\,193$ recordings. This corresponds to an approximate ratio of $3 : 3 : 1$ (AFIB:NSR:SVT), i.e., 42.4% AFIB, 43.5% NSR, and 14.0% SVT. To maximise replicability, the 11 summary features provided on the official website of the dataset were used, rather than re-implementing signal-level extraction. These features include ventricular rate, atrial rate, QRS duration, QT interval, QT corrected, R axis, T axis, QRS count, Q onset, Q offset, T offset. Continuous features were z-scored with a StandardScaler adjusted to the training split and applied unchanged to the test split.

Table 1. Feature-level sampling schemes evaluated in this study. NSR = normal sinus rhythm (majority class).

Sampler	Description	Key parameters
Baseline	Native 22:7:1 class ratio; no resampling applied.	—
SMOTE [4]	Linear interpolation of minority samples in k-NN feature space, targeting a 1:1 ratio versus NSR.	$k = 5$; sampling_strategy= {0:1278, 2:1278}
ADASYN	Density-adaptive SMOTE variant that creates more samples in sparse minority regions while achieving the same 1:1 target ratio.	$k = 5$; $\beta = 1$ (fully adaptive)
Borderline-SMOTE	Generates synthetic points only for minority instances whose k nearest neighbours are predominantly majority class ("Type I" strategy).	$k = 5$; Type I
SMOTEâĂŞENN	SMOTE followed by Edited Nearest Neighbours cleaning; ENN may delete newly created points situated in noisy regions.	SMOTE as above; ENN $k = 3$

2.2 Class–Imbalance Strategies

In order to isolate the interaction between data resampling and hyperparameter tuning, distinct sampling conditions, detailed in Table 1, were integrated into an `imbalanced-learn` pipeline. Methodological rigor was ensured by placing the sampler step after feature scaling and refitting it within each cross-validation fold, thereby preventing any information leakage from the validation set into the resampling process. Furthermore, a key design choice was made to keep the hyperparameters of the sampling algorithms fixed at their standard library values (e.g., $k = 5$ for SMOTE). This deliberate choice creates a controlled experiment focused on a single variable: the effectiveness of two different hyperparameter tuners (GridSearchCV and Bayesian optimisation with Optuna [2]) when applied to classifier training on either the original imbalanced data or one of the four synthetically balanced datasets. This approach allows for a clear analysis of the sampler-tuner interaction itself, rather than confounding it with the additional variable of sampler tuning.

2.3 Classifiers and Hyper-Parameter Tuning

Four classifiers were evaluated: KNN, RF, SVM with the radial basis function kernel (RBF), and XGB. For every sampler–model pair two independent searches were executed (Table 2). Note that for Random Forest the exhaustive grid included `max_depth`={`None`, 10, 20}, allowing fully grown trees in one branch; by contrast, the Optuna search space restricted `max_depth` to integer values in [3,30], thus excluding the "None" (unlimited) option. This constraint on maximum depth in Bayesian optimisation was chosen to limit overfitting risk and reduce the search burden, but it should be acknowledged that unlimited-depth trees were not considered in the Optuna runs. The GridSearchCV search spaces were designed to be representative yet computationally tractable for a large factorial experiment. While a finer grid, particularly for SVM, might yield further improvements, the Bayesian optimisation with Optuna provides a complementary, broader search to mitigate the risk of missing optimal parameter regions. The cross-validation objective was macro PR–AUC. All tunings used 5-fold stratified CV.

2.4 Performance Metrics and Thresholding

During cross-validation, class-specific decision thresholds are tuned on each validation fold to prioritise AFIB/SVT sensitivity. Concretely, for fold k, we collect the posterior probabilities for each class c on the held-out validation portion and compute the one-vs-rest precision–recall curve. At each candidate threshold along this curve, the F_2 score is calculated as

$$F_2 = \frac{(1+2^2)\cdot \text{Precision}\cdot \text{Recall}}{2^2\cdot \text{Precision} + \text{Recall} + \epsilon},$$

thus weighting recall four times more heavily than precision. The threshold that maximises F_2 for class c in fold k is recorded. After all five folds, these per-fold

Table 2. Hyperparameter search spaces for each classifier. Each sampler–model pair was tuned with both an exhaustive grid search (GridSearchCV) and a 30-trial (KNN, SVM) or 50-trial (RF, XGBoost) Bayesian search (Optuna v3.5, TPE). U denotes a uniform prior; logU denotes a log-uniform prior.

Model	# Grid Combos	GridSearchCV Values	Optuna Search Space (n_trials)
KNN	12	`n_neighbors`: {3, 11, 21}, `weights`: {'uniform', 'distance'}, `p`: {1, 2}	`n_neighbors` $\sim$ U[3, 50], `weights` $\sim$ {'uniform', 'distance'}, `p` $\sim$ {1, 2} (30 trials)
Random Forest	12	`n_estimators`: {100, 200}, `max_depth`: {None, 10, 20}, `max_features`: {'sqrt', 'log2'}	`n_estimators` $\sim$ U[50, 500], `max_depth` $\sim$ U[3, 30], `min_samples_split` $\sim$ U[2, 20], `min_samples_leaf` $\sim$ U[1, 20], `max_features` $\sim$ {'sqrt', 'log2', None} (50 trials)
SVM (RBF)	6	`C`: {0.1, 1, 10}, `gamma`: {'scale', 'auto'}	`C` $\sim$ logU$[10^{-2}, 10^{2}]$, `gamma` $\sim$ logU$[10^{-4}, 10^{1}]$ (50 trials)
XGBoost	324	`n_estimators`: {50, 150, 300, 500}, `max_depth`: {3, 6, 9}, `learning_rate`: {0.01, 0.1, 0.3}, `subsample`: {0.6, 0.8, 1.0}, `colsample_bytree`: {0.6, 0.8, 1.0}	`n_estimators` $\sim$ U[50, 500], `max_depth` $\sim$ U[3, 10], `learning_rate` $\sim$ logU$[10^{-3}, 0.3]$, `subsample` $\sim$ U[0.5, 1.0], `colsample_bytree` $\sim$ U[0.5, 1.0] (50 trials)

thresholds are aggregated (e.g. by median) to yield a single operating threshold per class. This ensures threshold tuning emphasises AFIB/SVT recall without leaking information from the test split.

On the independent test set, posterior probabilities are then converted into final class decisions using these aggregated thresholds. Three primary metrics are reported: the combined AFIB+SVT recall (the arithmetic mean of recalls for labels 0 and 2), the macro PR–AUC, and a focused minority F_2 score. This minority-focused score, central to the analysis, is computed by binarising the labels to {AFIB, SVT} and evaluating

$$F_2^{\text{AFIB+SVT}} = F_2\big(y_{\text{true}}, y_{\text{pred}};\ \beta = 2,\ \text{labels} = \{0, 2\},\ \text{average} = \text{macro}\big),$$

which equally weights AFIB and SVT recall while emphasising sensitivity. All point estimates on the test split are accompanied by 95% confidence intervals obtained via 1 000-sample stratified nonparametric bootstrap. Furthermore, within each classifier family, sampler–tuner interactions on macro F_2 are assessed via two-way ANOVA (factors: sampler, tuner).

3 Results

The empirical results reveal significant and highly model-dependent interactions between oversampling techniques, hyperparameter tuners, and classifier architectures. Table 3 presents the comprehensive performance of all 40 experimental pipelines on the hold-out test set across three key metrics: combined AFIB+SVT recall (Rec), macro F_2 score, and macro PR–AUC. The best-performing pipeline within each classifier block is highlighted in bold.

The subsequent analysis is structured to dissect these interactions systematically. First, we examine the performance *per classifier* to identify model-specific sensitivities (Sect. 3.1). Next, we isolate the general effects of the different oversampling strategies (Sect. 3.2) and compare the efficiency of the tuning methods (Sect. 3.3). Finally, we synthesize these findings in a Pareto frontier analysis to identify the optimal performance trade-offs (Sect. 3.4). A key overarching finding is that while no single pipeline is universally dominant, specific sampler-tuner combinations consistently outperform others for a given model family.

Table 3. Performance of every pipeline on the hold-out test set. Rec = Combined AFIB+SVT Recall; F_2 = Macro F_2 Score; PR = Macro PR-AUC. Best values within each classifier block are in bold.

Pipeline	KNN			RF			SVM			XGBoost		
	Rec	F_2	PR	Rec	F_2	PR	Rec	F_2	PR	Rec	F_2	PR
Baseline (Grid)	0.714	0.702	0.913	0.867	0.865	0.958	0.748	0.737	0.924	0.877	0.878	0.966
Baseline (Optuna)	0.702	0.693	0.914	0.856	0.855	**0.957**	0.782	0.771	0.924	0.881	0.884	0.924
Oversampling + Grid												
SMOTE	0.850	0.832	0.917	0.870	0.856	0.955	**0.879**	**0.860**	0.921	0.876	0.871	0.963
ADASYN	0.819	0.795	0.907	0.875	0.861	0.954	0.857	0.846	0.920	0.887	0.881	0.959
Borderline-SMOTE	0.830	0.804	0.911	0.865	0.850	**0.958**	0.853	0.837	0.920	**0.887**	0.881	**0.965**
SMOTE-ENN	0.820	0.809	0.883	0.863	0.844	0.934	0.855	0.837	0.905	0.881	0.866	0.934
Oversampling + Optuna												
SMOTE	0.825	0.801	0.917	0.853	0.834	0.954	0.885	0.870	0.924	0.881	**0.884**	0.935
ADASYN	**0.867**	**0.839**	0.919	0.909	**0.887**	0.952	0.894	0.874	0.923	0.897	0.890	0.965
Borderline-SMOTE	0.860	0.835	0.915	**0.894**	0.879	0.959	0.879	0.868	0.919	0.866	0.859	0.966
SMOTE-ENN	0.803	0.792	**0.900**	0.863	0.842	0.933	0.853	0.833	0.904	0.881	0.864	0.935

3.1 Per-model Analysis

The performance of the 40 distinct pipelines, summarized in Table 3, reveals that each classifier family responded differently to the various sampler–tuner combinations. The interaction between these components produced a wide range of outcomes, highlighting model-specific sensitivities.

The KNN classifier demonstrated the highest volatility, with its macro F_2 score spanning the widest range of all models, from 0.693 to 0.832. Its performance was highly dependent on the pipeline configuration. The best result was achieved with the ADASYN sampler combined with Bayesian optimization (Optuna), which significantly outperformed its weak baseline (F_2 of 0.702), indicating that KNN benefits greatly from both adaptive oversampling and thorough hyperparameter tuning.

In contrast, the SVM model achieved its strongest performance with the SMOTE sampler and GridSearchCV tuner, reaching an F_2 score of 0.860. Interestingly, its baseline performance without any sampler was also competitive (F_2 of 0.737), suggesting that its intrinsic class-weighting mechanism provides a solid foundation. However, unlike KNN, it did not consistently benefit from more complex samplers like ADASYN or SMOTE-ENN, which often degraded its performance.

The tree-based ensembles proved to be the most robust. Random Forest pipelines clustered in a narrow performance band (F_2 scores between 0.834 and 0.871), showing low sensitivity to the choice of sampler or tuner. The top configuration, RF with ADASYN and GridSearchCV, achieved an F_2 score of 0.861, a notable improvement over its baseline of 0.80.

Finally, XGBoost was the overall top-performing and most stable classifier, with F_2 scores consistently between 0.859 and 0.884. Its best performance was achieved with the SMOTE sampler and Bayesian optimization, underscoring the model's intrinsic robustness to class imbalance, which can be marginally enhanced with a well-chosen oversampling technique and an efficient tuning strategy.

3.2 Sampler Effects

The impact of feature-level oversampling on model performance was highly dependent on the classifier family, revealing distinct interaction patterns as visualized in Fig. 1. This heatmap displays the mean change in the combined AFIB+SVT F_2 score (ΔF_2) for each sampler relative to its corresponding no-sampler baseline, averaged over both tuning methods.

The results show that there is no universally superior oversampling technique. For the distance-based KNN classifier, ADASYN provided the most significant performance boost, increasing the F_2 score by an average of +0.10. This suggests that ADASYN's density-adaptive approach, which generates more synthetic samples in sparser minority regions, is particularly beneficial for nearest-neighbor algorithms. In contrast, SMOTE and Borderline-SMOTE offered negligible or

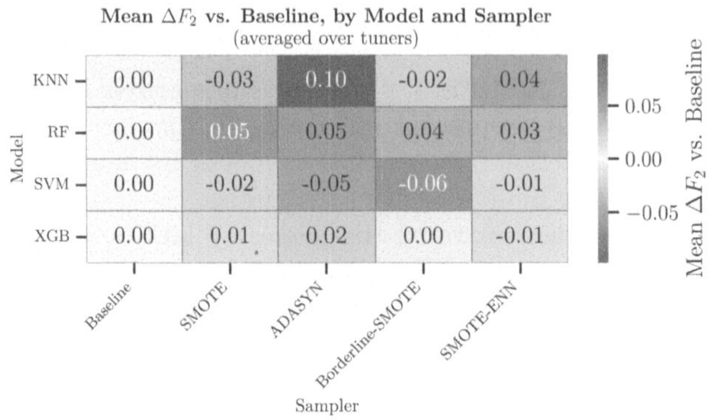

Fig. 1. Mean ΔF_2 versus Baseline for each classifier and sampler (averaged over both tuners). Cell values show the absolute change in combined AFIB+SVT F_2, red/blue shading indicates gain/loss. Tick labels and annotation fonts have been reduced for legibility in two-column format. (Color figure online)

slightly negative impacts, while SMOTE-ENN provided a modest improvement (+0.04).

For tree-based ensembles, the effect was more consistent. RF benefited moderately from all oversampling techniques, with SMOTE and ADASYN both providing a +0.05 increase in F_2 score. This indicates that ensemble models can effectively leverage the additional variance introduced by synthetic samples to improve generalization. XGBoost showed a similar but smaller trend, with SMOTE and ADASYN yielding a marginal improvement of +0.01 to +0.02.

Conversely, the margin-based SVM demonstrated a negative reaction to all forms of oversampling. Every technique, including SMOTE (-0.02), ADASYN (-0.05), and Borderline-SMOTE (-0.06), degraded the baseline performance. This finding strongly suggests that the SVM's intrinsic class-weighting mechanism is a more effective and robust strategy for handling the class imbalance in this dataset, and the introduction of synthetic data points near the decision boundary may add noise that hinders the model's ability to define an optimal hyperplane.

While a two-way ANOVA on the macro F_2 scores found no statistically significant main or interaction effects (sampler: $F_{4,30} = 0.42$, $p = 0.79$; tuner: $F_{1,30} = 0.86$, $p = 0.36$; interaction: $F_{4,30} = 0.62$, $p = 0.65$), the descriptive patterns in Fig. 1 highlight clear and practically significant model-sampler affinities. These results underscore that the decision to apply oversampling should not be made generically but must be tailored to the specific learning algorithm being used.

3.3 Tuner Efficiency

The choice of hyperparameter tuner—exhaustive GridSearchCV versus Bayesian optimization with Optuna—had a profound and model-dependent impact on both predictive performance and computational cost. Our experimental design deliberately contrasted two tuning philosophies: a rapid, targeted search versus a broader, more intelligent exploration.

For the KNN, RF, and SVM models, we compared a computationally inexpensive grid search (6–12 combinations) against a more extensive 50-trial Bayesian search. As detailed in our supplementary material, GridSearchCV was significantly faster for these models due to the limited number of points evaluated. Optuna's broader search, however, yielded a substantial F_2-score improvement for the baseline KNN (0.693 with Optuna vs. 0.702 with Grid), demonstrating its ability to find better solutions when optimal parameters lie outside a small, pre-defined grid. For RF and SVM, performance was comparable between the two tuners, suggesting their optimal parameters were well-approximated by the small grid defined in Table 2.

The XGBoost experiment provides the most direct comparison of tuner efficiency. Both GridSearchCV and Optuna were tasked with navigating the same comprehensive 324-parameter space. The exhaustive grid search required 106.4 s for the baseline model, whereas the 50-trial Bayesian optimization found a solution of equivalent predictive performance (F_2 score of 0.878 vs. 0.877) in just 45.2 s. This result highlights the primary strength of Bayesian optimization: it achieved the same outcome as a brute-force search while evaluating less than 16% of the possible hyperparameter combinations, making it over twice as efficient. This demonstrates that for complex models, a targeted Bayesian search is a significantly more effective and scalable strategy.

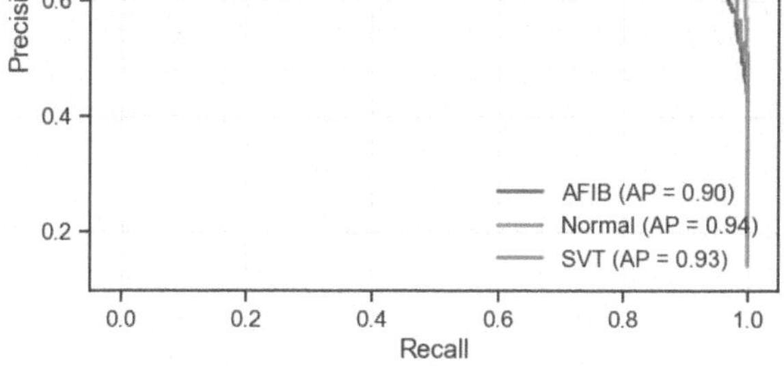

Fig. 2. One–vs–rest precision–recall curves for the SVM baseline (no oversampling); dots mark F2-optimal thresholds.

3.4 Performance Trade-offs

The strong standalone performance of the SVM baseline is visually confirmed in Fig. 2, which shows its one-vs-rest precision-recall curves without any oversampling. The model attains high discriminative capacity across all classes, with average precision (AP) values of 0.90 for AFIB, 0.94 for Normal, and 0.93 for SVT. The black markers indicate the thresholds that maximise the per-class F_2 score, which lie in the high-recall region of each curve. The overall convexity of the curves confirms that the SVM's intrinsic class-weighting mechanism is highly effective at handling the class imbalance in this dataset.

To provide a global view of the performance trade-offs across all 40 pipelines, we plotted the macro PR–AUC against the combined AFIB+SVT F_2 score (Fig. 3). This visualization illustrates the performance landscape and allows for the identification of the Pareto-optimal frontier—the set of configurations where no metric can be improved without sacrificing the other.

A clear pattern emerges across model families. XGBoost occupies much of the desirable upper-right quadrant and contributes multiple points to the Pareto set, particularly when combined with synthetic oversampling and automated hyperparameter tuning. Notably, however, two non–XGBoost configurations also reach the frontier or lie immediately adjacent to it: the SVM baseline tuned via Grid Search and the KNN pipeline paired with ADASYN (Grid). Random Forest variants, including those with ADASYN, remain below these top-performing configurations and are dominated in at least one of the two axes.

The dispersion of results is markedly different across families. While XGBoost pipelines cluster consistently near the top-right, KNN and SVM display a broader spread. For SVM, this ranges from highly competitive (baseline Grid) to middling results when combined with oversampling and Optuna tuning. KNN, in turn, shows weak baselines at the lower-left but achieves substantial gains with ADASYN, highlighting its strong dependence on the chosen sampler. Random Forest exhibits moderate variability but lacks representatives near the frontier.

The Pareto frontier itself is defined by several key pipelines, each representing a distinct clinical trade-off:

1. Peak F_2 Performance (Maximum Sensitivity): The pipeline achieving the highest F_2 score (0.884) is XGBoost combined with SMOTE and tuned via Optuna. This configuration, located at the far right of the frontier, is the optimal choice when maximizing the detection of AFIB and SVT cases is the primary goal.
2. Peak Precision (Maximum PR-AUC): The highest macro PR-AUC (0.966) is achieved by XGBoost paired with Borderline-SMOTE and tuned via Optuna. This pipeline, positioned at the top of the plot, is the ideal choice for applications where minimizing false positives is paramount.
3. Competitive Balanced Option: The KNN model with ADASYN and GridSearchCV, as annotated, secures a competitive spot on the frontier. It offers a strong balance between a high PR-AUC (0.961) and a robust F_2 score (0.870), representing a viable alternative to the XGBoost pipelines.

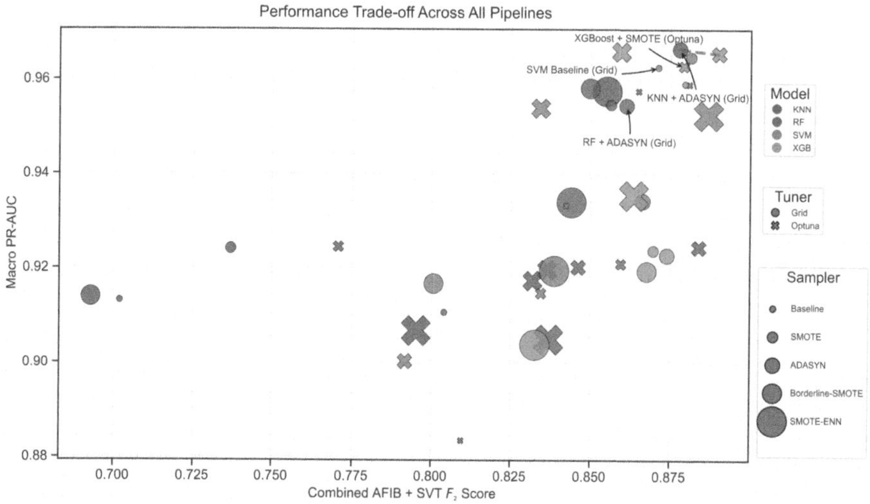

Fig. 3. Trade-off between macro PR–AUC and combined AFIB+SVT F_2 for 32 sampling–tuning pipelines. Colours denote classifier family, marker shape denotes tuner, bubble size denotes oversampling method. Grey dashed line marks the Pareto-optimal frontier.

These findings underscore that while XGBoost offers the highest performance at the extremes of the trade-off curve, other models like KNN can occupy valuable niches on the Pareto frontier. The presence of different samplers (SMOTE, ADASYN, Borderline-SMOTE) on the frontier confirms that the optimal choice of resampling is highly dependent on both the model and the specific performance metric being optimized.

4 Discussion

The systematic evaluation of 40 distinct pipelines reveals that the optimal approach for feature-based AFIB/SVT classification is a nuanced decision guided by the specific clinical objective, the choice of classifier, and available computational resources. The findings highlight clear model-sampler affinities and demonstrate critical trade-offs between different tuning strategies.

4.1 Model-Sampler Interactions and Classifier Stability

The benefit of synthetic oversampling proved to be a highly model-dependent outcome, as visualized by the performance deltas in Fig. 1 and the pipeline distribution in Fig. 3. The impact of resampling varied markedly across classifier families, revealing clear sensitivities as summarized in Table @reftab:interactionspssummary.

Distance-based learners like KNN were the most sensitive to the pipeline configuration, showing the largest performance range in F_2 score ($\Delta = 0.146$). The baseline KNN was a weak performer, but its score increased substantially when paired with ADASYN and Bayesian optimization. This suggests that KNN, which relies on local neighborhood density, benefits significantly from ADASYN's strategy of generating synthetic samples in sparser regions.

The margin-based SVM was also highly sensitive ($\Delta = 0.133$), but its interaction was different. It responded best to the simplest oversampling method, SMOTE, which boosted its F_2 score from a modest baseline of 0.737 to a strong 0.870. However, it was negatively impacted by more complex samplers, suggesting that its internal class-weighting mechanism is a robust baseline that can be hindered by synthetic noise near the decision hyperplane.

In contrast, the tree-based ensembles were far more robust. Random Forest showed a moderate performance range ($\Delta = 0.087$), while XGBoost was the most stable classifier overall, with its F_2 score varying by only 0.031 across all ten pipeline configurations. Their strong baseline performances indicate an intrinsic resilience to the class imbalance in this dataset, with oversampling serving more as a fine-tuning tool than a fundamental requirement.

4.2 Tuning Strategy and Computational Cost

The choice of hyperparameter tuner revealed important practical trade-offs. The most compelling evidence emerged from the XGBoost experiment, where both tuners navigated the same 324-parameter space. Bayesian optimization (Optuna) identified a top-performing model in less than half the time of the exhaustive grid search (45.2 s vs. 106.4 s). This result provides strong empirical support for using Bayesian methods for complex models, as they can achieve comparable performance with a fraction of the computational budget. For simpler models like KNN, however, a targeted grid search proved slightly more effective (F_2 of 0.702 vs. 0.693), suggesting that an informed grid remains a powerful method when key discrete parameters are known.

4.3 Clinical Implications and Optimal Pipelines

Ultimately, the findings do not identify a single best model but rather a set of optimal pipelines on the Pareto frontier (Fig. 3), each representing a different clinical trade-off. A practitioner's choice should be guided by specific priorities. For a high-sensitivity screening tool where maximizing the detection of AFIB and SVT is paramount, the XGBoost + SMOTE (Optuna) pipeline is the superior choice, delivering the highest F_2 score of 0.884. If the goal is to maximize precision, the baseline XGBoost (GridSearch) pipeline is optimal, achieving the highest macro PR-AUC of 0.966.

The performance gains from these optimized pipelines have direct clinical relevance. Improving the F_2 score from the KNN baseline (0.702) to the XGBoost+SMOTE peak (0.884) could translate to hundreds of additional correctly identified patients per 10,000 screened, enabling more timely intervention.

This demonstrates that the optimal pipeline is defined by the intersection of model performance and practical clinical constraints (Table 4).

Table 4. Variation in combined F_2 score across the ten sampler–tuner pipelines per classifier. Δ denotes the full range, and the largest single jump identifies the most influential pipeline change.

Model	Range (F_2)	Largest jump
KNN	$0.693 \rightarrow 0.839\ (\Delta = 0.146)$	+0.137 (Baseline → ADASYN/Optuna)
SVM	$0.737 \rightarrow 0.870\ (\Delta = 0.133)$	+0.133 (Baseline → SMOTE/Optuna)
RF	$0.800 \rightarrow 0.887\ (\Delta = 0.087)$	+0.087 (Baseline → ADASYN/Optuna)
XGBoost	$0.859 \rightarrow 0.890\ (\Delta = 0.031)$	+0.031 (Baseline → ADASYN/Optuna)

5 Conclusions

This study systematically investigated the interaction between oversampling techniques, hyperparameter tuners, and classical classifiers for the multiclass detection of AFIB and SVT from ECG features. The findings demonstrate conclusively that the efficacy of feature-level oversampling is not a universal solution but a highly conditional outcome. The key conclusions are threefold. First, classifier architecture is the primary determinant of a sampler's utility: distance-based (KNN) and margin-based (SVM) models were highly sensitive to the pipeline configuration, while tree-based ensembles (RF and XGBoost) proved far more robust. Second, no single oversampling technique was universally optimal, with ADASYN being most effective for KNN and SMOTE for SVM. Third, Bayesian optimization was a significantly more efficient tuning strategy than exhaustive grid search for complex models like XGBoost.

Ultimately, this work establishes that thoughtful pipeline engineering is a more reliable path to robust clinical performance than the indiscriminate application of oversampling. As illustrated by the Pareto frontier analysis, the optimal pipeline is a context-dependent trade-off between recall, precision, and computational cost.

The primary limitation of this work is its focus on a single feature set from the Chapman-Shaoxing dataset. While this ensures methodological clarity, the generalization of these specific interactions to other datasets or feature sets remains an open question. Future research should extend this framework by benchmarking these optimized classical pipelines against end-to-end deep learning models to quantify the trade-off between interpretability and raw performance. Furthermore, exploring signal-level data augmentation with techniques like Generative Adversarial Networks (GANs) [3,12] and applying formal interpretability methods like SHAP analysis to the best-performing classical models are promising directions for developing trustworthy and effective tools for clinical deployment.

Acknowledgements. The authors acknowledges the Universidad ECCI.

Funding Information. This research received no external funding.

Conflicts of Interest. The authors declare no conflicts of interest.

References

1. Acharya, U.R., et al.: A deep convolutional neural network model to classify heartbeats. Comput. Biol. Med. **89**, 389–396 (2017). https://doi.org/10.1016/j.compbiomed.2017.08.022
2. Akiba, T., Sano, S., Yanase, T., Ohta, T., Koyama, M.: Optuna: a next-generation hyperparameter optimization framework. In: Proceedings of the 25th ACM SIGKDD, pp. 2623–2631 (2019)
3. Berger, L., Haberbusch, M., Moscato, F.: Generative adversarial networks in electrocardiogram synthesis: recent developments and challenges. Artif. Intell. Med. **143**, 102632 (2023). https://doi.org/10.1016/j.artmed.2023.102632
4. Chawla, N.V., Bowyer, K.W., Hall, L.O., Kegelmeyer, W.P.: Smote: synthetic minority over-sampling technique. J. Artif. Intell. Res. **16**, 321–357 (2002)
5. Fernández, A., García, S., Herrera, F., Chawla, N.V.: SMOTE for learning from imbalanced data: progress and challenges, marking the 15-year anniversary. J. Artif. Int. Res. **61**(1), 863–905 (2018)
6. Hannun, A.R., Rajpurkar, P., Haghpanahi, M., et al.: Cardiologist-level arrhythmia detection and classification in ambulatory electrocardiograms using a deep neural network. Nat. Med. **25**(1), 65–69 (2019). https://doi.org/10.1038/s41591-018-0268-3
7. Ketu, S., Mishra, P.K.: Empirical analysis of machine learning algorithms on imbalance electrocardiogram based arrhythmia dataset for heart disease detection. Arab. J. Sci. Eng. , 1–23 (2021). https://doi.org/10.1007/s13369-021-05972-2
8. Kim, Y.K., Lee, M., Song, H.S., Lee, S.W.: Automatic cardiac arrhythmia classification using residual network combined with long short-term memory. IEEE Trans. Instrum. Meas. **71**, 1–17 (2022). https://doi.org/10.1109/TIM.2022.3181276
9. Martinez, H., Martínez, J.P., Laguna, P.: A wavelet-based QRST descriptor for arrhythmia classification. Comput. Cardiol. , 229–232 (2004)
10. Moody, G.B., Mark, R.G.: The MIT-BIH arrhythmia database on cd-rom and software for use with it. Comput. Cardiol., 185–188 (1990)
11. Ribeiro, A.L.D., et al.: Automatic diagnosis of the 12-lead ECG using a deep neural network. Nat. Commun. **11**, 1760 (2020). https://doi.org/10.1038/s41467-020-15432-4
12. Simone, L., Bacciu, D., Gervasi, V.: ECG synthesis for cardiac arrhythmias: integrating self-supervised learning and generative adversarial networks. Artif. Intell. Med. **167**, 103162 (2025). https://doi.org/10.1016/j.artmed.2025.103162
13. Siontis, K.C., Noseworthy, P.A., Attia, Z.I., Friedman, P.A.: Artificial intelligence-enhanced electrocardiography in cardiovascular disease management. Nat. Rev. Cardiol. **18**(7), 465–478 (2021). https://doi.org/10.1038/s41569-020-00503-2, publisher: Nature Publishing Group
14. Xiao, Q., et al.: Deep learning-based ECG arrhythmia classification: a systematic review. Appl. Sci. **13**(8), 4964 (2023). https://doi.org/10.3390/app13084964, number: 8 Publisher: Multidisciplinary Digital Publishing Institute

15. Yang, Y., Khorshidi, H.A., Aickelin, U.: A review on over-sampling techniques in classification of multi-class imbalanced datasets: insights for medical problems. Front. Digit. Health **6** (2024). https://doi.org/10.3389/fdgth.2024.1430245, publisher: Frontiers
16. Zheng, J., Guo, Q., Lyu, Y., et al.: A 12-lead ECG database for arrhythmia research. Sci. Data **7**, 48 (2020)

Unsupervised Clustering and Classification of Upper Limb EMG Signals During Functional Movements: A Data-Driven

L. F. Salazar Álvarez[1], D. Escobar-Saltarén[1], M. B. Salazar Sánchez[1](✉), and S. C. Henao-Aguirre[2]

[1] Systems Engineering Department, Engineer Faculty, Universidad de Antioquia, Calle 70 No. 52-21, A.A. 1226, Medellín, Colombia
bernarda.salazar@udea.edu.co

[2] School of Engineering and Sciencies-Campus Guadalajara, Tecnológico de Monterrey, Guadalajara, Mexico

Abstract. This study presents a comprehensive approach for the clustering and classification of upper-limb surface electromyography (sEMG) signals during functional reach and grasp movements. The methodology was applied to the NINAPRO DB4 dataset, which provides multichannel EMG recordings of 52 gestures. A four-stage pipeline was designed, including signal preprocessing, feature extraction, gesture selection via hierarchical clustering, and comparative model evaluation. Preprocessing involved a fourth-order low-pass filter (0.6 Hz) and Hilbert envelope transformation, effectively reducing noise and enhancing signal clarity. Feature extraction yielded 26 temporal and frequency-domain metrics, which were later refined using visual analysis, mutual information, principal component analysis, and decision tree importance scores. A final subset of five key features was selected for classification tasks. Gesture selection was performed through hierarchical clustering using Mahalanobis distance, resulting in six representative movements that balanced biomechanical diversity and computational efficiency. A 200 ms window was identified as optimal for temporal segmentation based on stability and physiological plausibility. Classifier models were evaluated in two stages. Automated comparison using PyCaret identified Extra Trees (ET) and Artificial Neural Networks (ANN) as top performers. Subsequent independent training confirmed their stability and generalization capacity, with ANN showing progressive learning and ET maintaining robust, consistent results. The findings support the implementation of adaptive, low-latency control strategies for myoelectric prostheses and provide a scalable pipeline for future real-time applications.

Keywords: prosthesis · classifier models · feature extraction · sEMG signals · cluster

W. J. Sarmiento and E. L. Sierra-Ballén (Eds.): CCC 2025, CCIS 2692, pp. 158–171, 2027.
https://doi.org/10.1007/978-3-032-28447-1_11

1 Introduction

Upper-limb amputations constitute a public-health problem, accounting for 4.5% of all reported amputations and exerting a disproportionate impact on the functional autonomy and quality of life of those affected [1]. The inability to perform everyday tasks as simple as reaching for objects has physical, emotional, psychological and social repercussions for the individual. Worldwide prevalence is estimated at 200–300 cases per 100000 inhabitants, with a relevant share involving the upper extremities.

In Colombia this figure warrants a distinct analysis because of the complex interplay of sociopolitical violence, occupational accidents and road traffic incidents—factors that raise the incidence of limb loss.

In recent years, engineering advances have driven the design of myoelectric prostheses with multiple degrees of freedom, integrated sensors and increasingly lightweight interfaces. Yet the continuing challenge is to achieve intuitive control that naturally replicates the complexity of human movement while ensuring comfort and ease of use.

Even with sophisticated solutions, contemporary abandonment rates remain close to 35% owing to issues such as weight, limited wrist dexterity, insufficient grip strength, heat build-up, high cost and—above all—the difficulty users face in perceiving the device as an extension of themselves [2].

Modern approaches process EMG in the time and frequency domains, using statistical features to capture different aspects of muscle activation. Feature selection—through principal-component analysis, decision trees or mutual information—reduces dimensionality and enhances the interpretability of classification models that discriminate movements.

These models learn the nonlinear relationship between the feature vector and specific gestures while striving to generalize to new instances and mitigate overfitting. To evaluate such methods, researchers rely on public datasets.

The NINAPRO DB4 (Non-Invasive Adaptive Hand Prosthetics) database used in this study provides multichannel sEMG recordings from 12 electrodes sampled at 2 kHz in healthy subjects performing 50 static and functional gestures [3]. Six repetitions per gesture, demographic metadata and carefully curated labels make it a reference standard for training and validating movement-recognition algorithms.

Framed within the research line of biological-signal characterization, the present study focuses on reach and grasp movements essential for fine manipulation and, consequently, patient independence. Our objectives are, first, to cluster grips based on sEMG dynamics, applying statistical criteria to select the most distinguishable classes via Mahalanobis distance; and second, to implement robust, adaptive and interpretable classification models that can underpin more precise and intuitive control systems.

Ultimately, improving motor-intent detection will not only enhance prosthetic functionality but also reduce abandonment rates, foster motor rehabilitation and restore the psychosocial well-being of individuals with upperlimb amputations.

2 Data Source

Using NINAPRO as the information source for this project, it is relevant to note that the complete database is divided into 10 subsets (DB1-DB10), each with its own acquisition protocol and number of recorded movements. This study focuses on DB4, which is designed for classification and regression tasks in gesture recognition. The selected subset contains recordings from 10 healthy volunteers (6 men and 4 women, with an average age of about 29 years) as shown in Fig. 1.

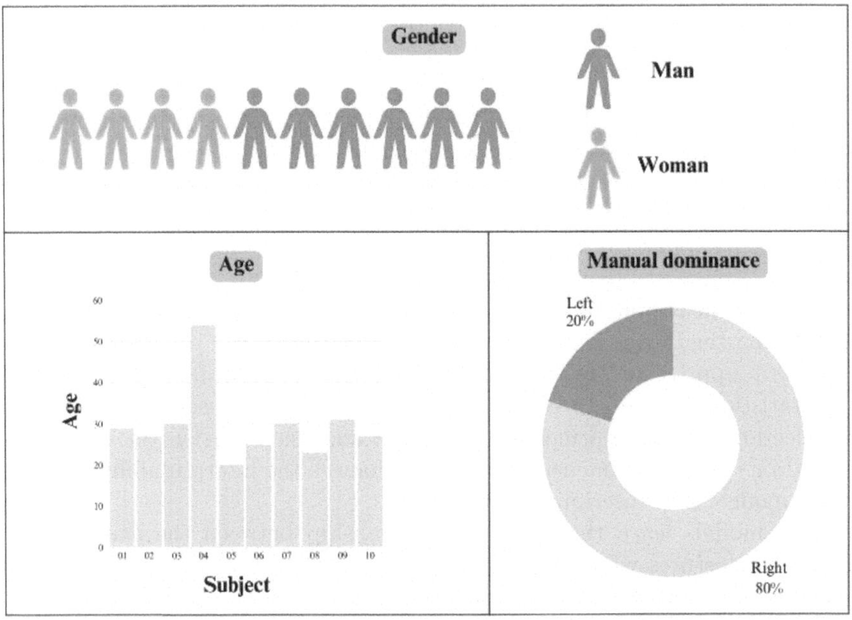

Fig. 1. Database - DB 4 demographic information

Their sEMG signals were captured with 12 Cometa electrodes placed around the forearm-from the distal biceps' region to the extensor and flexor musculature of the forearm (Fig. 2)-to cover the main muscle groups involved in reaching, gripping, and pronation-supination. Each channel was digitized at 2 kHz, enabling analyses in both the time and frequency domains.

The protocol comprises 52 distinct gestures, shown in Fig. 3. Each gesture is repeated six times per subject: 5 s of sustained contraction followed by roughly 3 s of rest. This design strikes a balance between minimal fatigue and enough training samples. The original files contain time-synchronized variables, making it straightforward to align muscle signals with protocol events-an essential feature for developing prostheticcontrol algorithms and functional-rehabilitation tools.

The present article presents the insights gained from characterizing sEMG recordings obtained during functional upper-limb movements, using the pub-

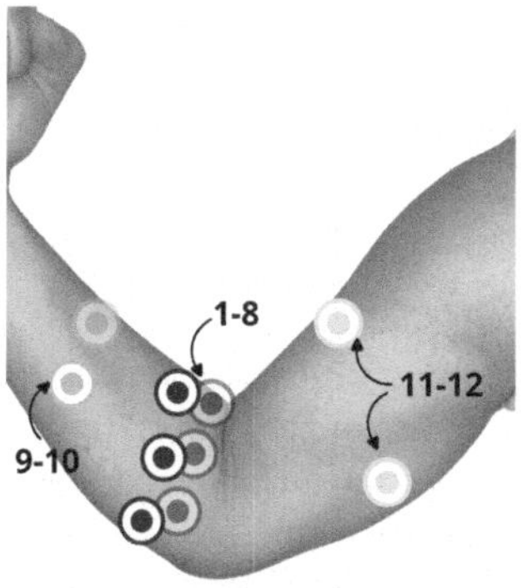

Fig. 2. Anatomical disposition of the electrodes

lic **NINAPRO DB4** database. The methodology is organized into four main stages: (A) signal preprocessing and transformation, (B) feature extraction, (C) movement selection by cluster analysis, and (D) classifier selection and validation. The overarching goal is to identify muscle-activation patterns with sufficient discriminative power to feed movement-classification algorithms for prosthetic control.

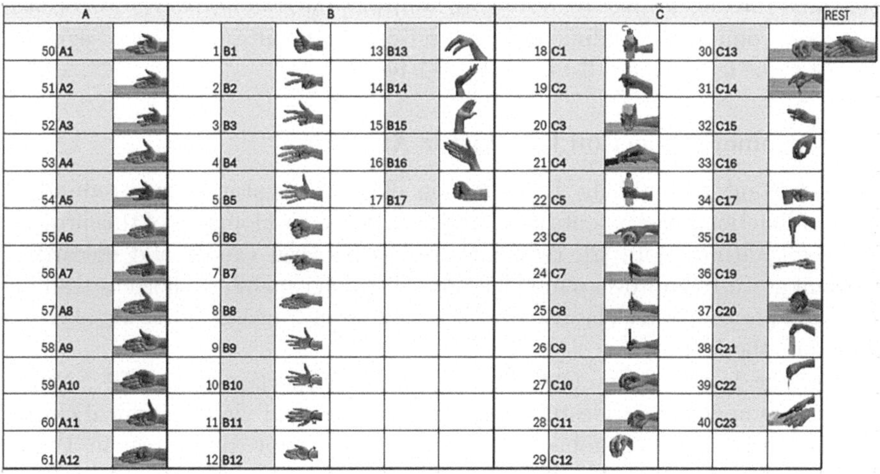

Fig. 3. Movements recorded in DB4 and their respective labeling

3 Methodology

3.1 Signal Preprocessing and Transformation

After downloading the raw DB4 files, a pipeline was applied to attenuate noise generated by motion artefacts, highlight active contraction phases, and ensure inter-subject comparability:

a) Electrical-noise filtering: mains interference (60 Hz) was removed.
b) Butterworth low-pass filter: a fourth-order digital filter with a 0.6 Hz cut-off frequency smoothed abrupt variations without affecting the low-frequency content relevant to muscle activation.
c) Envelope calculation (Hilbert): the Hilbert transform was used to obtain the analytic envelope of each channel, providing a clearer and more continuous representation of the signal.
d) Subject- and channel-wise normalization: each signal was scaled to reduce intrinsic variability between individuals.

3.2 Feature Extraction

Statistical features were extracted from the processed signals in both the time and frequency domains. Selected for their balance of low computational cost and high discriminative capacity reported in the literature, these metrics included mean absolute value (MAV), root mean square (RMS), integrated absolute value (IAV), and the wavelet-based mDWT. In addition, the standard deviation of each metric was computed, yielding an initial set of 26 attributes per signal segment. Some of these features are listed in Table 1 below.

3.3 Movement Selection by Cluster Analysis

It is important to highlight the selection of a representative, non-redundant subset of functional movements as a key phase in the development of this project. As several authors note [2], reducing overlap between classes and discarding irrelevant gestures improves a model's generalization capability, shortens training times, and prevents inter-class ambiguities that often translate into control errors during prosthesis use.

After the feature extraction, an exploratory inspection of the 52 Ninapro gestures was undertaken. Gesture similarity was assessed via hierarchical cluster analysis, with the Mahalanobis distance chosen as the proximity metric. Unlike Euclidean distance, Mahalanobis accounts for the variance-covariance structure of the features, penalizing highly correlated dimensions and thus providing a more robust criterion in the presence of internal redundancies [4]. The clustering strategy employed complete linkage, which defines inter-cluster distance as the largest Mahalanobis distance between any pair of members from different clusters. This method tends to yield compact, well-separated groups, desirable for minimizing overlap in feature space [4].

Results were visualized with a dendrogram, revealing six major clusters that satisfied two fundamental principles:

Table 1. Extracted features categorized by domain and functional grouping

Feature	Domain	Group
RMS (Root Mean Square)	Time Domain	Energy/Amplitude
MAV (Mean Absolute Value)		
IAV (Integrated Absolute Value)		
WL (Waveform Length)		Slope/Change
MAVS (Mean Absolute Value Slope)		Signal Pattern
ZC (Zero Crossing)		
SSC (Slope Sign Changes)		Variability/Dispersion
VAR (Variance)		
CoV (Coefficient of Variation)		
Autoregresive		Statistical
Kurt		
MNP (Mean Power)	Frequency Domain	Signal Pattern
MNF (Mean Frequency)		
mDWT (Marginal Discrete Wavelet)		

a) Maximum intra-cluster homogeneity, evidenced by reduced variances within feature vectors.
b) Minimum inter-cluster similarity.

Within each macro-cluster, the most representative movement was selected-defined as the gesture with the highest energy (based on RMS analysis) and the greatest potential for replicability in future studies.

3.4 Temporal Segmentation of the Signal

Once filtering, per-channel normalization and envelope calculation were completed, the sEMG signal underwent temporal segmentation-a critical phase for ensuring that the extracted features capture the true dynamics of each gesture without introducing biases from fatigue or overlap between consecutive movements. Following guidelines identified in the literature review, three non-overlapping rectangular window lengths were chosen- 100 ms, 200 ms and 300 ms -which also exhibit an inverse relationship with the system's computational demands. Windows were defined on the analytic envelope of each channel so that every segment contained an isolated portion of myoelectric activity associated either with the sustained-contraction phase or the corresponding rest period. To assess the reliability and replicability of the metrics, the 26 previously described features were extracted from each individual window, after which graphical representations-boxplots and violin plots-were constructed.

3.5 Feature Selection

Feature selection has a direct impact on model accuracy, generalization capacity, and computational efficiency. A hybrid scheme was designed that combines exploratory inspection, statistical indicators, and model-derived criteria to ensure that the final feature set captures muscle activity without unnecessary redundancies.

a) Visual exploratory análisis. Boxplots and violin plots were generated for each of the 26 features across the six selected gestures, making it easier to detect extreme biases that could undermine classifier robustness and to identify inter-class overlap, thus visually gauging the discriminative relevance of each feature.
b) Statistical dependence via Mutual Information (MI). MI was computed between every feature and the gesture label and then normalized to the $0-1$ range.
c) Multivariate structure with Principal Component Analysis (PCA). PCA was applied to the 12 features retained after the MI filter; components explaining $\geq 85\%$ of the cumulative variance were kept.
d) Variable importance in preliminary Decision Trees. A classification tree was trained, and the importance score of each feature within the model was extracted for further refinement.

3.6 Implementation of Classifier Models

In the present study, the recognition of the six selected gestures was carried out in two complementary phases, aimed at combining speed and precision, supporting clinical applications where a balance between accuracy and reduced calibration time is required. This multistage approach enabled both a systematic and automated comparison of various algorithms and a detailed evaluation of the top-performing classifiers:

a) PyCaret. To compare multiple classification algorithms under the same validation workflow in an automated manner, the PyCaret library used a free and open-source tool designed to simplify the automation of machine learning pipelines. This allowed for rapid experimentation with low computational cost. A total of 16 algorithms were evaluated, from which the top three were selected for the next phase.
b) Independent Testing. The models with the best preliminary performance were then trained independently using cross-validation and hyperparameter tuning, allowing for a more detailed analysis of their behavior. This process included manual hyperparameter adjustments and fold-byfold evaluation to assess model stability and generalization capability.

4 Results and Discussion

4.1 Exploratory Analysis and Signal Preprocessing

In the present study, the recognition of muscle gestures was based on a preliminary analysis of surface electromyography (sEMG) signals extracted from the target database. These signals exhibited typical characteristics of muscle biopotentials, such as low amplitude, the presence of motion artifacts, and electrical interference from the environment. It is important to note that, through visual inspection, high inter-channel variability was observed. This can be attributed to the anatomical placement of electrodes over different muscle groups in the forearm (Fig. 2), which exhibit specific activations depending on the performed movement. To address the limitations caused by high inter-channel variability, artifacts, and noise present in the signals, a twostep preprocessing procedure was implemented: (1) a low-pass filter with a cutoff frequency of 0.6 Hz, aimed at attenuating high-frequency components that could hinder the processing; and (2) the application of the Hilbert envelope, which enhances the instantaneous amplitude of the signal and provides a smoothed representation of muscle contraction, thus improving both temporal segmentation and feature extraction.

Figure 4 shows a comparison between the raw signal (blue) and its corresponding processed signal (orange) for channel 10 of a randomly selected subject. It can be clearly seen how the noise has been effectively suppressed, while the characteristic peaks of muscle activation are preserved, thereby facilitating the physiological interpretation of the signal.

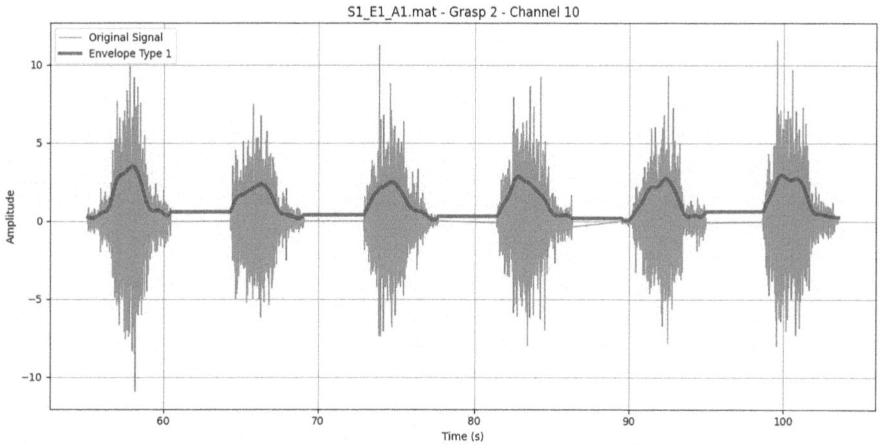

Fig. 4. Signal (blue) and processed signal (orange) for channel 10 of a particular subject. Comparison of the original signal with the signal smoothed by low-pass filter and Hilbert envelope, showing noise reduction. (Color figure online)

4.2 Cluster Analysis and Movement Selection

The selection of functional movements to be included in the classification model is a critical step that enables more reliable performance. This decision was based on the large number of gestures available in the NINAPRO DB4 database (over 50 types), which posed a challenge due to the presence of gesture classes that could be statistically similar or even redundant-introducing ambiguity and affecting the model's generalization capability.

To address this, a complete-linkage hierarchical clustering approach was applied, using Mahalanobis distance as the dissimilarity metric. The result of this analysis was graphically represented through a dendrogram (Fig. 5), which shows the hierarchical grouping of the various muscle gestures. In this visualization, gestures with high statistical similarity are clustered at lower levels, whereas those with greater dissimilarity merge at higher levels. This representation facilitated the identification of naturally differentiable movement subsets (Fig. 6), allowing the selection of those with greater discriminative capacity and high energy levels.

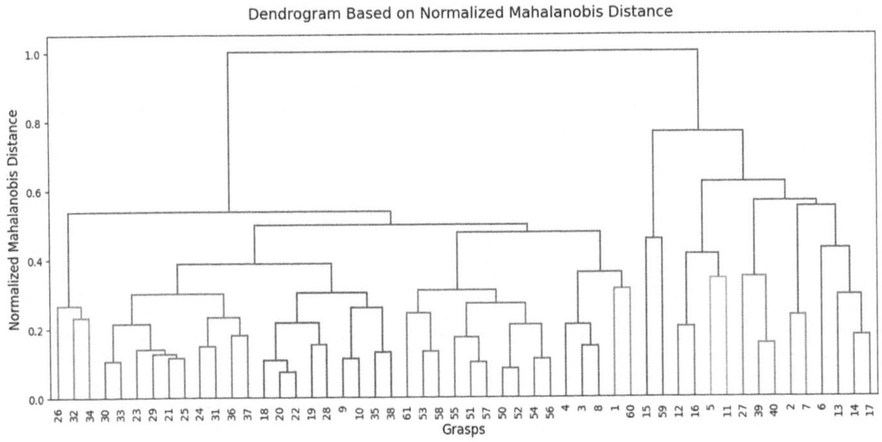

Fig. 5. Dendrogram based on Mahalanobis distance, hierarchical grouping of muscular gestures according to their multivariate statistical distance. Allows selection of differentiable movements.

Based on this representation, a final subset consisting of six movements (including the resting state) was selected. These gestures were considered sufficiently representative, distinguishable, and functionally relevant. This selection optimized the balance between biomechanical diversity and simplification of the classification scenario, reducing both the computational load and the complexity of the model.

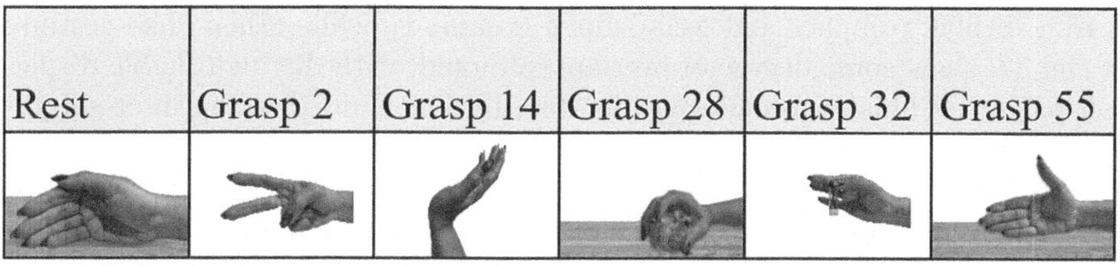

Fig. 6. Selected gestures based on dendrogram analysis. They represent a functionally diverse and statistically distinguishable subset within the original set of movements from the DB4 dataset.

4.3 Temporary Window Size Evaluation

The selection of the window size for this study was carried out by evaluating three window durations- 100 ms, 200 ms, and 300 ms -without overlapping, applying the same set of previously extracted features to each configuration. It was observed that 100 ms windows, although offering high temporal resolution, exhibited high variability in the extracted features, as well as a marked presence of outliers. This instability can hinder class discrimination and lead to an overestimation of noise within the model. In contrast, 300 ms windows significantly smoothed the signal and showed greater consistency across subjects and repetitions, but at the expense of reduced sensitivity to transient muscle activation events, negatively impacting the system's dynamic responsiveness.

On the other hand, the 200 ms window emerged as the most suitable option, providing a balance between resolution and stability. It allowed for the capture of relevant muscle activation patterns with lower susceptibility to noise, resulting in better inter-class discrimination in later processing stages. Moreover, this decision is supported by physiological literature showing that the central nervous system can organize posture-specific muscle activation patterns within time frames close to 200 ms, as reported by studies on motor control and neurophysiological reaction times [5].

4.4 Feature Selection and Debugging

The initial feature extraction phase resulted in a total of 26 features for each window, repetition, movement, and subject. These included metrics from the time domain, frequency domain, and dispersion measures. Subsequently, various techniques were applied to refine and select the most informative features: exploratory visual analysis, mutual information calculation, principal component analysis (PCA), and decision tree training. These tools made it possible to identify features with low discriminative power, redundancy, or statistically negligible contribution to the model.

PCA analysis was particularly useful for examining the internal structure of the dataset. Figure 7 shows a representation in the space of the first two principal components, where class separation can be observed. Some classes, such as rest, display compact and well-defined groupings, while others, like gestures 28 and 32, show some degree of overlap—although with distinguishable dispersion patterns. This non-homogeneous distribution highlights the inter-subject variability inherent to EMG signals. As a result of the refinement process, all standard deviation features, six additional low-relevance features, and channels 11 and 12 were eliminated due to their consistently low importance across all applied approaches. The final feature set consisted of: IAV, MAV, MAVS, RMS, and mDWT, applied across the 10 most relevant channels.

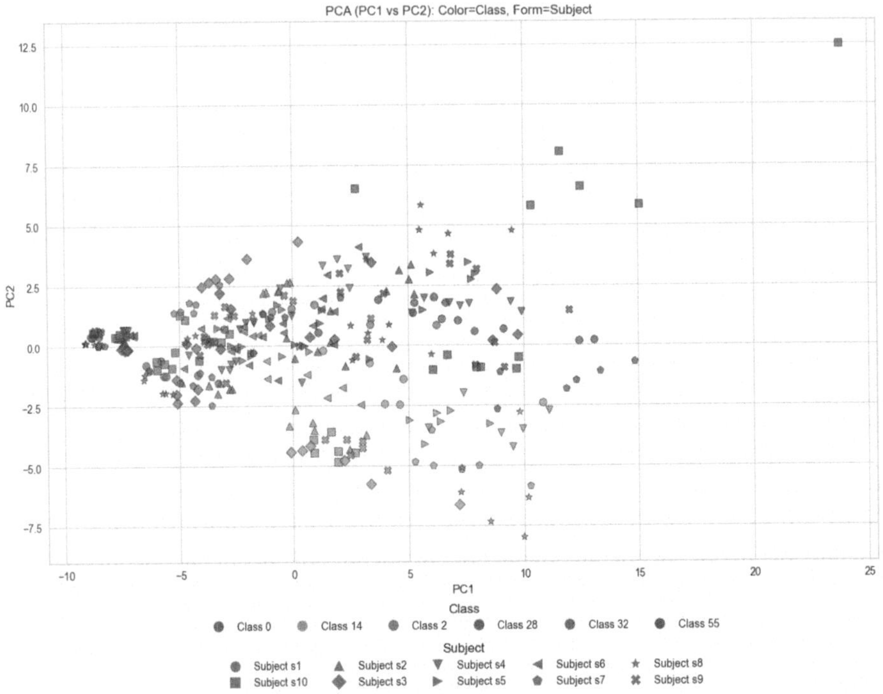

Fig. 7. Representation of EMG signals in the space of the first two principal components (PCA).

4.5 Comparison of Classification Models

The classification model comparison stage represented a vital component in the development of this research, allowing for the identification of the most effective approaches for the task of multiclass movement recognition based on sEMG signals. This phase was divided into two main stages: an initial evaluation using the PyCaret library, and a subsequent implementation involving independent hyperparameter tuning and controlled crossvalidation. In the first stage, PyCaret facilitated the automated comparison of multiple algorithms under standardized conditions, streamlining the analysis without compromising methodological rigor. Models such as K-Nearest Neighbors (KNN), Support Vector Machines (SVM), logistic regression, decision trees, Extra Trees Classifier (ET), and Artificial Neural Networks (ANN) were evaluated, among others. The metrics used included accuracy, recall, F1-score, precision, and Cohen's Kappa. The results showed that ANN and ET models outperformed the others, achieving accuracy close to 95%, along with high and consistent values across all evaluated metrics (see Table 2). This preliminary performance positioned both models as leading candidates for more in-depth validation.

Table 2. Performance of RNA and ET models evaluated by PyCaret.

Model	Accuracy	Recall	F1	Prec.	Kappa
RNA	95.2%	95.2%	95.7%	95.2%	94.3%
ET	94.4%	94.4%	94.9%	94.4%	93.3%

In the second phase, the selected models were trained independently, allowing for manual hyperparameter tuning and more detailed cross-validation. This stage aimed to verify the robustness observed in PyCaret within a less automated environment. The results confirmed the stability of both the ANN and ET models, each maintaining high metric values with minimal differences compared to the previous results (see Table 3). At this point, ANN showed a slight improvement in the F1-score, while ET maintained consistent performance, reaffirming its reliability.

Table 3. Performance of RNA and ET models evaluated independently

Model	Accuracy	Recall	F1	Prec.	Kappa
RNA	94.7%	94.7%	95%	94.7%	93.6%
ET	94.4%	94.4%	94.9%	94.4%	93.3%

During the analysis of the learning curves, both models demonstrated stable behavior. In the case of the ET model, the validation curve remained very close to the training curve, indicating a low tendency toward overfitting and good generalization. ANN, on the other hand, showed progressive improvement as the size of the training set increased, suggesting effective and gradual learning-typical of well-configured neural network architectures. This trend was confirmed by the confusion matrices, which reflected a low error rate and few misclassifications between similar gestures. Additionally, models such as KNN and SVM were also compared, but they showed lower performance, particularly in multiclass tasks with high dimensionality and inter-subject variability. This reinforces the idea that models like ET, due to their ability to handle nonlinear relationships and low risk of overfitting, and ANN, due to their capacity to learn complex representations, are the most suitable for this type of problem.

5 Conclusion

This study proposed and validated a comprehensive approach for the characterization, selection, and classification of upper limb sEMG signals during functional reach and grasp movements, within the framework of developing more adaptive control systems for myoelectric prostheses. Using the public NINAPRO DB4 database as the source of information, a step-by-step workflow was designed, covering the transformation of raw signals through to the comparison of classification models. One of the main contributions of this work lies in the preprocessing strategy applied, based on the use of a fourth-order low-pass filter (0.6 Hz) and the analytic envelope obtained via the Hilbert transform. This combination enabled the attenuation of high-frequency noise and provided a smoother, more continuous representation of muscle activations, improving not only the temporal segmentation of gestures but also the overall quality of the extracted features. This methodology proved particularly useful in prototypical instrumentation contexts, as evidenced by preliminary testing with a custom-designed acquisition armband. The movement selection stage, using hierarchical cluster analysis with Mahalanobis distance, introduced an objective, data-driven perspective to reduce inter-class ambiguity. This technique allowed the 52 original gestures to be grouped into homogeneous macro-clusters and led to the selection of six representative gestures (including rest), optimizing the balance between biomechanical diversity and computational complexity. This selection was also based on estimated muscle energy (via RMS) and the gestures' potential for replicability.

Regarding temporal segmentation, both empirical and theoretical evidence confirmed that the 200 ms window represents an optimal compromise between temporal resolution and statistical stability of the features. In addition to improving inter-class discrimination, this choice was supported by neurophysiological studies demonstrating the central nervous system's ability to organize specific muscle activation patterns within such time framesparticularly during postural or reactive tasks. Feature extraction and subsequent refinement were also central to the study. Starting from an initial set of 26 statistical attributes from the time and frequency domains, a hybrid selection strategy was designed using visual criteria (boxplots and violin plots), statistical dependencies (mutual information), multivariate analysis (PCA), and variable importance scores from decision trees. This combination enabled the identification and removal of redundant or low-discriminative features, resulting in a reduced and efficient feature set consisting of IAV, MAV, MAVS, RMS, and mDWT, distributed across 10 channels selected for their high inter-subject relevance.

On the other hand, the staged implementation and comparative evaluation of classification models constituted a significant phase of the project. In the first stage, PyCaret was used to automatically explore the performance of 16 algorithms under standardized conditions. This evaluation revealed that the Extra Trees (ET) and Artificial Neural Networks (ANN) models delivered the best results in terms of accuracy, F1-score, precision, recall, and Cohen's kappa coefficient. In the second stage, both models were independently trained and validated, with manual hyperparameter tuning and fold-based cross-validation. Their stability and generalization capacity were confirmed through learning curves and confusion matrices. ANN showed progressive improvement with increased training data, while ET maintained consistent performance-ideal for real-time implementation. Alternative models such as KNN and SVM were also evaluated but demonstrated lower performance, particularly in multiclass tasks with high dimensionality and inter-subject variability. Despite these advances, the study also revealed persistent challenges, such as inter-subject variability even after normalization, and the need to incorporate dimensionless features or adaptive scaling techniques. Future lines of research should focus on clinical validation with real users, integration into real-time embedded systems, and the fusion of multimodal signals (EMG, accelerometry, force sensors) to enhance the contextual richness and functionality of myoelectric control systems.

Acknowledgments. The authors express deep gratitude to the University of Antioquia, especially the Department of Bioengineering, for providing the academic and scientific environment necessary for the completion of this work. Thanks, we are also due to the Intelligent Information Systems Lab (In2Lab-UdeA) for their technical support and for providing access to the computational resources essential for the analysis of EMG signals. Special appreciation goes to her research colleague for his continuous assistance in data processing and model validation, and to the professor who contributed during the statistical analysis stage, for his guidance in the use of multivariate and machine learning techniques. Finally, the author acknowledges all individuals and institutions who, directly or indirectly, contributed to the development of this work, making possible the integration of engineering, data science, and functional rehabilitation.

References

1. Chapurri, D., Felipe, A., Vélez, M., Patricia, C., López, G., Marcela, P.: Medellín, 201AD. Medellín, Univesidad CES (2014). http://hdl.handle.net/10946/2656
2. Stival, F.S., Michieletto, M., Cognolato, E., Pagello, H., Müller, Atzori, M.: A quantitative taxonomy of human hand grasps. J. Neuroeng. Rehabil. **16**(26) (2019). https://doi.org/10.1186/s12984-019-0488-x
3. University of Applied Sciences Western Switzerland - Valais, NINAPRO. https://ninapro.hevs.ch/
4. Hernández Arellano, F.M.: El Concepto de Distancia y su Aplicación en Estadística Multivariada. Datos Diagnosticos y Tendencias (1970)
5. Bobbert, M.F., Koopman, A.S.: Humans need only 200 ms to generate posture-specific muscle activation patterns for successful vertical jumps in reaction to an auditory trigger. Front. Sports Active Living **5**(26) (2023). https://doi.org/10.3389/fspor.2023.1123335

Short Papers and Workshops

Towards a Model–Driven Engineering Approach for Trusted Execution Environments

Juan Jáuregui[1], Jose Bocanegra[2](✉), Regis Schuch[3], Sandro Sawicki[3], Fabricia Roos-Frantz[3], and Rafael Z. Frantz[3]

[1] Department of Systems and Computing Engineering, Universidad de Los Andes, Bogotá, Colombia
j.jauregui@uniandes.edu.co
[2] Faculty of Engineering, Universidad Distrital Francisco José de Caldas, Bogotá, Colombia
jjbocanegrag@udistrital.edu.co
[3] Unijuí University, Ijuí, Brazil
{regis.schuch,sawicki,frfrantz,rzfrantz}@unijui.edu.br

Abstract. As cities evolve, the integration of diverse digital services becomes critical. Traditional integration platforms such as Apache Camel or MuleSoft usually lack mechanisms to guarantee the privacy and integrity of the data exchanged by digital services. Trusted Execution Environments (TEEs) offer secure compartments in memory for the execution of code and processing of sensitive data, providing stronger protection against attacks. Among these TEEs is the Morello Board, developed by ARM, which has capabilities for compartmentalized execution using the CHERI architecture. Despite its potential, development for the Morello platform is hindered by the complexity of writing secure code in C language. To overcome this development barrier, we propose a Model–Driven Engineering approach that automates the generation of C code for TEEs, enabling the creation of secure execution scaffolding compatible with the Morello Board. This approach includes the definition of a metamodel, a high–level abstraction model, and a code generation process. The generated code runs on the Morello Board. Furthermore, we demonstrate the feasibility of this approach using a case study focused on the integration of secure digital services in the context of smart cities.

Keywords: Trusted Execution Environment · Morello Board · CHERI · Model-Driven Engineering · Code Generation · Digital Cities · Secure Integration

1 Introduction

Cities have become increasingly digitalized, and the need for robust integration of services is more urgent than ever. Smart cities rely on the seamless interaction of diverse digital services, including healthcare, transportation, messaging, and

W. J. Sarmiento and E. L. Sierra-Ballén (Eds.): CCC 2025, CCIS 2692, pp. 175–184, 2027.
https://doi.org/10.1007/978-3-032-28447-1_12

many more. This interaction is commonly implemented through integration processes, which are software entities that retrieve, combine, and process data from multiple digital services to enable new functionalities or coordinated operations. Some platforms like Apache Camel [1] and MuleSoft [9] have become widely adopted for addressing integration challenges in these domains. These tools enable the orchestration of data and services from multiple sources, supporting interoperability across systems. Apache Camel provides a powerful routing and mediation engine that allows developers to connect various systems using a declarative Domain–Specific Language (DSL) in Java, XML, or YAML. On the other hand, MuleSoft enables developers to connect applications, data, and devices across cloud and on-premise environments. It provides tools for building APIs and orchestrating services through various connectors. However, these types of platforms lack support to protect and process sensitive data exchanged by digital services. In a smart city integration context, data protection and privacy have become very important factors to ensure the safe and trustworthy operation of interconnected services. Thus, it has motivated us to explore alternative solutions. There are sensitive scenarios such as digital identity management, health data exchange, among others, where exposure to critical information can have serious consequences. These limitations have motivated the exploration of secure execution environments that can provide stronger isolation and security guarantees.

Trusted Execution Environments (TEEs) [11] have emerged as a promising solution to this problem. TEEs allow to create isolated compartments within hardware platforms that are protected from external interventions, even from complex software, such as operating systems. Well–known TEE implementations include Intel SGX [8], ARM TrustZone [4], and AWS Nitro Enclaves [5], which are widely used in commercial and research settings to ensure secure execution and data protection. One of the most promising developments in this field is the Morello Board [3], a research platform based on the Capability Hardware Enhanced RISC Instructions (CHERI) architecture [12]. Morello introduces memory protection and hardware compartmentalization, making it suitable for executing security sensitive code.

Despite these advances, programming for TEEs such as Morello remains a challenge. Developers are required to write secure, low–level C code and to manage complex memory settings. This introduces restrictions for teams working in high–level developing modeling environments. For individuals who are not experienced developers, writing secure C code is especially difficult, particularly when relying on TEE libraries that are often complex and tightly coupled to specific hardware platforms. For teams accustomed to working in visual or model–driven tools that abstract away low–level details, the jump to manual memory management can be overwhelming. As a result, the potential of TEEs often remains inaccessible to those without deep systems programming expertise.

In this paper, we propose a Model–Driven Engineering (MDE) approach to facilitate the development of secure orchestration logic targeting the Morello Board. By defining a high–level metamodel and graphical modeling language using Sirius [10], and generating secure C code using Acceleo [6], we enable the automatic production of code that can run on TEEs like the Morello Board. Our approach connects secure hardware platforms with modern model-driven development practices.

To clarify the scope of our proposal, rather than generating the integration process itself, our approach focuses on producing the structure of the execution environment required to securely deploy and launch such a process. This includes the generation of the launcher code, and the scaffolding necessary for execution within a TEE such as the Morello Board. By doing so, we provide a foundation that ensures the process handling sensitive data can be executed securely and reliably, while maintaining compatibility with hardware-enforced isolation mechanisms. The separation between the process logic and its secure execution environment allows for greater modularity and reduces the complexity of developing secure systems from scratch. This bridge between high-level modeling and secure low-level execution reduces the entry barrier for TEE deployment, making it feasible for broader adoption in privacy-sensitive domains such as smart cities.

The remainder of this paper is organized as follows. Section 2 provides background on TEEs, the Morello Architecture, and challenges associated with secure C development. Section 3 presents the motivation and problem statement, focusing on the gap between high-level modeling and low-level secure implementation. Additionally, it introduces our MDE approach, detailing the proposed metamodel, graphical modeling environment built with Sirius, and the code generation strategy using Acceleo. Section 4 demonstrates the application of our approach through a case study targeting the Morello Board, illustrating how secure C code is automatically generated from models. Finally, Sect. 5 concludes the paper and outlines directions for future work.

2 Related Work

The development of secure applications for TEEs has increased in recent years. There are multiple frameworks such as Intel SGX [8] and ARM TrustZone [4] that have laid the foundation for isolating sensitive code and data from the rest of the system. These platforms have made big efforts to ease the development process, including SDKs that provide low-level APIs for trusted and untrusted communication. However, these approaches require expertise in low-level programming and security.

To overcome these limitations, other techniques have been explored in the context of secure system development. SysML-Sec, proposed by Apvrille and Roudier [2], integrates security requirements directly into system models using SysML, enabling early verification and formal validation of secure embedded designs. Similarly, Hamid et al. [7] present an MDE approach that incorporates reusable security and dependability patterns to guide the design of trusted embedded systems in domains like automotive and railways. These patterns encapsulate best practices to address common vulnerabilities.

However, despite these advances, there is a lack of approaches that specifically target the automatic generation of low–level secure code for TEE like the Morello Board from high–level graphical models. Our work addresses this gap by bridging MDE with trusted hardware through a fully generative workflow.

3 Our Proposal

In order to mitigate the complexity of secure software development for the Morello Board, we propose a model–driven approach that facilitates the automatic generation of C code compatible with this platform. Our solution leverages Eclipse Sirius for graphical modeling and Acceleo for model-to-text transformation.

To implement our model–driven approach, we follow a structured methodology composed of three key steps. First, we define a metamodel that formally specifies the concepts and relationships relevant to secure integration within TEEs. This metamodel acts as the foundation of our approach. Second, we create a model that graphically instantiates the metamodel, representing a concrete solution for secure integration. This step enables domain experts to design systems visually without requiring low-level code knowledge. Finally, we apply Acceleo for code generation, transforming the model into C code that is syntactically and semantically aligned with the Morello platform requirements.

3.1 Metamodel Definition

The first step of our methodology involves defining a metamodel (which is depicted in Fig. 1) that abstracts the core components involved in the integration of secure and non-secure processes in a TEE environment. The metamodel, implemented using Eclipse Ecore, is composed of six main classes:

- *IntegrationSolution:* serves as the root concept of the model and represents a complete integration scenario. This element acts as the container or canvas upon which the rest of the architecture is built.
- *Tee* represents the Trusted Execution Environment under analysis. It encapsulates secure execution features and provides the context in which secure components reside.
- *Launcher:* models the processes that are executed outside the secure memory compartment. These processes are typically responsible for initiating or communicating with secure processes.
- *MemoryCompartment:* Represents the isolated memory region offered by the TEE where secure processes are executed. This element is crucial for the enforcement of confidentiality and integrity.
- *IntegrationProcess:* Denotes a secure process that is executed within the memory compartment. These processes must be written securely and are subject to strict execution constraints.
- *Port:* Represents the logical ports used for data exchange between secure and non–secure domains. Ports are also used for interaction with external services.

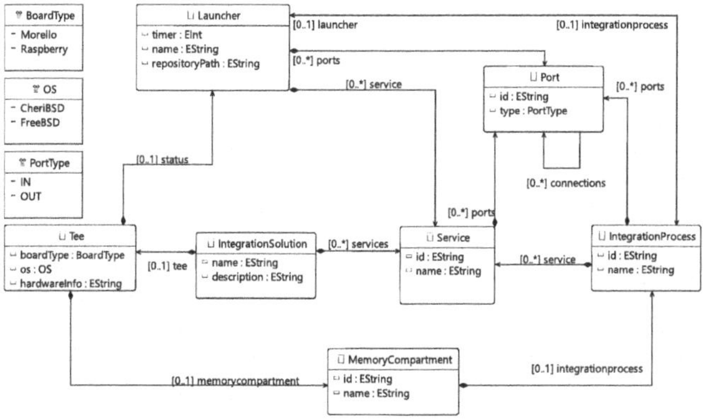

Fig. 1. Metamodel for secure integration in a TEE context.

3.2 Model Definition

Based on the defined metamodel, we used Eclipse Sirius to construct a graphical modeling environment. Sirius allowed us to create a visual editor in which the *IntegrationSolution* serves as the canvas. Each instance of the *Tee*, *Launcher*, *MemoryCompartment*, *IntegrationProcess*, and *Port* classes is visually represented and positioned within the solution model according to its execution domain.

- Tee components are graphically displayed with nested compartments for MemoryCompartment and IntegrationProcess elements. This visual structure reflects the internal organization of a TEE. The nesting indicates that both the memory spaces and integration logic are encapsulated within the TEE, reinforcing the concept of secure isolation. Graphically, this helps the modeler understand which components are protected and executed within the secure domain.
- Launcher elements are placed outside the Tee, representing the boundary between secure and non-secure execution. In the model, Launcher elements serve as interfaces or initiators for TEE-based operations. Placing them outside the TEE visually conveys that they belong to the normal (non-secure) world. This separation is key to understanding the interaction between trusted and untrusted components and emphasizes the controlled entry points into the TEE.
- Port elements are attached to IntegrationProcess and Launcher components, visually signifying the communication flow. Ports act as connectors between components. By attaching them to both IntegrationProcess elements (inside the TEE) and Launcher elements (outside), the model highlights how data or commands flow across the secure boundary. This visual linking helps express the allowed and designed communication paths.

3.3 Code Generation with Acceleo

To connect high–level modeling and executable code for the Morello platform, we employed Acceleo, an Eclipse-based model–to–text transformation engine. We developed a transformation script defined in an .mtl file, which takes Sirius model as input and generates a C code file tailored for secure execution on the Morello Board. The structure of the generated file is directly informed by the structure and relationships defined in the metamodel. In Table 1 we show the mapping of the metamodel elements to the C code generation templates.

Table 1. Mapping of metamodel concepts to generated C code structures.

Metamodel Concept	Generated C Code Construct
IntegrationSolution	Top-level trigger for code generation. Names output files.
TEE	Organizes the model structure; does not generate code.
Launcher	Creates the non-secure launcher .c file with main() function.
MemoryCompartment	Generates a typedef struct and its variable instance.
IntegrationProcess	Adds comments linking processes to compartments.
Port	Generates handler functions for each port.

The IntegrationSolution serves as the top-level controller for the entire code generation process. It initiates the creation of the key .c files and defines the naming used throughout generated filenames and in-code comments, ensuring consistency. The TEE acts as a logical container within the model. Although it does not generate code directly, it plays an important organizational role by grouping related elements. The MemoryCompartment is responsible for generating a typedef struct definition representing the memory layout for a particular compartment. Additionally, it declares a corresponding variable instance of that structure, providing a concrete memory representation used by other parts of the generated code. The Launcher element maps directly to the creation of the non-secure launcher file. It includes the implementation of the main() function, marking the entry point of the non-secure application and initializing necessary components. The IntegrationProcess does not generate code constructs like structs or functions itself. Instead, it contributes to the generation process by adding informative comments in the code, indicating which processes are logically associated with each memory compartment. Finally, each Port element results in the generation of a handler function. These functions are created empty, ready to be implemented to handle specific types of data or events related to their port type.

4 Case of Study

To validate the applicability and benefits of our model-driven approach for secure software generation on TEEs, we developed a case study that simulates the integration of several digital services commonly used in smart cities. The objective of this scenario is to securely coordinate interactions between a messaging platform, a transportation service, and a retail system, while ensuring confidentiality and integrity of user data throughout the process.

4.1 Scenario Description

In this case, three different services participate in the integration process: (i) Store Service: A digital marketplace where users can purchase products; (ii) Transport Service: A transportation platform offering promotional discounts; (iii) Messaging Service: A communication system through which users are notified of discounts and provided with redemption codes.

The promotional workflow operates in a particular way. Initially, when a user makes a purchase of, for example, $100 or more through the store service, they become eligible for a discount voucher applicable to the transportation service. Then, this discount code is generated securely, based on purchase details and user identity, and must not be exposed. The code is transmitted to the user through the messaging service, which is responsible for securely notifying users.

This scenario requires the secure handling of several pieces of sensitive information such as: user identifiers, personal data (names, addresses, contact information, etc.), purchase records and transaction amounts, encrypted discount tokens and eligibility rules.

This type of multi–service integration introduces some security concerns that must be addressed in order to avoid risks. First, sensitive information must be protected during transit and storage, which means that data confidentiality is crucial. Second, the logic that validates purchases and generates discount tokens must be executed in a protected environment in order to avoid tampering. Third, only authorized services should be allowed to give or retrieve discount tokens. Finally, each services operates in its own domain, so interactions must occur through secure interfaces to prevent unauthorized access. Figure 2 presents the graphical representation of this integration scenario, showing the architecture of secure service orchestration within a TEE.

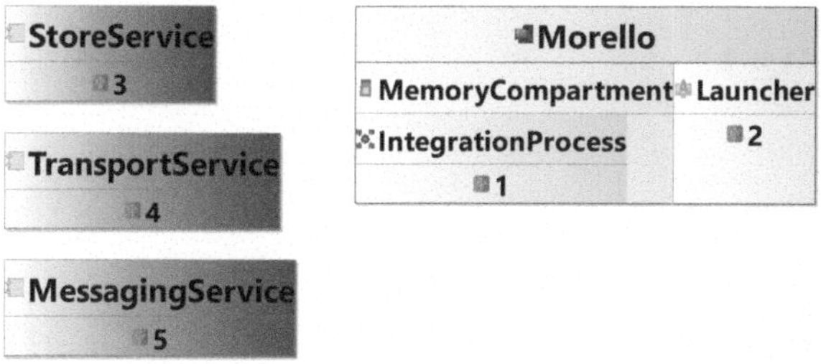

Fig. 2. Graphical representation of the integration model using Sirius for the case of study.

The model visually represents a secure integration architecture within a TEE for a system composed of three distinct services: Store, Transport, and Messaging. Traditional integration platforms, such as Camel or Mule, lack built-in support for these type of security at execution level. This motivates the need for TEEs, where critical operations can be isolated from the rest of the system.

4.2 Code Generation and Evaluation

```c
/* ---------------------------------------------- */
/*                NON-SECURE LAUNCHER              */
/* Generated from IntegrationSolution: IntegrationSolution */
/* ---------------------------------------------- */

#include <stdio.h>

/* Integration Process */
/* Integration Process: Secure Integration Process */

int main() {
    printf("Launching secure application...\n");

    // TODO: Initialize communication with secure app

    return 0;
}

/* End of Launcher */
```

Listing 1.1. Non-secure Launcher Code.

```c
/* ---------------------------------------------- */
/*              SECURE APPLICATION                 */
/* Generated from IntegrationSolution: IntegrationSolution */
/* ----------------------------------------- -----*/

#include <stdio.h>

/* Memory Compartment: SecureProgram */

typedef struct {
    int id;
  char name[50];
} SecureProgram_t;

/* Example variable for memory compartment */
SecureProgram_t secureProgram = { 1, "SecureProgram" };


```

```
/* Port: IN (ID: 1) */
void handle_IN(void) {
    // TODO: Implement Port IN
}

/* End of Secure App */
```

Listing 1.2. Secure Application Code.

To validate our model-driven approach, we automatically generated the implementation code for the elements proposed in the model. The code adheres to the functional and security requirements derived from the model, ensuring proper isolation of sensitive logic within the TEE. In Listings 1.1 and 1.2, there is a template of each file generated: one is a secure code and the other one is a non-secure code. The elements in each file are described in Sect. 3.3. Listing 1.1 shows the non-secure code that contains the `main()` function, and Listing 1.2 shows the secure code that contains the memory compartment structure and the port handlers as stated before.

The full source code of the generated implementation is available in our GitHub repository[1]. To assess the maintainability and quality of the generated code, we used SonarQube for static analysis. The analysis reports metrics such as code smells and duplicated blocks. Some relevant code metrics from the generated system are as follows: total lines of code: 45; bugs: 0. These results confirm that our generated code is not only functionally correct and secure by design, but also meets high standards of code quality and maintainability.

5 Conclusions and Future Work

The increasing demand for secure, integrated services in digital cities highlights the limitations of traditional integration platforms that lack native support for data privacy and execution integrity. TEEs, offer a promising alternative by isolating sensitive processes within protected memory compartments. However, using TEEs like the Morello Board, which is based in the CHERI architecture, requires a deep understanding of low-level programming and security concepts in the C language. This paper addresses that challenge through an innovative model–driven approach for the automatic generation of secure C code, reducing the need for manual development. By designing a metamodel that captures the essential elements of secure integration, and implementing it through Sirius and Acceleo, we enable developers to graphically define secure orchestration logic and generate the corresponding TEE code. This code is compatible and executable in the Morello Board as we demonstrated in the case of study presented before. The proposed approach reduces development complexity, enhances maintainability, and facilitates the adoption of TEEs in applications where security is crucial. The

[1] https://anonymous.4open.science/r/mde4tee-F801/README.md.

graphical abstraction layer provides reasoning about secure workflows promoting the separation of concerns.

Future research and development directions include extending the metamodel to support more complex data types and a broader applicability in digital cities. Also, it would be interesting to adapt the code generation templates for other TEE platforms such as ARM TrustZone and Intel SGX to promote platform independence. Additionally, the integration of an automated verification of the generated code to ensure memory safety and absence of vulnerabilities would be of great importance.

Acknowledgements. This research is partially funded by the Coordination for the Brazilian Improvement of Higher Education Personnel (CAPES) and the Brazilian National Council for Scientific and Technological Development (CNPq) under the project grants 311011/2022-5, 309425/2023-9, and 402915/2023-2; and by de Universidad Distrital Francisco José de Caldas under the PR2-2025 call.

References

1. Apache Software Foundation: Apache camel – integration framework based on enterprise integration patterns. https://camel.apache.org/. Accessed May 2025
2. Apvrille, L., Roudier, Y.: SysML-Sec: a model driven approach for securing embedded systems. In: 12th International Workshop on Modeling in Software Engineering, pp. 29–35. IEEE (2015). https://arxiv.org/pdf/1404.1985
3. ARM: Morello program – arm® (2019). https://www.arm.com/architecture/cpu/morello. Accessed May 2025
4. ARM: ARM TrustZone for Cortex-A (2025). https://www.arm.com/technologies/trustzone-for-cortex-a. Accessed May 2025
5. AWS: AWS Nitro Enclaves (2025). https://aws.amazon.com/ec2/nitro/nitro-enclaves/?nc1=h_ls. Accessed May 2025
6. Eclipse Foundation: Acceleo - code generator. https://www.eclipse.org/acceleo/. Accessed May 2025
7. Hamid, B., Geisel, J., Ziani, A., Bruel, J.-M., Perez, J.: Model-driven engineering for trusted embedded systems based on security and dependability patterns. In: Khendek, F., Toeroe, M., Gherbi, A., Reed, R. (eds.) SDL 2013. LNCS, vol. 7916, pp. 72–90. Springer, Heidelberg (2013). https://doi.org/10.1007/978-3-642-38911-5_5
8. Intel: Intel Software Guard Extensions (Intel SGX) Developer Guide (2025). https://www.intel.com/content/www/us/en/content-details/671334/intel-software-guard-extensions-intel-sgx-developer-guide.html. Accessed May 2025
9. MuleSoft Inc.: Mulesoft – anypoint platform for APIs and integrations. https://www.mulesoft.com/. Accessed May 2025
10. Obeo: Sirius - the open source graphical modeling workbench. https://www.eclipse.org/sirius/. Accessed May 2025
11. Sabt, M., Achemlal, M., Bouabdallah, A.: Trusted execution environment: what it is, and what it is not. In: 2015 IEEE Trustcom/BigDataSE/ISPA, vol. 1, pp. 57–64. IEEE (2015). https://doi.org/10.1109/Trustcom.2015.357
12. Watson, R.: Capability hardware enhanced RISC instructions (CHERI) (2019). https://www.cl.cam.ac.uk/research/security/ctsrd/cheri/. Accessed May 2025

Towards an Efficient Solution for Enterprise Application Integration Processes Deployed in Trusted Execution Environments

Ernesto Perez[1], Jose Bocanegra[2(✉)], Regis Schuch[3], Eldair Dornelles[3], Rafael Z. Frantz[3], Sandro Sawicki[3], and Fabricia Roos-Frantz[3]

[1] Department of Systems and Computing Engineering, Universidad de Los Andes, Bogotá, Colombia
ec.perez@uniandes.edu.co
[2] Faculty of Engineering, Universidad Distrital Francisco José de Caldas, Bogotá, Colombia
jjbocanegrag@udistrital.edu.co
[3] Unijuí University, Ijuí, Brazil
{regis.schuch,eldair.dornelles,rzfrantz,sawicki,frfrantz}@unijui.edu.br

Abstract. Enterprise Application Integration is a research domain within software engineering that addresses the challenge of integrating autonomous business systems. In this context, ensuring the confidentiality and integrity of data exchanged between applications constitutes a complex task. Several authors have proposed the utilisation of Trusted Execution Environments (TEEs) as a means to guarantee the integrity, isolation, reliability, and security of critical operations on the data processed and transmitted by applications, thereby preventing exfiltration and unauthorised access. However, this approach may adversely affect application efficiency due to substantial resource consumption. In this article, we propose an efficient solution for Enterprise Application Integration processes deployed within Trusted Execution Environments.

Keywords: Application Integration · Trusted Execution Environments

1 Introduction

Enterprise Application Integration (EAI) is a subfield of software engineering dedicated to providing methodologies, techniques, and tools for developing integration processes that orchestrate enterprise applications. An integration process connects heterogeneous applications, enabling data sharing and functional interoperability [10]. EAI solutions are commonly developed using platforms that offer resources to design, implement, test, and execute integration processes [7]. These processes are software artefacts composed of a well-defined workflow of atomic tasks, such as data transformation and validation, message enrichment

W. J. Sarmiento and E. L. Sierra-Ballén (Eds.): CCC 2025, CCIS 2692, pp. 185–195, 2027.
https://doi.org/10.1007/978-3-032-28447-1_13

with third-party data, and routing to target systems, in accordance with specific business rules [8]. These rules define the logic and conditions that govern the flow of data and tasks, ensuring that the process operates efficiently and aligns with organisational objectives and constraints.

Figure 1 illustrates a typical integration process that connects two applications (`App 1` and `App 2`). The workflow comprises two atomic tasks (`T1, T2`) connected by slots (`S1, S2, S3`) and two communication ports (`P1, P2`) that facilitate interaction with the integrated applications. The slots act as buffers that decouple tasks and connect the output of one task to the input of the next, thereby ensuring the continuous flow of data within the workflow. Ports abstract the technical details required to implement inbound and outbound communication channels, enabling data to be read from and written to the integration process and the applications. These communication ports are typically configured to periodically poll or push data to external applications. For instance, they may poll `App 1` for new data or push processed data to `App 2` at regular intervals. However, such periodic execution often encounters scenarios where no data is available for processing in the application or within the final slot (e.g., `S3`) of the workflow, resulting in unnecessary resource consumption.

An integration process is often exposed to various risks, particularly those concerning the security and privacy of data, as information transmitted through communication ports may be improperly exfiltrated. At present, there is a lack of robust technical guarantees to prevent such exfiltration and to establish a trustworthy environment for the development and execution of integration processes. In recent work, [12] proposed the use of Trusted Execution Environments (TEEs) to mitigate exfiltration risks and safeguard sensitive data, including credit card numbers, personal details, financial records, vehicle information, medical test results, academic records, among others.

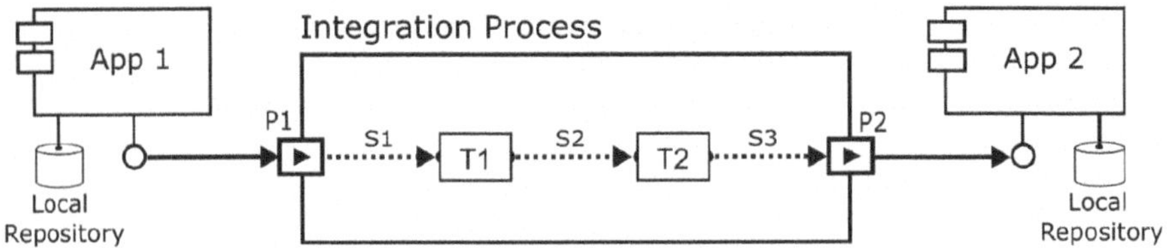

Fig. 1. Conceptual view of a typical integration process.

A Trusted Execution Environment (TEE) is a secure execution context that establishes protected memory compartments in which code handling sensitive data is executed [11]. Unlike conventional execution environments, a TEE enforces a hardware-based boundary that prevents unauthorised processes—including the operating system—from accessing the memory regions where sensitive operations are performed. When an application, or a portion of its code, runs within such a compartment, both the data and the operations are shielded from external observation, tampering, and exfiltration.

Memory compartmentalisation, enabled by TEEs, enhances overall system security by logically isolating different components of a system—such as tasks or software modules—into distinct memory compartments. Each compartment is governed by specific permissions that precisely define accessible memory regions, thereby restricting access strictly to what is necessary for the execution of a given task [2]. For instance, a digital payment service can be partitioned into compartments to separately manage payment and credit card data, user accounts, and stock control. This separation ensures that, even if one component—such as stock control—is compromised, the remaining components remain secure, thereby mitigating the risk of a full-system breach.

Several technologies have emerged to support the development of software systems requiring trusted execution environments. These include AMD SEV [1], AWS Nitro Enclaves [5], Intel SGX [9], ARM TrustZone [4], and, more recently, the Morello Board [3]. While each of these technologies implements TEEs with varying mechanisms for memory isolation and access control, they share the common objective of establishing a trusted boundary within memory compartments to protect sensitive operations and data.

We conducted a series of experiments and observed that executing code within memory compartments on a Trusted Execution Environment (TEE) results in reduced efficiency, as such execution demands greater computational resources. These include increased processing time for memory allocation, data read/write operations, and memory deallocation, all of which contribute to longer execution times compared to code executed outside of memory compartments. The inefficiencies arise from several factors, including context switching between secure and non-secure operations, unnecessary poll/push operations, cryptographic processing required to ensure data security and privacy, and the overhead associated with compartmentalised memory management. Consequently, prolonged or continuous execution within TEEs can place significant strain on system resources, leading to performance degradation in scenarios that require frequent or time-sensitive processing.

In this paper, we propose an efficient solution for the development and execution of integration processes deployed within Trusted Execution Environments (TEEs). Our approach significantly enhances process efficiency by decoupling tasks that require secure handling of sensitive data from those that do not, thereby restricting resource-intensive operations to instances where they are strictly necessary. The integration process is executed exclusively in response to events that necessitate secure processing, effectively reducing the overhead associated with continuously running processes within the TEE.

By streamlining cryptographic operations and implementing adaptive compartment management, the proposed solution ensures efficient utilisation of computational resources without compromising security guarantees. This strategy benefits both execution-time applications—where low-latency interactions are critical—and large-scale integration scenarios, such as those found in smart cities, where multiple digital services must interoperate.

We conducted an experiment comparing two configurations: (1) complete execution of the integration process, including communication ports, within the memory compartment; and (2) relocation of the communication ports to an external component termed the Communication Gateway. The Communication Gateway manages non-critical data exchanges outside the compartment via secure communication channels, such as SSL or VPN, and triggers secure tasks within the TEE only when necessary. The experimental results demonstrate that offloading the communication ports significantly reduces computational overhead and optimises resource usage, without compromising the security of sensitive data. This approach underscores the importance of decoupling communication mechanisms from the memory compartment to improve efficiency, thereby enabling large-scale integration processes to be both feasible and scalable in environments with stringent performance and security requirements.

The remainder of this paper is structured as follows. Section 2 presents the current approach to executing EAI processes within Trusted Execution Environments. Section 3 describes the proposed solution. In Sect. 4, we provide experimental data comparing the efficiency of the two approaches. Finally, Sect. 5 concludes the paper and outlines directions for future work.

2 Current Approach

The proposal to utilise Trusted Execution Environments (TEEs) to protect integration processes from data exfiltration was introduced by [12]. The authors implemented an integration process operating within a TEE, specifically targeting the development of software systems supported by ARM Morello Board technology. The conceptual model, illustrated in Fig. 2, depicts the execution of integration processes within a TEE using hardware-supported secure memory compartments. The double lines in the figure indicate the isolation and protection guarantees provided for sensitive data. In this model, the integration process interacts with external applications through poll and push operations. The poll operation retrieves data from `App 1` via an API designed for programmatic interaction, while the push operation transmits the processed data from the integration process to `App 2`. These operations are managed by communication ports that operate within the secure memory compartment.

The periodic polling mechanism commonly employed by communication ports results in frequent checks for new data, even when none is available, thereby leading to unnecessary resource consumption. Each time the integration process performs a poll or push operation, several resource-intensive tasks are executed within the memory compartment. These include: (i) Memory allocation, which reserves a secure memory block within the compartment to store the polled data; (ii) Data writing, which involves recording the retrieved data into the allocated memory block; (iii) Data reading, where tasks such as `T1` and `T2` access the data for processing; and (iv) Memory deallocation, which securely releases the memory block after the data has been processed and transferred between tasks,

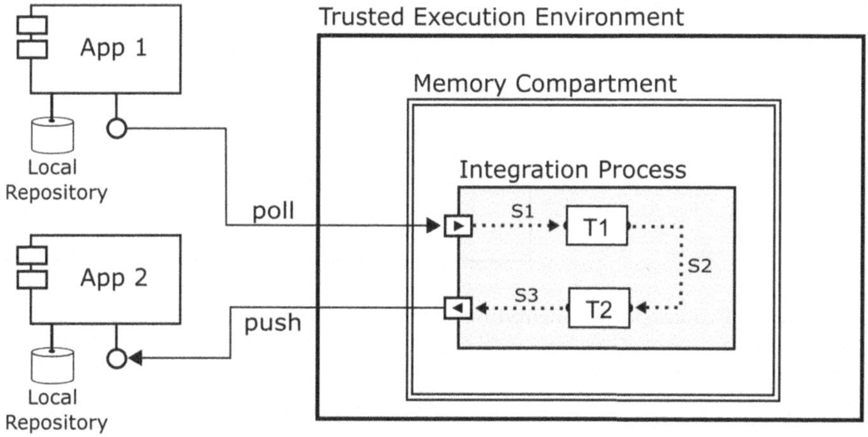

Fig. 2. Conceptual view of an integration process in a memory compartment.

ensuring that no residual data remains accessible. This repeated cycle of operations, particularly when no new data is available, contributes significantly to the inefficiency of continuous polling within TEEs.

Context switching between secure and non-secure states, combined with continuous integrity and permission checks, further contributes to performance degradation. Each interaction with external applications necessitates state transitions, data integrity verification, and the enforcement of access permissions, all of which consume substantial computational resources. As a result, the continuous execution of read and write operations within the memory compartment leads to significant performance overhead, particularly in scenarios that require frequent or real-time data exchanges. This inefficiency becomes a critical bottleneck when attempting to scale integration processes in environments that demand low latency and large-scale interactions.

3 Proposed Solution

We propose decoupling the tasks executed within the memory compartment of the Trusted Execution Environment (TEE) to optimise computational resource usage and minimise unnecessary overhead. In this approach, communication ports operate outside the memory compartment, in a faster and less resource-intensive environment. The integration process is executed only when an event necessitating secure processing occurs, and tasks are triggered solely when data is available in their respective input slots. Figure 3 illustrates the main components of the proposed solution.

A key component of the proposed architecture is the `Communication Gateway`, which operates outside the memory compartment but remains within the Trusted Execution Environment. The Communication Gateway continuously monitors external applications (such as `App 1` and `App 2`) by periodically polling

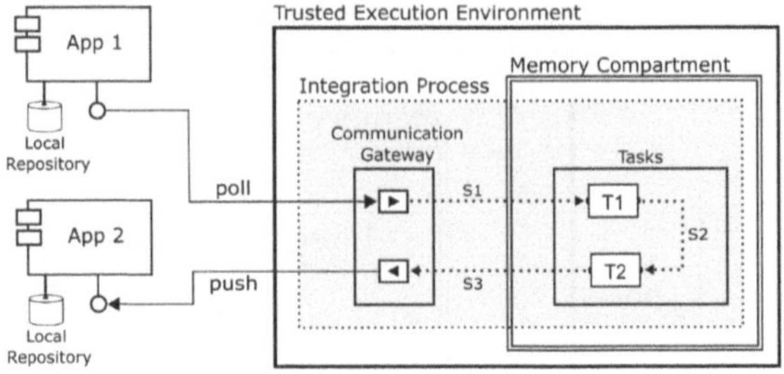

Fig. 3. Proposed implementation of the integration process in a TEE.

or pushing for events. It also manages the communication ports used to interface with both external applications and internal tasks within the integration process. When an event requiring secure processing is detected, the data polled from `App 1` is first verified by the Communication Gateway. Upon successful verification, the gateway triggers the execution of tasks within the TEE's memory compartment by transmitting the relevant data through the communication ports. The tasks inside the compartment then process the data sequentially. To ensure secure data transmission, the Communication Gateway encrypts the polled data before sending it to the internal tasks. These tasks subsequently decrypt the data using a private key, thereby ensuring that sensitive information is processed securely within the trusted environment.

In the proposed architecture, the only component that operates continuously is the `Communication Gateway`, which handles non-critical operations outside the memory compartment, such as reading from and writing to external applications. The tasks within the memory compartment are activated only when necessary, thereby reducing the computational cost and overhead associated with continuously running processes. This design effectively confines resource-intensive operations to instances where they are strictly required, enhancing the overall performance and scalability of the integration process without compromising security guarantees. The interaction between components proceeds as follows: the `Communication Gateway` polls data from `App 1` using a poll operation and performs validation; upon successful validation, it triggers the execution of tasks within the memory compartment; the tasks process the data sequentially, passing intermediate outputs through slots `S1`, `S2`, and `S3`; once processing is complete, the final output is pushed to `App 2` via a push operation. This approach decouples non-critical communication from the memory compartment, thereby optimising resource usage and improving scalability, while ensuring that sensitive data remains protected throughout execution.

4 Experiments

We conducted two experiments within the development environment of the Morello Board to evaluate the impact of memory compartmentalisation on the performance of the integration process. The objective was to measure the time required to perform operations such as memory allocation, writing, reading, and deallocation. Each experiment was repeated 100 times, and the results were analysed in accordance with the Central Limit Theorem for each tested configuration [6]. In the first experiment, we employed the scenario described in a previous section, in which the entire integration process was executed within the memory compartment of the Morello Board. In the second experiment, we applied our proposed approach, detailed in Sect. 3, which decouples the tasks executed within the memory compartment. Memory blocks ranging from 100 MB to 1 GB—both inside and outside the memory compartment—were evaluated. The experiments were conducted on the Morello Board, equipped with the Research Morello SoC r0p0 processor, four CPU cores, and 16 GB of DDR4 RAM. The operating system used was CheriBSD 24.5.

The collected data include the execution times, measured in milliseconds, for memory allocation, writing, reading, and deallocation operations. Allocation refers to the time required to reserve a memory block; writing denotes the time taken to populate the memory block with data; reading corresponds to the time needed to access the entire contents of the memory block; and freeing indicates the time required to release the memory block. These execution times were recorded in CSV files, from which the means and standard deviations across 100 repetitions for each block size were calculated. Each value presented in the tables represents the *mean execution time* (in milliseconds) and its *standard deviation*. For example, a value presented as 2 ± 4.77 indicates that 2 is the mean, and 4.77 is the standard deviation, reflecting the variability across repetitions. Table 1 presents the data collected when the integration process was executed outside the memory compartment, while Table 2 presents the data collected when it was executed within the compartment.

Table 1. Running outside a compartment, including mean and standard deviation.

Block Size (MB)	Allocation Time (ms)	Write Time (ms)	Read Time (ms)	Free Time (ms)
100	2 ± 4.77	282,584 ± 13.86	282,581 ± 12.79	6 ± 4.52
100	4 ± 4.19	565,164 ± 17.12	565,163 ± 18.85	10 ± 4.03
300	4 ± 1.77	847,755 ± 21.18	847,752 ± 64.89	13 ± 3.66
400	5 ± 3.09	1,130,330 ± 21.00	1,130,328 ± 28.20	14 ± 2.27
500	5 ± 3.07	1,412,907 ± 31.49	1,412,903 ± 28.92	15 ± 2.37
600	5 ± 1.56	1,695,493 ± 32.97	1,695,493 ± 30.19	16 ± 1.28
700	5 ± 1.52	1,978,083 ± 52.24	1,978,098 ± 79.47	17 ± 0.86
800	5 ± 1.73	2,260,662 ± 41.09	2,260,660 ± 53.11	18 ± 0.62
900	5 ± 0.54	2,543,249 ± 47.19	2,543,234 ± 42.16	18 ± 0.97
1000	5 ± 0.50	2,825,823 ± 47.72	2,825,818 ± 41.68	18 ± 0.64

A comparison of Tables 1 and 2 reveals that executing code within the memory compartment incurs higher computational costs across all operations, with

Table 2. Running inside a compartment, including mean and standard deviation.

Block Size (MB)	Allocation Time (ms)	Write Time (ms)	Read Time (ms)	Free Time (ms)
100	93 ± 171.27	283,239 ± 58.31	283,133 ± 28.83	89 ± 180.05
200	98 ± 221.17	566,458 ± 82.10	566,269 ± 65.02	214 ± 397.35
300	99 ± 295.44	849,705 ± 131.43	849,396 ± 87.16	222 ± 452.92
400	127 ± 430.92	1,132,983 ± 189.58	1,132,550 ± 106.44	430 ± 788.02
500	159 ± 599.09	1,416,190 ± 189.97	1,415,698 ± 123.68	217 ± 420.54
600	151 ± 648.00	1,699,454 ± 255.41	1,698,795 ± 174.82	439 ± 921.59
700	195 ± 880.05	1,982,654 ± 245.07	1,981,909 ± 122.70	453 ± 979.92
800	216 ± 1,084.49	2,265,901 ± 235.38	2,265,075 ± 139.94	818 ± 1,513.98
900	288 ± 1,536.92	2,549,115 ± 258.37	2,548,205 ± 196.83	816 ± 1,579.74
1000	248 ± 1,543.50	2,832,372 ± 337.74	2,831,332 ± 167.56	444 ± 1,003.29

memory allocation and deallocation being particularly affected. For example, allocating 100 MB of memory takes approximately 2 ms outside the compartment, whereas the same operation requires 90 ms inside. Both tables exhibit a linear increase in write and read times as the memory block size increases; however, execution within the compartment consistently results in higher latencies. The free operation shows the most pronounced difference, reaching 6 ms outside the compartment compared to 89 ms inside when allocating 100 MB of memory.

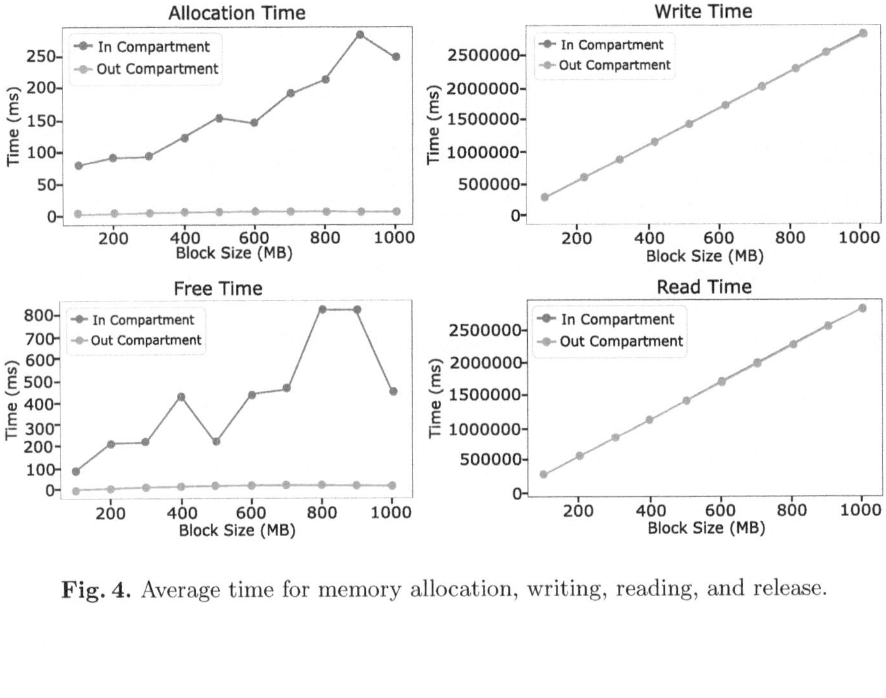

Fig. 4. Average time for memory allocation, writing, reading, and release.

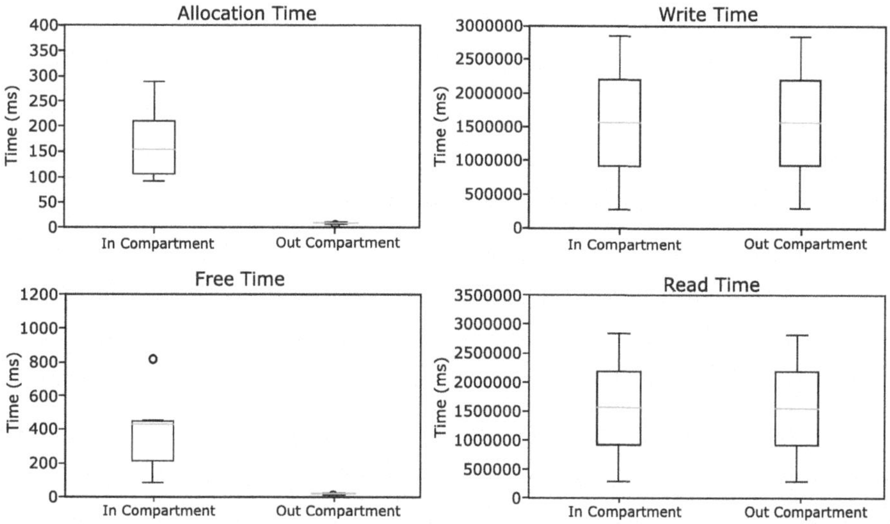

Fig. 5. Execution time dispersion for allocate, write, read, and free operations.

The results indicate that executions within the memory compartment exhibit higher latency and significant unpredictability, particularly in memory allocation and deallocation operations. The graphs in Fig. 4 highlight the linear behaviour of write and read times, while also emphasising the increasing overhead incurred inside the compartment as block sizes grow. Meanwhile, the boxplot in Fig. 5 illustrates considerable dispersion, with long tails and outliers especially evident in the allocation and free operations. In contrast, execution times outside the compartment remain stable and predictable. This unpredictability within the memory compartment can be attributed to the overhead introduced by TEE control and security mechanisms, which impose additional computational costs and variability on memory management.

5 Conclusions and Future Work

Enterprise Application Integration typically operates with sensitive data in conventional (untrusted) environments, thereby exposing it to security risks such as data exfiltration and unauthorised access. Trusted Execution Environments (TEEs) have emerged as a promising solution to address these challenges by providing support for the secure execution of integration processes. However, executing integration processes within TEEs can significantly impact system efficiency due to their high resource consumption. The approach presented in this paper offers a viable solution that both protects sensitive data during processing and reduces computational resource usage. Experimental results from a case study conducted on the Morello Board TEE validate the applicability and effectiveness of the proposed method. As future work, we intend to extend the case study to other trusted environment technologies, including AWS Nitro Enclaves, Intel SGX, AMD SEV, and ARM TrustZone. This will allow us to evaluate the adaptability of the approach across diverse technological contexts and enable a comparative analysis of these technologies for secure software development.

Information about Artifact Availability

The experiment's implementation and results are available in the repository: https://github.com/gca-research-group/in-out-tee-integration-case-study/

Conflicts of Interest

All authors declare that they have no conflicts of interest to disclose.

Acknowledgements. This research is partially funded by the Coordination for the Brazilian Improvement of Higher Education Personnel (CAPES) and the Brazilian National Council for Scientific and Technological Development (CNPq) under the project grants 311011/2022-5, 309425/2023-9, and 402915/2023-2; and by de Universidad Distrital Francisco José de Caldas under the PR2-2025 call.

References

1. AMD: AMD Secure Encrypted Virtualization (SEV) (2025). https://www.amd.com/en/developer/sev.html. Accessed 21 Jan 2025
2. ARM: Morello prototype architecture (2022). https://developer.arm.com/documentation/. Accessed 21 Jan 2025
3. ARM: Arm Morello Program (2025). https://www.arm.com/architecture/cpu/morello. Accessed 21 Jan 2025
4. ARM: Arm TrustZone for Cortex-A (2025). https://www.arm.com/technologies/trustzone-for-cortex-a. Accessed 21 Jan 2025
5. AWS: AWS Nitro Enclaves (2025). https://aws.amazon.com/ec2/nitro. Accessed 21 Jan 2025
6. Dudley, R.M.: Uniform Central Limit Theorems, vol. 142. Cambridge University Press (2014)
7. Frantz, R.Z., Corchuelo, R., Basto-Fernandes, V., Rosa-Sequeira, F., Roos-Frantz, F., Arjona, J.L.: A cloud-based integration platform for enterprise application integration: a model-driven engineering approach. Softw. Pract. Exp. **51**(4), 824–847 (2020)
8. Hohpe, G., Woolf, B.: Enterprise Integration Patterns: Designing, Building, and Deploying Messaging Solutions. Addison-Wesley Professional (2004)
9. Intel: Intel®Software Guard Extensions (Intel®SGX) (2025). https://www.intel.com/content/www/us/en/products/docs/accelerator-engines/software-guard-extensions.html. Accessed 21 Jan 2025
10. Ritter, D., Nordvall Forsberg, F., Rinderle-Ma, S.: Responsible composition and optimization of integration processes under correctness preserving guarantees. Inf. Syst., 102400 (2024)

11. Sabt, M., Achemlal, M., Bouabdallah, A.: Trusted execution environment: what it is, and what it is not. IEEE Trustcom/BigDataSE/Ispa **1**, 57–64 (2015)
12. Schuch, R., Frantz, R.Z., Bocanegra, J., Roos-Frantz, F., Sawicki, S., Molina-Jiménez, C.: Digital services integration in smart cities: a trusted execution environment based solution. In: Congresso Ibero-Americano em Engenharia de Software, pp. 437–438 (2024)

CookieSentinella: A Chrome Extension to Prevent Cookie Theft and Cloning in Educational Platforms with a Focus on Usability

Mateo Jesús Cadena Cabrera(✉), Miguel Ángel Llanten Llanten, Marta Cecilia Camacho Ojeda, and John Jairo Perafán Ruiz

Institucion Universitaria Colegio Mayor del Cauca, Carrera 7 # 2-41 Edificio Bicentenario, Popayán, Colombia
mcadena@unimayor.edu.co

Abstract. This paper introduces CookieSentinella, a Google Chrome extension designed to prevent the theft and cloning of authentication cookies in Moodle-based educational platforms. The proposed solution combines real-time detection and cookie context isolation techniques with a usability-focused design, featuring a user-friendly interface for non-expert users. CookieSentinella was developed using an iterative approach that combines PTES guidelines with Scrum methodology. The initial version focuses on modifying authentication cookie attributes. Consequently, if a cookie is used maliciously, it will fail to produce the expected results, as its content will have been altered beforehand. Unlike existing solutions, this tool aims not only to effectively counter cookie data theft but also to incorporate usability principles that reduce the complexity of its installation and use.

Keywords: cookies · usability · browser extension · data privacy · cookie cloning · cybersecurity · session hijacking · session protection

1 Introduction

Authentication cookie vulnerabilities represent one of the most common yet underexplored security issues. Cybercriminals can employ various techniques to steal these cookies and impersonate legitimate users [7]. Authentication cookies are critical for ensuring traceability and security in session management, particularly in social media and institutional contexts such as universities, where academic, administrative, and personal information can be easily compromised. These cookies allow servers to identify users without requiring credentials for each access [9]. However, due to their functionality and structure, cookies are frequent targets of cyberattacks. Research by [4] analyzed multiple websites and platforms, finding that 60% exhibited vulnerabilities in unprotected cookies, posing a significant issue for higher education institutions [12].

W. J. Sarmiento and E. L. Sierra-Ballén (Eds.): CCC 2025, CCIS 2692, pp. 196–205, 2027.
https://doi.org/10.1007/978-3-032-28447-1_14

According to [6], cybersecurity poses significant challenges in digital education, necessitating concrete actions to ensure information protection and user privacy. Consequently, technologies are needed to enhance security on virtual educational platforms while promoting cybersecurity awareness and training for both students and educators [6].

Usability principles play a crucial role, as they enable security tools, such as browser extensions, to be accessible to all user types. Effective usability not only enhances the user experience but also facilitates safer decision-making, minimizes errors, and encourages the adoption of protective tools [11]. In contexts where users are the most vulnerable link, usability-centered design becomes an essential component of cybersecurity. Browser extensions with a focus on security and usability offer a viable and efficient solution [11]. Recent studies on the effectiveness of extensions against threats like phishing [10] demonstrate that well-designed tools with real-time detection and comprehensible alerts can significantly reduce risks if integrated intuitively and non-intrusively. However, many extensions fail due to overly technical terminology, which hinders their use by average users [10].

In this context, we propose CookieSentinella, a Chrome extension that detects and prevents cookie theft and cloning in real time on educational platforms like Moodle. With a usability-focused design, the extension not only protects user sessions but also ensures an intuitive experience, fostering the adoption of cybersecurity tools in academic environments. To better understand these vulnerabilities, the following section presents the foundational concepts framing this research.

2 Theoretical Framework

To conduct this research, it is necessary to clarify key concepts, which are briefly presented in this section.

Cookies are a commonly used tool in web applications to store user state information and maintain active sessions [12]. However, improper use of cookies can expose applications to various security vulnerabilities [12]. To mitigate misuse, tools like **Cookie Editor extensions** allow users to view and manipulate cookies to prevent third-party exploitation; however, these same tools can be used in attacks based on *JavaScript Injection* APIs. One such attack is **session hijacking**, where an adversary steals session cookies to impersonate an authenticated user [12]. In response, systems like **client-side intrusion detection** have been developed to detect anomalous activities from the client side (browser).

While information security is critical, designing tools to prevent attacks without prioritizing ease of use can be a limitation. **Usability** is a software quality characteristic that evaluates how easy and efficient it is for users to interact with a system or product. According to ISO 9241-11, usability is defined as "the extent to which a product can be used by specified users to achieve specified goals with effectiveness, efficiency, and satisfaction in a specified context of use" [8]. Usability influences **user satisfaction**, encompassing the level of comfort, trust, or enjoyment a user experiences when interacting with a system. It relates to how users feel during and after use, whether the tool is clear and helpful, and whether they would reuse or recommend it [8].

2.1 Background

Cybersecurity aims to protect individuals' and institutions' data within information systems in the internet environment [7].

Web-based systems allow users to authenticate their identity through various methods, depending on the required security level. Some applications require only a username and password, while others involve additional verification through external devices, such as fingerprints, SMS codes, or authentication keys. However, the number of attempted attacks on applications continues to grow, targeting data theft, identity theft, or service disruptions [7]. These attacks can result in financial damage to individuals, institutions, businesses, and even governments [7].

Virtual education experienced rapid growth following the COVID-19 pandemic, with online learning platforms and videoconferencing applications becoming widely used for online or hybrid education. Consequently, the risk of attacks such as identity spoofing, unauthorized data access, and personal data theft has significantly increased [1]. According to [1], ensuring cybersecurity in remote activities remains a constant challenge for higher education institutions. It is necessary to control information access, develop access policies for applications and stored data, and implement mechanisms to minimize unauthorized access or data breaches.

According to [3], the use of cloud computing (CC) in educational environments offers numerous benefits, such as low-cost services for academics and students. However, its widespread use presents significant adoption challenges, as most existing systems do not adequately address reliability, security, privacy, or integrity.

Information security in higher education institutions is a pressing concern due to their vulnerability to threats, stemming from the complexity of their IT environments, diverse user base, and sensitive data. Poor password management practices, such as using easily guessable passwords, storing credentials in browsers, or maintaining credentials for extended periods, can expose institutional systems to password-cracking vulnerabilities [2].

Studies like [11] have analyzed browser extensions like CookieBlock, which block third-party cookies to protect privacy, but lack a specific focus on securing active sessions in educational platforms. CookieSentinella stands out by combining real-time detection with a usability-centered design, addressing both security and user experience.

Various studies have highlighted the need to rethink the relationship between cybersecurity and usability. [5] introduced the "3U" model (user, usage, and usability), arguing that security solutions neglecting the user as the central focus tend to fail due to technical complexity and lack of comprehension, leading to human errors or bypassing protective measures. Similarly, [11] demonstrated through a lab study with CookieBlock that while privacy tools can be technically effective, their adoption and proper use are compromised by issues with comprehension, configuration, or interaction. Many users lack clear mental models of how these extensions function, leading to frustration and usage errors. These findings align with research on cybersecurity in education, underscoring the scope of the problem. Both studies agree that usable cybersecurity must offer intuitive interfaces, clear feedback, and a balance between protection and user experience, especially in contexts where privacy is constantly at risk, such as with web cookies. Building on these identified limitations, the design of CookieSentinella is proposed.

3 Design of the CookieSentinella Extension

The project methodology integrates PTES (Penetration Testing Execution Standard) guidelines, Scrum, and tools for validating functionality and usability to develop a browser extension. The proposed extension, named CookieSentinella, aims to protect active user sessions on Moodle-based educational platforms by detecting and preventing cookie theft and cloning in real time, mitigating attacks such as those using Cookie Editor extensions or JavaScript injection APIs. Additionally, the extension is designed to be accessible to users without advanced digital security knowledge through a user-friendly interface that prioritizes comprehension and user control. This usability-focused approach ensures not only tool adoption but also a progressive improvement in users' digital security culture. CookieSentinella is developed using the Chrome Web Extensions API and JavaScript, with a modular design comprising three main components: a real-time detection module, a cookie context isolation module (cookie partitioning), and a graphical interface built with HTML5 and CSS3. The detection module monitors cookie requests on Moodle platforms, analyzing changes in the browser's digital fingerprint (e.g., HTTP headers, user agent, or environment properties) to identify cloning or unauthorized access attempts. The isolation module implements cookie partitioning, ensuring session cookies are valid only in the user's original browser context, automatically invalidating any reuse attempts in other environments. The graphical interface, designed for intuitiveness, employs a minimalist design with clear notifications and accessible controls, avoiding complex technical terminology. The workflow of CookieSentinella is as follows:

- **Monitoring:** Intercepts requests using `chrome.webRequest` and flags anomalous patterns (e.g., unauthorized domains, injection attempts).
- **Detection:** Compares the browser's fingerprint with the one recorded at login, alerting on discrepancies.
- **Protection:** Applies cookie partitioning, blocks reuse in other environments without invalidating the legitimate session, and notifies non-intrusively.
- **Interaction:** Allows enabling/disabling protection and clearing cookies, with preferences stored in `chrome.storage.local`.

Unlike traditional approaches that invalidate sessions, CookieSentinella keeps the legitimate session active, allowing uninterrupted user interaction. The extension employs detection and cookie partitioning techniques, ensuring that attempts to reuse cookies in other environments are automatically invalidated. It relies on the cookie partitioning principle, ensuring that cookies visible to an attacker lack useful authentication information. Due to this logical separation, genuine cookies are valid only in the original browser, and any reuse attempt in another environment is automatically invalid. The development of CookieSentinella follows an iterative and incremental approach. Each cycle introduces new functionalities, such as improved attack detection or usability refinements, based on tests in the Moodle web environment. Broader usability evaluations with students and educators of varying academic levels are planned to enhance both security and user experience. In conclusion, CookieSentinella offers an innovative solution addressing the identified challenges.

To illustrate CookieSentinella's internal workings, As shown in Fig. 1 presents a diagram clearly and structurally describing the extension's architecture. This diagram illustrates the three core modules: the detection module, responsible for intercepting and analyzing incoming cookie requests; the isolation module, tasked with applying cookie partitioning and blocking cloning attempts; and the user interface, enabling straightforward management by end users. The three modules are embedded within the Chrome browser hosting the cookie. The figure also depicts the cookie's interactions with the Moodle platform.

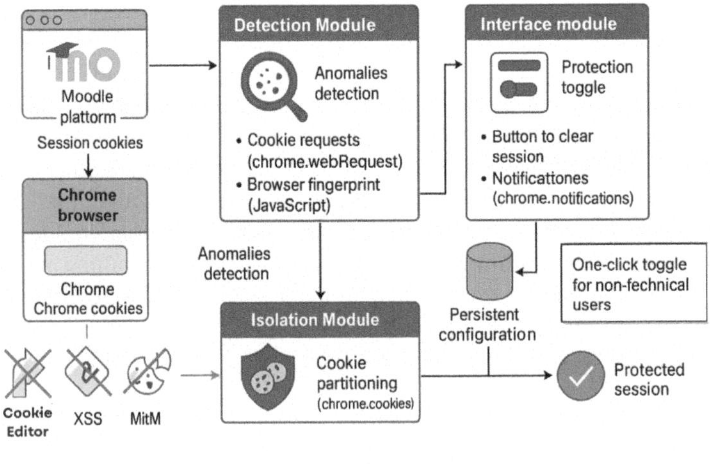

Fig. 1. CookieSentinella Architecture.

The lack of control over the backend and servers of any website posed a critical challenge in developing CookieSentinella. Operating solely on the client side with the Chrome API makes it difficult to modify session logic and server-generated cookie attributes, complicating the prevention of attacks like XSS and *session hijacking*. This limitation, common in many web environments, explains the scarcity of similar extensions. CookieSentinella's innovation lies in implementing cookie partitioning from the browser, achieving autonomous cookie isolation and blocking unauthorized access.

Figure 2 depicts CookieSentinella's main interface, where users can enable or disable protection and clear session cookies with a single click. The minimalist design features a bottom options menu with clear, understandable buttons, allowing non-technical users to interact without needing to leave the extension or open another browser. The figure highlights the interface's simplicity, with a central control panel and visual security alerts. With the design established, a testing process was planned to evaluate both technical effectiveness and usability.

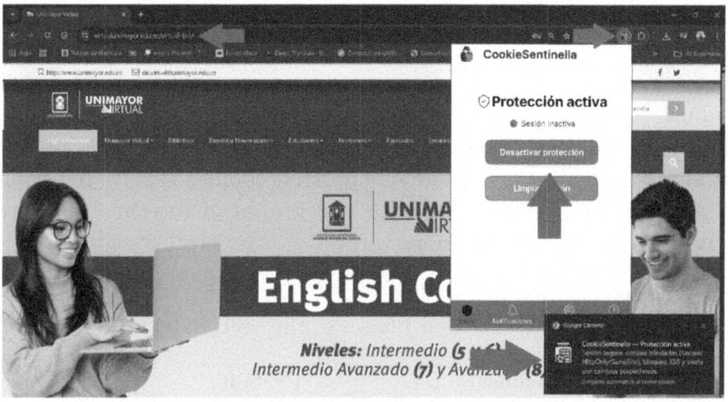

Fig. 2. CookieSentinella Interface.

4 Results

Initial tests were conducted by the research team on 15 Moodle-based educational platforms using the Chrome browser, with tools like Burp Suite, EditThisCookie, and Cookie-Editor to simulate attacks and gather evidence, as all platforms use the same cookie structure. The goal was to verify CookieSentinella's functionality, ensuring the extension modified cookie attributes (e.g., SameSite, HttpOnly, HostOnly, Secure) to invalidate extracted cookies, simulating a data theft attack. In these cases, attempts to use modified cookies for fraudulent access failed, resulting in HTTP 401/403 errors. Additionally, XSS attacks were simulated using Burp Suite to inject malicious scripts (`<script>document.cookie</script>`) into HTTP requests to Moodle, confirming that the detection module, implemented with `chrome.webRequest`, blocked suspicious requests, and generated real-time alerts.

Table 1. Evaluation Phases

Identified Extensions	Burp Suite	EditThisCookie	Cookie Editor	CookieSentinella
Performed Attacks	1	2	4	2
Moodle Platforms (ES) Declaring `MoodleSession`	6	6	6	6
Cookies with Modified Attributes	2	2	2	3
Tools	1	0	2	0

To evaluate the extension's effectiveness, specific metrics were considered: for the detection module, the rate of accurate alerts versus false positives; for the isolation module, the percentage of exported cookies that were invalid when reused in another environment; and for the graphical interface, user response time and notification comprehension.

Table 2. Evaluation Phases of the Web Extension

	Technical	Usability
Objective	Verify the extension's ability to detect and prevent attacks.	Evaluate ease of use and comprehension for average users.
Method	- Attack simulation: XSS, MitM, session hijacking. - Controlled tests with sniffers and cookie editors. - Log review.	- Guided tasks: installation, configuration, alert response. - SUS questionnaires. - Direct observation of use.
Indicators	- Successful attack blocking. - Real-time alerts generated. - Anomalous event logging.	- Average task completion time. - Task success rate. - Comprehension level. - SUS score.
Tools	- EditThisCookie - Cookie-Editor - Scripts	- SUS questionnaire - Forms - UX lab

The aggregated values in Table 1 support the results, showing that attempts with extracted cookies were rejected, and alerts were issued in real time without interrupting the legitimate session.

In parallel, Table 2 organizes the evaluation plan into two complementary areas—technical and usability—and succinctly outlines the objectives, methods, indicators, and tools considered.

For future validation phases, acceptance criteria yet to be implemented are proposed, which are detailed in the *Future Work* section to clarify that they correspond to proposed analyses, not validated results.

5 Future Work

As part of the ongoing development of the tool and its impact on user-centered cybersecurity, several future research directions are identified to strengthen and expand the extension's scope:

1. **Continuous Usability Improvement:** Conduct iterative usability testing with diverse user profiles to further simplify interaction and ensure digital inclusion, considering accessibility, language, and devices.
2. **Multiplatform and Cross-Browser Compatibility:** Extend the extension to work not only on Google Chrome but also on browsers like Firefox, Edge, and Brave, enabling broader reach and adoption by diverse users.

Additionally, proposed acceptance criteria for future validation phases are included to measure the extension's effectiveness in both cybersecurity (Table 3) and usability (Table 4).

Table 3. Proposed Acceptance Criteria – Cybersecurity

Criterion	Expected Target
Detection of cookie cloning attempts	$\geq$ 60% of cases
Notifications issued for detected cloning or theft attempts	(Number of notifications / number of detections) = 0.8
Detection of XSS script attacks	$\geq$ 70% of users
Detection of browser fingerprint changes	$\geq$ 70% of users

For CookieSentinella's usability evaluation, a workshop with real Moodle users at UnimayoVirtual is planned. The selection will use stratified sampling by role (40% students, 40% educators, 20% usability experts) and experience (novices <1 year; experienced $\geq$1 year), with AB/BA counterbalancing of conditions (baseline without extension vs. CookieSentinella use) and brief neutral familiarization. The planned sample size is $n = 16$ for a population of 3160. Tasks will include login, navigation, and an induced alert scenario. Metrics will cover success rate, efficiency, and effectiveness. Table 4 outlines the proposed usability acceptance criteria.

Table 4. Proposed Acceptance Criteria – Usability

Criterion	Expected Target
Installation and configuration without assistance	$\geq$ 80% of users
Comprehension of security notifications	$\geq$ 85% of users
Average SUS score	$\geq$ 70/100
Positive feedback in survey	$\geq$ 80%

6 Conclusions

CookieSentinella represents an innovative solution for securing sessions on educational platforms, integrating real-time cookie cloning detection with an intuitive design that facilitates adoption by non-technical users. Preliminary results demonstrate that the extension significantly reduces identity theft risks while ensuring a seamless user experience. Thus, it contributes to strengthening cybersecurity in virtual education and addresses the need for higher education institutions to have effective solutions to mitigate threats and minimize their potential impacts.

Acknowledgments. This work was funded by the project ID IUCMC 2025.SII.0080, under the 2025 research group project call of the Institución Universitaria Colegio Mayor del Cauca.

References

1. Alexei, A., Alexei, A.: Cyber security threat analysis in higher education institutions as a result of distance learning. Inter. J. Sci. Technol. Res. **10**, 128–133 (2021)
2. Bwiino, K., Kituyi, G., Nkamwesiga, L., Nyamadi, M.: Information security in higher education institutions: a systematic literature review. ORSEA J., 302–320 (2025). https://doi.org/10.56279/orseaj.C2024.18
3. Elgelany, A., Alghabban, W.: Cloud computing: Empirical studies in higher education - a literature review. Int. J. Adv. Comput. Sci. Appl. **8**(10) (2017)
4. Gonzalez Brito, H.R., Zuniga Solorzano, M.E.: Study of the impact of cookies on the security of web applications. In: Proceedings of the National Cybersecurity Conference, pp. 1–20. Universidad de las Ciencias Informaticas, La Habana (2017)

5. Grobler, M., Van Heerden, R., Jansen, R., Van Niekerk, B.: User, usage and usability: redefining human centric cyber security. In: Proceedings of the 16th International Conference on Cyber Warfare and Security (ICCWS 2021), pp. 132–139. Academic Conferences International, Reading, UK (2021). https://doi.org/10.34190/ICCWS.21.086
6. Guana-Moya, J.: The importance of cybersecurity in digital education: challenges and solutions. RECIMUNDO **7**(1), 609–616 (2023). https://doi.org/10.26820/recimundo/7.(1).enero.2023.609-616
7. Gugala, L., Laba, K., Dul, M.: Protecting web applications from authentication attacks. Adv. Web Develop. J. **1**(4), 26–42 (2023). https://doi.org/10.5281/zenodo.10049992
8. International Organization for Standardization (ISO): Iso 9241-11:2018 – ergonomics of human-system interaction – part 11: Usability: Definitions and concepts (2018)
9. Londono Congote, A.: Personal data processing through web cookies: Analysis under colombian data protection legislation. Universidad de los Andes, Bogota, Independent Academic Article (2021)
10. Olusegun, J.: Case studies on the effectiveness of anti-phishing browser extensions in real-world scenarios, article (January 2025)
11. Schoni, L., Kubicek, K., Zimmermann, V.: Block cookies, not websites: analysing mental models and usability of the privacy-preserving browser extension cookieblock. Proc. Priv. Enhancing Technol. **2024**(1), 192–216 (2024). https://doi.org/10.56553/popets-2024-0012
12. Zambrano Rendon, A.D., Cedeno Valarezo, L.C., Avellan Vera, D.A., Herrera Molina, J.E., Cedeno Zambrano, K.J.: Cookie vulnerabilities in web applications: social media and streaming. Informatica y Sistemas: Revista de Tecnologias de la Informatica y las Comunicaciones **7**(1), 40–52 (2023). https://doi.org/10.33936/isrtic.v7i1.5792

Interaction and Popularity Based Metrics for Estimation of Interpersonal Influence on Social Media

Juan José Viáfara[1(✉)], Robinson Andrey Duque[1], Juan Francisco Díaz[1], and Carlos Alberto Ramírez[2]

[1] Universidad del Valle, Cali, Colombia
{viafara.juan,robinson.duque,juanfco.diaz}@correounivalle.edu.co
[2] Pontificia Universidad Javeriana - Cali, Cali, Colombia
carlosalbertoramirez@javerianacali.edu.co

Abstract. This paper presents two complementary metrics (interaction-based and popularity-based) to estimate interpersonal influence among users on the X platform, which can be particularly used under restricted data conditions. The first metric is based on explicit user interactions, such as replies, mentions, and retweets, which are interpreted as direct signs of influence. The second metric is built from this former measure by incorporating the popularity of the influential user, calculated from the volume of received interactions.

The evaluation of the two proposed measures of interpersonal influence was carried out using data from the pension reform in Colombia carried out in 2024. The analysis is focused on the correlation between the proposed metrics and three baseline indicators of influence in social networks (i.e., retweets, mentions, betweenness centratlity). The interaction-based metric achieved a Spearman correlation of 80% with retweet counts. Meanwhile, the popularity-based metric showed a Spearman correlation of 75% with betweenness centrality. The results support the usefulness of both metrics in capturing different aspects of interpersonal influence.

Keywords: Influence estimation · Interpersonal influence metrics · Digital influence measurement

1 Introduction

This study aims to estimate interpersonal influence within digital social networks, using data from the X platform (formerly Twitter). Interpersonal influence is defined as the ability of one user to affect the beliefs, attitudes, or behaviors of others through social interactions [10]. On platforms like X, this influence is often expressed through explicit interactions such as replies, mentions, and retweets [11]. Previous research has also highlighted that influence is determined by factors such as familiarity, perceived similarity, and trust in the sender of the message [9].

W. J. Sarmiento and E. L. Sierra-Ballén (Eds.): CCC 2025, CCIS 2692, pp. 206–216, 2027.
https://doi.org/10.1007/978-3-032-28447-1_15

Recent restrictions in X's data access policies have reduced the availability of information for academic use. In addition to this limitation, applying opinion dynamics models to real-world scenarios requires the ability to estimate interpersonal influence from observable interactions. Although models such as French-DeGroot, Friedkin-Johnsen, and the Expressed-Private Opinion framework offer valuable theoretical insights [2], they typically rely on predefined influence weights. Without empirical measures of influence, these models cannot be grounded in actual social media behavior. Developing such measures is therefore essential to bridge the gap between theoretical modeling and practical analysis of opinion evolution.

Since 2023, X has imposed more restrictive and expensive data access policies for academic research, introducing a $100 per month API plan for 10,000 publications, which doubled to $200 for 15,000 publications by 2025. This shift from free large-scale datasets highlights the growing need to develop alternative approaches that can operate effectively with smaller datasets. In response, this study proposes a method for estimating interpersonal influence using limited data, using established metrics from previous research to ensure analytical validity.

The remainder of this article is structured as follows: Sect. 2 introduces previous work related to interpersonal influence, then Sect. 3 presents the methodology used to access data for a case study and introduces interaction-based and popularity-based metrics to estimate influence with limited data. Additionally, Sect. 4 presents experimental results and finally Sect. 5 presents conclusions and open research lines.

2 Related Work

A review of the literature on influence estimation in digital social networks revealed two main strategies: the first one is focused on estimating the social influence of each user by the estimation of its overall impact on the network as a whole. Although this first approach estimates a single influence value for each user, the second one emphasizes on interpersonal influence, that is, estimates the directional influence on each pair of users of the network.

Within the social influence strategy, the most common methods rely on structural metrics such as degree centrality, betweenness centrality, eigenvector centrality, or PageRank [7,11]. These approaches seek to identify the most relevant users based on the structural characteristics of the network. Studies such as those by Ahmed et al. [1] and Chakraborty et al. [3] have applied these techniques to analyze public actors in various contexts. Although effective in identifying broadly influential users, these methods do not account for how much one specific user influences another.

In contrast, the interpersonal influence strategy assigns a specific value to each user-to-user connection, evaluating the strength of the relationship in terms of influence. This more granular approach allows researchers to examine not just who is influential but also over whom. A widely used method in this line is to analyze explicit interactions between users, such as retweets, mentions, or replies [1,6,11–13]. These actions are often interpreted as evidence of resonance or alignment with the content shared by another user. For example, Venkatesan et al. [12] and Yuan et al. [13] used this approach and even incorporated multiple types of interactions to gain a broader perspective.

Additionally, some methods have combined direct interactions with measures of the popularity or authority of the influencing user. This means that not all interactions are weighted equally, an interaction with a highly visible or authoritative user is considered stronger evidence of influence. Zhai et al. [14] proposed such an approach, aiming to capture not only the act of influencing but also its potential reach based on the sender's status within the network.

3 Data Access and Influence Metrics

This section presents two complementary measures designed to estimate interpersonal influence on X under limited data conditions. These measures aim to capture different but related dimensions of influence observed in social media platforms. The first is focused on direct interactions, while the second introduces an adjustment based.

3.1 Data Collection

To evaluate the proposed measures of interpersonal influence, a dataset of 10,001 Spanish-language tweets on Colombia's pension reform was collected from X between June 22 and July 19, 2024. Each week, 2,500 posts matching the term *"reforma pensional"* were sampled, with daily quota adjusted to match platform activity over the previous seven days using X's API. The result is a structured dataset in `.csv` format containing the necessary information to apply the proposed influence estimation measures. Table 1 lists the main fields included in the dataset.

Table 1. Fields from the dataset used for evaluating interpersonal influence.

Field	Data Type	Description
`text`	String	First 180 characters of the tweet text
`created_at`	String	Date of tweet creation
`public_metrics`	Object	Interaction metrics (retweets, likes, replies, etc.)
`entities`	Object	Entities in the tweet (e.g., user mentions)
`author_username`	String	Username of the tweet author
`ref_type`	Enum<String>	Type of referenced tweet (reply, retweet, etc.)
`ref_author`	String	Username of the referenced tweet's author
`ref_text`	String	First 180 characters of the referenced tweet
`ref_note_tweet`	String	Full text of the referenced tweet, if longer than 180 characters

The dataset includes both the textual content and metadata associated with each tweet, such as interactions (retweets, replies, mentions and quotes) and details about the users involved. Each row represents a single post on X, identified by a unique ID. Although the `geo` field is present, it remains empty across all records due to the access limitations of the API. Several fields are stored as objects or key-value arrays because they encapsulate additional structured data such as `public_metrics` and `context_annotations`, which are useful for deeper interaction analysis.

The quality of the dataset is considered high, as the data was collected directly from X by the project, without intermediaries or external transformations. The dataset includes 953 originals (9.5 %), 161 quotes (1.6 %), 1,013 replies (10.1 %) and 7,874 retweets (78.7 %), with only minor omissions—29 originals and 4 replies without entities, plus one quote and 10 replies missing reference text. During preprocessing, unused fields were dropped, nulls handled, and all required data retained for influence

3.2 Interpersonal Influence Based on Interactions

On social media platforms like s often reflect underlying dynamics of influence. When a user mentions, replies to, quotes, or retweets another (as described in Table 2), a relationship is established in which the first user —whether intentionally or not— acknowledges and amplifies the content of the second. These interactions are not only acts of dissemination but also potential signs of how one user's ideas may influence another's behavior or perspective.

Table 2. Types of interactions used to estimate interpersonal influence

Interaction Type	Description
Mentions	Occur s another by writing "@" followed by the username in the body of the post.
Retweets	Occur when a user shares someone else's post with their own followers.
Replies	Occur when a user responds to another's tweet. The reply becomes part of the original tweet's thread.
Quotes	A type of retweet where the user adds a comment while sharing the original post, creating a separate conversation.

Taking into account the interactions described in Table 2, the influence of one user over another one can be estimated by the number of interactions recorded between them. The more frequently a user interacts with another, the higher the probability that the second user's content or views are impacting their activity on the platform. This relationship is formally modeled through the normalized matrix M, where each cell M_{ij} represents the level (in the range $[0, 1]$) of interaction from user j toward user i, and therefore the influence that user i exerts over user j. The proposed interpersonal influence measure is defined as follows:

$$\forall i, j \in U, \quad M_{ij} = \frac{\tilde{M}_{ij} - \min(\tilde{M})}{\max(\tilde{M}) - \min(\tilde{M})} \tag{1}$$

where:

- U: set of users in the analyzed dataset,
- $M_{ij} \in [0, 1]$: interpersonal influence based on interactions from user i to user j,
- $I_{ij} \in \mathbb{Z}_{\geq 0}$: number of interactions (detailed in Table 2) from user j to user i,
- $\tilde{M}_{ij} \in \mathbb{R}_{\geq 0}$: logarithmic transformation of I_{ij}, defined as $\tilde{M}_{ij} = \log(1 + I_{ij})$ to reduce the impact of outliers.

It is worth noting that the matrix M is not symmetric, as it is based on direct interactions from one user to another one. This means that the value M_{ij} —determined by the interactions that user i receives from user j— can differ from M_{ji}, especially when user j has not received interactions from user i in the dataset. As a result, the heuristic captures both the strength and direction of user-to-user interactions, where $M_{ij} = 0$ implies that no influence from i to j was recorded, while $M_{ij} = 1$ reflects the highest level of influence observed using this heuristic.

Prior work shows that direct interactions such as retweets, mentions, and replies are effective for mapping and quantifying influence relationships on social platforms [4,13], particularly in X, where empirical studies link higher interaction rates to greater influence in political and social contexts [1,11,12]. The measure implemented in this paper builds on this evidence and is adapted to the interaction types and user data available on the X platform.

3.3 Popularity-Based Influence

Users with higher numbers of followers or received interactions (likes, replies, retweets, etc.) tend to have greater capacity to amplify and shape narratives on social platforms. This capacity is not only related to the number of direct interactions with other users, but also to their level of recognition on the platform. The rationale for this comes from the authority bias [8], which suggests that individuals are more likely to accept, internalize, or spread information coming from figures perceived as influential or with recognized status within a group.

From this perspective, a user who receives a large number of interactions has a higher probability of influencing those who mention, reply to, or retweet them, compared to another user with the same number of direct interactions but lower overall recognition. Based on this idea, we adjust the interaction-based influence measure by incorporating a popularity factor, so that the influence a user exerts depends not only on how often they are interacted with by a specific person, but also on how broadly recognized they are on the platform. This results in a more realistic model of how actors amplify messages and shape discussions in digital environments.

Due to recent policy changes on X, access to data on user relationships, such as who follows whom, has become restricted and costly. As a result, this research does not include follow relations. Therefore, user popularity is estimated solely based on the number of interactions their posts receive. The popularity of user i is defined as the total number of interactions received across all their posts:

$$\forall i \in U, \quad P_i = \sum_{t \in T_i} (r_t + q_t + rp_t + l_t) \tag{2}$$

where T_i is the set of posts made by user i, and r_t, q_t, rp_t, and l_t are the retweets, quotes, replies, and likes received by post t, respectively. The popularity-based influence of user i on user j is computed by multiplying the popularity score P_i by the number of interactions I_{ij} from user j to user i. Then, a log transformation is applied to reduce the impact of outliers, this also avoids obtaining negligible influence values, which, with current limited access, is now the norm. After that, the results are normalized to the range $[0, 1]$:

$$\forall i, j \in U, \quad IP_{ij} = \frac{\tilde{IP}_{ij} - \min(\tilde{IP})}{\max(\tilde{IP}) - \min(\tilde{IP})} \tag{3}$$

with the following definitions:

- U: set of users in the dataset,
- $IP_{ij} \in [0, 1]$: normalized popularity-based interpersonal influence from user i to j,
- $\overline{IP}_{ij} = I_{ij} \cdot P_i$: unnormalized popularity-based influence,
- $\tilde{IP}_{ij} = \log(1 + \overline{IP}_{ij})$: log-transformed influence to reduce outliers.

This heuristic builds on prior research that has used interaction volume as a proxy for user influence. For instance, Riquelme and González-Cantergiani [11] emphasized the relevance of received interactions in measuring influence on X. Similarly, other studies have used degree centrality, mentions, and retweets to model influence relationships, arguing that users with higher levels of interaction are better positioned to amplify narratives [1,5].

4 Experiments

As explained in previous sections, the evaluation of the two proposed measures of interpersonal influence was carried out using data collected by the PROMUEVA research project, which focused on public discussions on X about the pension reform in Colombia. This dataset serves as a case study to analyze the behavior of the proposed metrics and compare them with widely used baseline measures.

To quantify the similarity between the proposed measures and the reference metrics, Pearson and Spearman correlation coefficients were computed. Pearson correlation assesses the linear relationship between two variables, indicating whether higher influence values in one metric are associated with higher values in the other. In contrast, Spearman correlation evaluates the consistency in the relative ordering of values, verifying whether users identified as more influential by one measure hold similar ranks in the other, regardless of the exact values.

Table 3 shows the correlations between the proposed interpersonal influence measures and standard reference metrics: retweets, mentions, and betweenness centrality. In the following sections, we focus only on conceptually relevant comparisons. The interaction-based measure is compared with retweets and mentions, as all three focus on direct interactions. The popularity-based measure is contrasted with betweenness centrality, since both capture a user's recognition or importance in the network.

Table 3. Correlations between the proposed interpersonal influence measures and baseline metrics.

Proposed Measure	Retweets		Mentions		Betweenness Centrality	
	Pearson	Spearman	Pearson	Spearman	Pearson	Spearman
Interactions	79%	80%	29%	26%	47%	62%
Popularity	69%	69%	13%	15%	49%	75%

4.1 Evaluation of Interpersonal Influence Based on Interactions

This analysis is focused on comparing this measure with two common indicators of interaction on X: retweets and mentions. The interaction-based metric shows a strong correlation between the measure and retweet counts, with a Pearson coefficient of 79.2% and a Spearman coefficient of 80.7%. This indicates that the proposed metric reflects patterns similar to those captured by retweets, which represent a direct form of content amplification. This strong relationship may be influenced by the distribution of tweet types in the dataset, where more than 78% of the analyzed tweets were retweets.

In contrast, the correlation with mentions was significantly lower (Pearson = 29%, Spearman = 26%), suggesting that this type of interaction does not align as closely with the notion of influence captured by the combined measure. Although 12,796 direct mentions were recorded in the dataset, the low correlation indicates that, mentions alone do not accurately reflect interpersonal influence under the proposed approach. This result can be attributed to the fact that tagging another user may have various purposes (such as critique, visibility, or mere reference) and may lack the directional clarity needed to infer influence reliably.

Figure 1 shows a dense graph built with the interaction-based heuristic. Node size reflects the number of users influenced, while light edge colors indicate generally low influence values.

4.2 Evaluation of Interpersonal Influence Based on Popularity

In this analysis, Betweenness centrality is used as the reference metric for comparison. In this case, betweenness centrality replaces the popularity (P_i) as

Fig. 1. Graph generated using the interaction-based influence measure. For privacy reasons, user identifiers have been omitted.

the adjustment factor in the estimation of influence based on user interactions. Betweenness centrality, widely used in network analysis, quantifies the importance of a node based on its ability to act as a bridge between others, facilitating the flow of information across the network.

When computing the correlations between this measure and the Betweenness-based influence, a Spearman correlation of 75% and a Pearson correlation of 49.2% were obtained. The Spearman correlation indicates that, although the user rankings are not identical across both metrics, there is a meaningful correspondence. This relatively high correlation suggests that, in general, higher popularity (as measured by centrality) is associated with higher values assigned by the proposed measure and successfully captures the structural characteristics in the network. This may be due to users occupying structurally important positions in the network while not necessarily receiving many direct interactions (i.e. a relatively low P_i). The Pearson correlation is more moderate (49%), indicating that while the ranking of influential users is similar, the exact influence scores differ.

Figure 2 shows a sparser graph, as users without interaction metrics lose their ability to influence other users under the popularity-based metric. Despite the lower edge density, the influence values are noticeably higher, which is reflected in the stronger edge color intensity. This shows that integrating popularity can sharpen the model's sensitivity to latent structural influence, even under constrained data conditions.

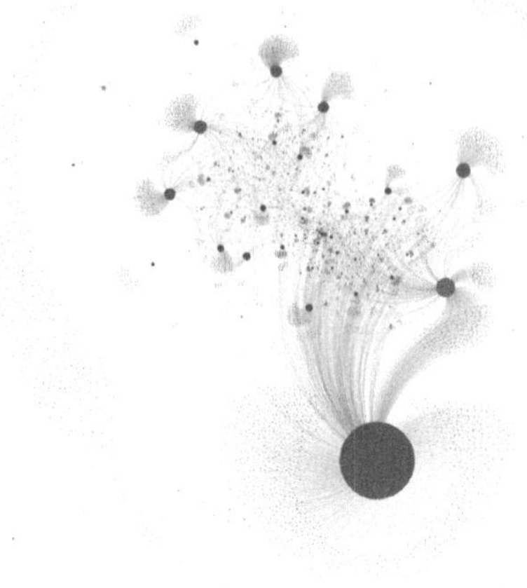

Fig. 2. Graph using the popularity-based influence measure.

5 Conclusions

This study introduced two complementary metrics for estimating interpersonal influence under data-restricted conditions on the X platform that offer opportunities to analyze influence in social networks: one based on observed user behaviors, and the other on social prominence. The **Interaction-based** measure shows high correlation with established indicators of influence such as retweets, while offering a more comprehensive view by accounting for various types of interactions on the platform, which is especially valuable when working with relatively limited data and its effectiveness is particularly notable given the dominance of retweets in the dataset. The **popularity-based** measure extends the interaction-based approach by incorporating a cumulative recognition of users. Its strong correlation with betweenness centrality suggests that it captures users' structural importance in the network beyond direct interactions and might help to identify users who play a bridging role in information flows.

Although the measures are context-independent, the conclusions note the need to test in other domains (e.g., sports, health) with different distributions of interaction types. Future work might include sentiment or qualitative analysis (criticism, visibility, or mere reference) to contextualize influence within issue-specific narratives. Although mentions were included as part of the study, their low correlation with influence suggests they serve diverse functions (critique, visibility, or simply reference) that can be studied using stance detection models to enrich the interaction matrix with qualitative polarity, not just frequency.

Acknowledgments. This work is supported by the Colombian Minciencias project PROMUEVA, BPINă2021000100160.

References

1. Ahmed, W., Aiyenitaju, O., Chadwick, S., Hardey, M., Fenton, A.: The influence of joe wicks on physical activity during the covid-19 pandemic: thematic, location, and social network analysis of X data. J. Med. Internet Res. **26**(1), 22 (2024). https://doi.org/10.2196/49921
2. Anderson, B.D.O., Ye, M.: Recent advances in the modelling and analysis of opinion dynamics on influence networks. Int. J. Autom. Comput. **16**(2), 129–149 (2019). https://doi.org/10.1007/s11633-019-1169-8
3. Chakraborty, A., Badyal, N., Sharma, A., Mukherjee, N.: A novel centrality-based measure for election network analysis. In: 2022 IEEE Silchar Subsection Conference (SILCON 2022). IEEE, Silchar (2022). https://doi.org/10.1109/SILCON55242.2022.10028812
4. Gaumont, N., Panahi, M., Chavalarias, D.: Reconstruction of the socio-semantic dynamics of political activist Twitter networks–method and application to the 2017 French presidential election. PLoS ONE **13**(9), e0201879 (2018). https://doi.org/10.1371/journal.pone.0201879
5. Lara-Navarra, P., López-Borrull, A., Sánchez-Navarro, J., Yànez, P.: Measuring the influence of users on social networks: social engagement proposal. Profesional de La Información **27**(4), 899–908 (2018). https://doi.org/10.3145/epi.2018.jul.18
6. Li, X., Cheng, S., Chen, W., Jiang, F.: Novel user influence measurement based on user interaction in microblog. In: 2013 IEEE/ACM International Conference on Advances in Social Networks Analysis and Mining, pp. 615–619. ACM, Niagara Falls (2013). https://doi.org/10.1145/2492517.2492635
7. Logan, A.P., LaCasse, P.M., Lunday, B.J.: Social network analysis of twitter interactions: a directed multilayer network approach. Soc. Netw. Anal. Min. **13**(1), 65 (2023). https://doi.org/10.1007/s13278-023-01063-2
8. Milgram, S.: Behavioral study of obedience. Psychol. Sci. Public Interest **67**(4), 371–378 (1963). https://doi.org/10.1037/h0040525
9. Neubaum, G., Krämer, N.C.: Opinion climates in social media: blending mass and interpersonal communication. Hum. Commun. Res. **43**(4), 464–476 (2017). https://doi.org/10.1111/hcre.12118
10. Rashotte, L.: Social Influence. In: The Blackwell Encyclopedia of Sociology. Wiley (2007). https://doi.org/10.1002/9781405165518.wbeoss154
11. Riquelme, F., González-Cantergiani, P.: Measuring user influence on twitter: a survey. Inf. Process. Manage. **52**(5), 949–975 (2016). https://doi.org/10.1016/j.ipm.2016.04.003
12. Venkatesan, S., Valecha, R., Yaraghi, N., Oh, O., Rao, H.R.: Influence in social media: an investigation of tweets spanning the 2011 egyptian revolution. MIS Quarterly 45(4), 1679–1714 (2021). https://doi.org/10.25300/MISQ/2021/15297

13. Yuan, W., Zhang, Z., Chang, D.: A user influence network construction approach based on web mining and social network analysis. In: 2023 IEEE International Conference on Industrial Engineering and Engineering Management (IEEM), pp. 1123–1127. IEEE, Singapore (2023). https://doi.org/10.1109/IEEM58616.2023.10406593
14. Zhai, Y., Li, X., Chen, J., Fan, X., Cheung, W.K.: A novel topical authority-based microblog ranking. In: Chen, L., Jia, Y., Sellis, T., Liu, G. (eds.) APWeb 2014. LNCS, vol. 8709, pp. 105–116. Springer, Cham (2014). https://doi.org/10.1007/978-3-319-11116-2_10

Representation of Knowledge in a Specific Domain Model for Educational Data Mining

Emilcy Hernández-Leal[1,2](✉) and Néstor Duque-Méndez[2]

[1] Universidad de Medellín, Medellín, Colombia
ejhernandez@udemedellin.edu.co
[2] Universidad Nacional de Colombia, Manizales, Colombia

Abstract. The main objective of the use of data analysis techniques to support educational processes, as in other areas, is to improve decision-making and planning in this field. Information and communication technologies play a crucial role in this analysis process. Data mining offers a way to address the needs related to the management of academic data, which are generated both during the teaching-learning process and in associated administrative aspects. The structure and access to this educational data varies depending on the educational level and type of institution. In this article, characterizations and reviews on domain-specific approaches, data mining, and educational data sources are taken to propose a knowledge representation component in a domain-specific model that supports the application of data mining in the educational context. Additionally, its validation is done by presenting a series of activities to be considered for its application, such as the review of previous work for problems such as academic performance and school dropout and the construction of a guiding knowledge base based on questions and rules. It is concluded that a relevant and structured knowledge base translated into a set of educational data mining application rules can provide significant benefits, such as automation, efficiency, informed decision-making, and continuous improvement.

Keywords: Data analysis · Data mining · Educational data · Knowledge representation · Specific domain

1 Introduction

One of the fields of application of data mining is education, hence the emergence of educational data mining, better known as EDM (Educational Data Mining). EDM emerges as a paradigm that is oriented toward the design of models, tasks, methods, and algorithms that will be used in the exploration of educational data [13]. It can be said that EDM emerges from the combination of three main areas: computer science, education, and statistics; likewise, by intercepting these areas, three approaches closely related to EDM stand out: computer-based education, data mining, and machine learning, and learning analytics [15].

W. J. Sarmiento and E. L. Sierra-Ballén (Eds.): CCC 2025, CCIS 2692, pp. 217–227, 2027.
https://doi.org/10.1007/978-3-032-28447-1_16

The process followed for educational data mining is similar to the one developed in general data mining, but the beginning in an educational environment and the importance of ending with interpretation and evaluation are highlighted, since from it the need to return in hypothesis formation or any other of the previous stages may arise.

Among the objectives of educational data mining, the following stand out: to analyze the performance and behavior of each student; to analyze learning processes in various environments and with different teaching methods; to evaluate the quality of educational institutions; to generate patterns of access and permanence in the educational system; to provide different means to analyze logs in virtual educational environments; among others.

Now, when talking about a specific domain model for EDM, it is worth highlighting first the definition of a model, and this can be given as a representation of a real object or situation; therefore, the act of modeling implies the understanding of a process or phenomenon and the interactions between the parts of it with a possible environment or system. The principles of modeling allow for the creation of models according to the branch of science they are trying to explain. The way of constructing models of analysis is studied in philosophy of science, general systems theory, and scientific visualization. In almost all explanations of phenomena, one model or another can be applied, but it is necessary to adjust the model to be used so that the result is as accurate as possible [10].

In data mining, the definition of model is usually associated with supervised learning techniques, there are classification models, regression models, association models, and frequency models, among others. The concept is linked to the way of extracting knowledge from data or reality and taking it to a repeatable schema. In the domain-specific approach, the model is associated with the particular representation that can be made of a domain based on the knowledge that one has of it.

A domain model in problem solving and software engineering is a conceptual model of all the issues related to a specific problem. Describes the various entities, their attributes, roles, and relationships, as well as the constraints that govern the problem domain. The domain model is created to represent the vocabulary and key concepts of the problem domain. The domain model also identifies the relationships between all entities within the scope of the problem domain and commonly identifies their attributes. The domain model provides a structural view of the domain that can be complemented with other dynamic views, such as the use case model.

After these annotations, it is clarified that this article takes characterizations and reviews on domain-specific approaches, data mining, and educational data sources to propose a knowledge representation component in a domain-specific model that supports the application of data mining in the educational context. For this purpose, Sect. 2 presents the conceptual approach of the model, Sect. 3 the validation of the component, Sect. 4 the discussion and contributions of such a component for an EDM model, and finally, Sect. 5, the conclusions and lines of future work.

2 Conceptual Approach to the Component

A domain representation component can be the main source of differentiation of the models or methodologies traditionally used in data mining, even in educational data mining. It may be thought that domain knowledge is eventually always used in models; however, there is not always a strategy and a guideline to guide which domain elements should be taken into account and how they can be articulated with the other stages of the analysis process. Consequently, a domain representation component consists of metamodels that define various concepts that can be used to specify aspects of educational data analysis problems.

This component may include abstractions to represent various educational scenarios, such as dropping out or academic performance. Abstractions for grades, pass-fail status, or any other characteristics that interested users would like to analyze also need to be represented. The metamodel must also incorporate abstractions to represent the data formats and operations that the model must support. Figure 1 presents the knowledge representation component in the framework of a domain-specific model for EDM.

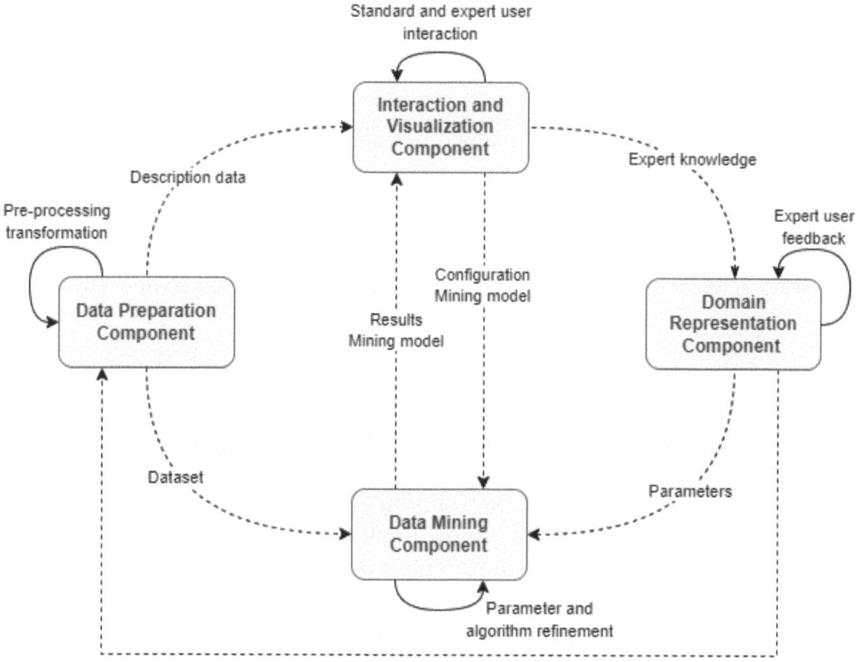

Fig. 1. Domain-specific model for Educational Data Mining.

Now, the principles of the conception of the domain representation component start from the main sources of knowledge supply, followed by the products and results; and finally, the impact delivered by the component should be highlighted. In Fig. 2 the principles are schematized, indicating in greater detail what each aspect refers to.

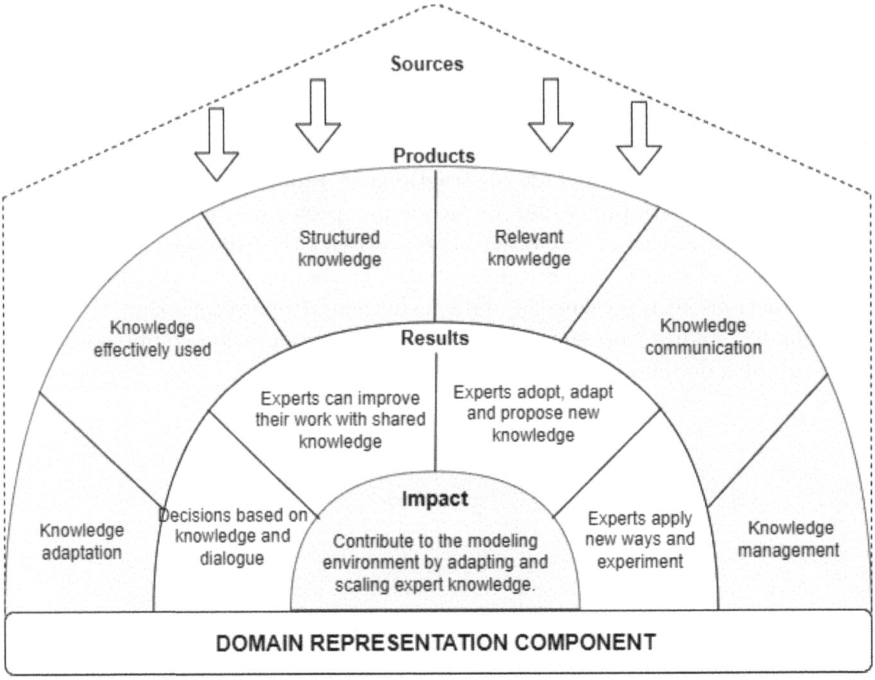

Fig. 2. Principles of the domain knowledge representation component.

The details of the interactions that occur within this component are presented in Fig. 3, starting from the problem domain, which is obtained from the experts and the educational context and which is materialized in elements such as knowledge, legislation, and educational standards, taxonomies, curricula, data mining techniques previously tested in the solution of educational problems together with the parameters defined for these, feedback from the actors of the context, among others. All of the above enters an extraction or inference phase, which allows obtaining from this accumulation of knowledge a series of solutions, explanations, rules, and instructions that translate into a structured and relevant knowledge of the data domain and of the problems adjacent to it. The above becomes part of a knowledge base that represents the domain.

A domain knowledge representation component can be seen as an approximation to an expert system, given that it has sources, that arise from a particular context and with the feedback of experts can be taken to a structured knowledge base in the form of rules that can contribute to the use of the model by standard

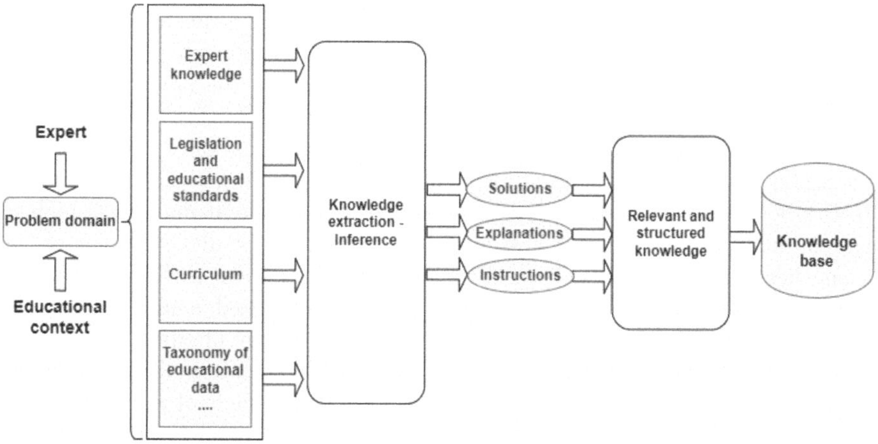

Fig. 3. Detail of the domain knowledge representation component.

users and to achieve that the application of mining techniques is oriented and facilitates the construction and interpretation of mining models.

3 Component Validation

Taking into account the basic components within a specific domain model and in particular the knowledge that transits through them, there is a need to represent it. To this end, the starting point is the generation of questions such as: what possible taxonomies or categorizations of educational data can be found, how to evaluate educational data in its original state, and can these educational data or their sources include metadata?

A taxonomy is a hierarchical classification system used to organize and categorize data according to their nature and characteristics. A taxonomy of educational data can be of great help in the representation of domain-specific knowledge [9]. In the context of educational data mining, a taxonomy can help define the variables and indicators of interest in a clear and consistent manner, which in turn facilitates the selection and application of appropriate data mining techniques for analysis. In addition, it can help identify relationships and patterns among the data, which in turn can lead to a better understanding and description of domain knowledge.

For example, an educational data taxonomy can classify data by student type (e.g., gender, age, grade level), school type (public or private), course type (math, science, languages, etc.), and other factors relevant to the education domain. By classifying data in this way, patterns and relationships between different types of data can be identified, which in turn can help develop more accurate prediction and classification models.

The use of taxonomy is considered to be of value since this instrument allows the classification of knowledge by means of a tree-like form where the more general terms are located at the root and the more specific terms at the branches and leaves. Additionally, in the validation of this component, the taxonomy facilitates the establishment of a system for classifying the specificity of the data as the structure of the tree is traversed, showing the categories of data of interest and the relationships between them. An educational data taxonomy can be a useful tool for the representation of domain-specific knowledge as it can help to organize and classify the data in a clear and coherent manner, which in turn can facilitate the analysis and interpretation of the results.

Another relevant aspect in the validation of the component is the construction of the knowledge base, for which different means can be used, some of which are educational standards and legislation, curricula, taxonomy of educational data, and of course, the knowledge of experts. Therefore, in order to build a knowledge base at the educational data level, activities such as a review of previous works in EDM, analysis of EDM algorithms and their parameters, and construction of guiding questions and support rules can be carried out.

To exemplify the first activity, a compilation of previous works is presented, including a classification of the techniques that report the best results for school dropout (see Table 1) and academic achievement (see Table 2).

Table 1. Best techniques for dealing with dropout.

Reference	Best technique	AUC	% Correctly classified
[7]	Random Forest	0.97	96.78
[8]	J48	-	95.43
[14]	Random Forest	0.88	85.50
[11]	Decision Tree	-	82.00
[18]	Random Forest	0.89	-
[6]	J48	0.79	80.23

For the case of attrition in Table 1, Random Forest is found to be the outstanding technique, as well as the J48 algorithm. In conclusion, decision trees stand out as the classification technique with the best results, compared to Bayesian algorithms, neural networks, support vector machines, and regression.

In the case of academic performance, Table 2 shows that decision trees are also the technique that stands out with the best results and algorithms with Random Forest and J48. It should also be noted as a particular characteristic of the data in this domain that the data always tend to be unbalanced, i.e., in these two problems there will always be more students who do not fail and who do not drop out than those who pass and drop out respectively, unless it is an exceptional case.

Another key activity for the representation of domain knowledge can be presented by means of a series of guiding questions, which help different types of users (experts and non-experts) to guide the application of EDM in the solution

Table 2. Best techniques to address academic performance

Reference	Best technique	AUC	% correctly classified
[3]	Naive Bayes	-	71
[1]	LMT	-	71.08
[12]	Decision Trees	-	0.89 (R2)
[16]	Neural networks	-	75.28
[5]	J48	-	66
[2]	Random Forest	-	69.35
[17]	J48	-	66.23
[4]	SVC	0.61	66.24

of problems such as school dropout and achievement. These guiding questions can help to develop a strategy to address problems using educational data mining and in particular a specific domain model that involves a knowledge representation component, thus involving a particular reflection of the context and knowledge implied by the nature of the data analyzed.

In this order of ideas, once the domain knowledge has been extracted, a transformation of this knowledge in terms of solutions, explanations, and instructions can be performed, allowing the relevant knowledge structured in a set of rules that help to guide the data mining process that connects in the next component. In fact, these rules can be considered as hypotheses to be tested.

The following are some rules that can serve as a guide for academic performance and dropout issues. These rules could even be graded according to specific parts of the aspect involved. For the academic performance issue:

Family status rule: The socioeconomic conditions of a student's family and social environment have a broad influence on his or her school results and therefore on his or her academic performance.

Family support rule: If a student has the active support of his or her family, in terms of monitoring and supervision, he or she is more likely to perform better academically.

Teacher feedback rule: If a student receives regular and constructive feedback from teachers, he or she is more likely to experience improvements in academic performance.

Previous grades rule: If a student has earned outstanding grades in previous courses or related subjects, he/she is more likely to perform well in subsequent courses.

Participation in extracurricular activities rule: If a student regularly participates in extracurricular activities, he/she is more likely to perform well academically.

Study Time Rule: If a student devotes a set minimum number of hours per day to study outside of school hours, he/she is more likely to perform well academically.

Attendance rule: If a student has an attendance percentage below a set figure, he/she is more likely to perform poorly academically.

It is also important to keep in mind that these rules must be adapted and adjusted according to the specific characteristics of the institutions and students being analyzed, as well as the sources and data available. Regarding another problem, dropout, it could be considered:

Irregular attendance rule: If a student has an attendance rate below a certain figure over an extended period of time, he or she is more likely to be at risk of dropping out.

Low academic achievement rule: If a student has a grade point average below the minimum passing grade according to the scale used by the institution, he/she is more likely to be at risk of dropping out.

Repetition rule: If a student has repeated more than two school years, he/she is more likely to be at risk of dropping out of school.

Absence of family support rule: If a student comes from a disadvantaged family background, with a lack of emotional, economic, or educational support, he/she is more likely to be at risk of dropping out of school.

Demotivation rule: If a student exhibits a general attitude of disinterest in education, lack of participation in school activities, and lack of clear academic goals, he or she is more likely to be at risk of dropping out of school.

Socioeconomic factors rule: If a student belongs to a low socioeconomic group, with economic difficulties and lack of access to additional educational resources, he/she is more likely to be at risk of dropping out of school.

It should be noted that these rules are examples and it is important to adapt them to the specific characteristics of the institutions and students being analyzed. In addition, dropout, like the prediction of academic performance, is a complex phenomenon that can be influenced by multiple interrelated factors.

4 Discussion and Contributions

How to discover the nature and perform knowledge representation from domain-specific data is a challenging and significant task. Giving this answer requires a large amount of relevant data, analysis methods, and even the approach of new theories. Nevertheless, it starts with efforts and proposals such as the component presented in this paper and its application. In order to make a contribution to this open research space and to fill knowledge gaps in a cumulative way, the work of building a knowledge base for the approach of a specific domain model of educational data mining and its application strategy includes the particularities of the data and the problems that must be faced in the processes of analysis of phenomena in the educational field.

It is considered a complex task given the dynamic nature to which educational environments are subject. Methods and models that work well in one context may not be applicable in another. This makes it difficult to generalize and validate models in different educational settings; likewise, educational data may be subject to variability and bias due to factors such as student diversity, different levels of granularity, and heterogeneous data sources. This variability can hinder the validation of models and affect their performance for different groups of students.

Another limitation corresponds to the absence of previous work reporting results of studies using data mining with data from different educational levels such as primary, secondary, and high school, with data from a traditional educational environment, not assisted by virtual support platforms. With regard to studies with higher education data, an opportunity is identified to further explore this data source and typology and to review options in terms of transfer of learning between educational levels and comparison of results to identify the causes and how to represent this in the construction of the mining models.

The guiding rules, presented in the previous section, justify the construction of an expert system that allows some inferences to be made in the application of educational data mining. Such a system can contribute to the automation of the educational data analysis process, allowing complex and repetitive tasks to be performed in a shorter time and with less margin of error than if they were done manually. In addition, with an expert system, it would be possible to identify more quickly and accurately the trends and issues to be covered and how to address them in terms of suggested techniques and parameterizations.

In the same sense, the use of an expert system in this domain can help decision-makers, such as heads of institutions or teachers, to obtain valuable and relevant information to better apply analysis approaches without being experts in data science since the system can provide recommendations based on rules previously established with the knowledge gathered from the domain. Educational data mining supported by an expert system would help to continuously monitor and evaluate the effectiveness of different interventions and analyses.

5 Conclusions and Future Work

In conclusion, a relevant and structured knowledge base brought into a set of educational data mining application rules can provide significant benefits, such as automation, efficiency, informed decision-making, and continuous improvement.

The inclusion of a knowledge representation component within a domain-specific model can help non-expert users, and support experts as well, in the educational data analysis process, even when not improving the results of the mining models in terms of traditional classification metrics, but responding to the need to carry out more conscious analysis processes and in which the results can be leveraged, either at institutional or regional level, in the construction of strategies, policies, action or improvement plans, among others. Given that many efforts are needed in the specific domain to extract meaningful information from the interactions and results of the students.

In future work, we propose to refine the conception of the component in terms of its capacity to assume the multiple data granularities that can occur for this particular domain and to collect the rules proposed as a closure of the construction of the knowledge base in an expert system that allows the user to enter the data available and the problem to be analyzed and to receive a recommendation of the technique, algorithms, hyperparameters, possible hypotheses and route to be taken.

References

1. Ayala Franco, E., López Martínez, R.E., Menéndez Domínguez, V.H.: Modelos predictivos de riesgo académico en carreras de computación con minería de datos educativos. Revista de Educación a Distancia (RED) **21**(66), 1–36 (2021). https://doi.org/10.6018/RED.463561
2. Candia Oviedo, D.I.: Predicción del rendimiento académico de los estudiantes de la UNSAAC a partir de sus datos de ingreso utilizando algoritmos de aprendizaje automático. Ph.D. thesis (2019). http://repositorio.unsaac.edu.pe/handle/20.500.12918/4120
3. Chiok, M., Higinio, C.: Predicción del rendimiento académico aplicando técnicas de minería de datos. Anales Científicos **78**(1), 26–33 (2017). https://dialnet.unirioja.es/servlet/articulo?codigo=6171237
4. Contreras, L.E., Fuentes, H.J., Rodríguez, J.I.: Predicción del rendimiento académico como indicador de éxito/fracaso de los estudiantes de ingeniería, mediante aprendizaje automático. Formación universitaria **13**(5), 233–246 (2020). https://doi.org/10.4067/S0718-50062020000500233
5. Díaz-Landa, B., Meleán-Romero, R., Marín-Rodriguez, W.: Rendimiento académico de estudiantes en Educación Superior: predicciones de factores influyentes a partir de árboles de decisión. Revista de Estudios Interdisciplinarios en Ciencias Sociales **23**(3), 616–639 (2021). https://dialnet.unirioja.es/servlet/articulo?codigo=8133982
6. Eckert, K., Suénaga, R.: Análisis de Deserción-Permanencia de Estudiantes Universitarios Utilizando Técnica de Clasificación en Minería de Datos. Formación universitaria **8**(5), 3–12 (2015). https://doi.org/10.4067/S0718-50062015000500002
7. Flores, V., Heras, S., Julian, V.: Comparison of Predictive Models with Balanced Classes Using the SMOTE Method for the Forecast of Student Dropout in Higher Education. Electronics **11**(3), 457 (2022). https://doi.org/10.3390/ELECTRONICS11030457
8. Hernández-Leal, E.J., Quintero-Lorza, D.P., Escobar-Naranjo, J.C., Ramírez-Gómez, J.S., Duque-Méndez, N.D.: Educational data mining for the analysis of student desertion. In: LALA 2018. Guayaquil, Ecuador (2018)
9. Lu, W., Zhou, Y., Yu, J., Jia, C.: Concept extraction and prerequisite relation learning from educational data. In: Proceedings of the AAAI Conference on Artificial Intelligence, vol. 33, no. 01, pp. 9678–9685 (2019). https://doi.org/10.1609/AAAI.V33I01.33019678. https://ojs.aaai.org/index.php/AAAI/article/view/5033
10. Maldonado, C.E.: Tipología de modelos científicos de explicación. Ciencia y complejidad. Sociología y Tecnociencia **7**(2), 58–72 (2017). https://doi.org/10.24197/st.2.2017.58-72
11. Miranda, M., Guzmán, J.: Análisis de la Deserción de Estudiantes Universitarios usando Técnicas de Minería de Datos. Formación universitaria **10**(3), 61–68 (2017). https://doi.org/10.4067/S0718-50062017000300007
12. Orozco Iguasnia, W.A., Villao Balón, A.J., Orozco Iguasnia, J., Villarroel Sánchez, M.V.: Aplicación de técnicas de minería de datos para predecir el desempeño

académico de los estudiantes de la escuela 'Lic. Angélica Villón L.'. Revista Científica y Tecnológica UPSE **8**(2), 68–75 (2021). https://doi.org/10.26423/RCTU.V8I2.637
13. Peña-Ayala, A.: Educational data mining: a survey and a data mining-based analysis of recent works. Expert Syst. Appl. **41**(4), 1432–1462 (2014)
14. Rodriguez Maya, N.E., Jimenez Alfaro, A.J., Reyes Hernandez, L.A., Suarez Carranza, B.A., Ruiz Garduno, J.K.: Data mining: a scholar dropout predictive model. In: Humanitarian Technology Conference (MHTC), pp. 89–93. IEEE, Puebla, Mexico (2017). https://doi.org/10.1109/MHTC.2017.8006421
15. Romero, C., Ventura, S.: Educational data mining and learning analytics: an updated survey. Wiley Interdisc. Rev. Data Min. Knowl. Discovery **10**(3), e1355 (2020). https://doi.org/10.1002/widm.1355
16. Salgado Reyes, N., Beltrán Morales, J., Guaña Moya, J., Escobar Teran, C., Nicolalde Rodriguez, D., Chafla Altamirano, G.: Modelo para predecir el rendimiento académico basado en redes neuronales y analítica de aprendizaje. RISTI **17**, 258–266 (2019)
17. Timarán-Pereira, R., Caicedo-Zambrano, J., Hidalgo-Troya, A.: Árboles de decisión para predecir factores asociados al desempeño académico de estudiantes de bachillerato en las pruebas Saber 11°. REVISTA DE INVESTIGACIÓN, DESARROLLO E INNOVACIÓN **9**(2), 363–378 (2019). https://doi.org/10.19053/20278306.V9.N2.2019.9184
18. Vargas-Cubero, A.L., Villalobos-Torres, G.: El uso de plataformas virtuales y su impacto en el proceso de aprendizaje en las asignaturas de las carreras de Criminología y Ciencias Policiales, de la Universidad Estatal a Distancia de Costa Rica. Revista Electronica Educare **22**(1), 20–39 (2018). https://doi.org/10.15359/ree.22-1.2

In-Silico Modeling of Abnormal Growth in Lung Tissue Using Fractals

Jose Santiago Duran Roa[1], María Isabel Romero Rodriguez[2](✉), Laura Manuela Orjuela Hernández[3], and Juan Camilo Vargas Pino[4]

[1] Ingeniería Biomédica, Facultad de Ingeniería, Cajicá, Colombia
[2] Departamento de Matemáticas, Facultad de Ciencias Básicas y Aplicadas, Bogotá, Colombia
maria.romeror@unimilitar.edu.co
[3] Ingeniería Mecatrónica, Facultad de Ingeniería, Bogotá, Colombia
[4] Ingeniería Biomédica, Facultad de Ingeniería, Universidad Militar Nueva Granada, Km 7 Cajicá–Zipaquirá, Cajicá 250240, Cundinamarca, Colombia

Abstract. This study presents a computational model for simulating tumor growth in pulmonary tissue using a stochastic process inspired by Diffusion-Limited Aggregation (DLA). The simulation carries out a binary matrix that evolves iteratively from user-defined seed points, representing tumor cells. Growth occurs through random walks and local adhesion rules. A fractal structure, resembling the bronchial tree, is generated and converted into a binary mask that serves as a spatial guide. To simulate a broader biological influence, the mask is expanding through binary dilation. Results demonstrate that proximity to the fractal structure significantly enhances tumor growth, producing denser and more anisotropic patterns. This suggests that the microenvironment, modeled through a fractal structure, plays a role in directing tumor expansion. Future comparisons with simulations lacking such spatial structures could further clarify the effect of environmental heterogeneity on growth dynamics. The model highlights the utility of combining fractal geometry and stochastic simulation to investigate tumor–environment interactions. Its flexibility allows extensions to simulate angiogenesis, local invasion, or evaluate responses to targeted therapies, offering a valuable tool for computational oncology and biomathematics.

Keywords: Tumor growth · Fractal geometry · Diffusion-Limited Aggregation (DLA)

1 Introduction

According to data from the Global Cancer Observatory, lung cancer is the deadliest cancer type in the world, with approximately 1.8 million deaths yearly [5]. Thanks to the current medical advancements against this disease, it has been identified alterations within the structure of the lung caused by oncological effects [7,15], it is known that although tumor growth initially follows a

W. J. Sarmiento and E. L. Sierra-Ballén (Eds.): CCC 2025, CCIS 2692, pp. 228–238, 2027.
https://doi.org/10.1007/978-3-032-28447-1_17

centrifugal pattern, it later evolves into a complex and unpredictable behavior [25]. There is still no study that fully describes the complexity of tumor structure, despite observing possible common characteristics between patient tissues, which suggests certain similarities [8].

A broad scientific literature exists around the usage of cancer mathematical models, incorporating the existence of a branch of mathematical oncology [19]. In this field, mathematics can be a tool to answer questions about cancer and inspire new theories based on biological processes, providing an opportunity to understand cancer behavior and obtain more effective treatments [2], however, this approach may be limited by a lack of biologically and clinically supported data to validate the mathematical models developed [18].

Fractal implementation within oncology studies provide quantitative and scalable measures that characterize biological structures that cannot be captured properly using classic Euclidian geometry. Fractals are mathematical constructions that exhibit autosimilitude in multiple scales and can quantify their complexity through parameters such as lacunarity and fractal dimension (FD), which allow for a more objective and reproducible analysis of cancer morphology [8]. Fractal algorithms implemented for cancer detection and analysis gained relevance during the last decade with the current technological improvements that allow faster computational processing, developing diagnostic applications based on pattern recognition, texture analysis, and tissue segmentation.

The most significant advances in oncology studies are the detection of tumors and their progression monitoring: A prominent example is the work of [20], who identified the edges of microcalcifications associated with breast cancer using segmentation techniques in X-ray images. Through fractal algorithms, images are divided into discrete regions to locate microcalcifications. Similarly, [13] implemented fractal analysis methods in patients with prostate cancer using images obtained with intraperitoneal magnetic resonance, highlighting fractal dimension (FD) as a quantitative descriptor to characterize tumor progression and allowing specialists to make more informed decisions about treatment according to the tumor stage. Recent studies have highlighted the characterization of tumor shape and spatial features using agent-based models (ABM), which allow the prediction of tumor morphology and dynamics considering factors such as cell migration, density, phenotypic ratios, and interactions with the extracellular matrix (ECM) that influence tumor shape and progression [3]. Additionally, the inclusion of machine learning methods enables the use of in silico models to generate substantial volumes of data, addressing the challenges and high costs associated with obtaining biological and medical data. ML approaches allow the calibration of ABM parameters and the updating of cellular behavior rules facilitating a more realistic reproduction of biological dynamics [3].

In the lung structure, the bronchial and vascular trees and the alveolar surface show fractal properties using techniques such as histological sections and bronchial molds [14]. This type of model describes the behavior of lung cancer in various research projects, an example of which is the work of [23] in which, using the fractal properties of lung tissue, chronic obstructive pulmonary disease

(COPD) is computationally described; another example is that of [16] through fractal analysis of computer images, describing lung behavior under the influence of COVID-19.

The current fractal-lung study integrates spectral image analysis, capturing the spatial heterogeneity of tissues and the complex dynamics of tumor progression and immune response [8]. Recent studies such as [12] have shown that techniques such as optical microscopy and partial waveform spectroscopy (PWS) allow the characterization of fractal and multifractal properties in cancer tissue samples, quantifying the structural disorder induced by the disease. [21] applied multifractal analysis to digital pathology images from squamous cell carcinoma (SCC) cases, revealing significant differences between the multifractal spectra of benign and malignant tissues. This approach allowed the quantification of alterations in the spatial distribution of cell nuclei, providing robust indicators of tumor heterogeneity. For a simplified approach to modeling lung tissue using fractals, the growth of tree-like or canopy fractals can be considered. Such structures have been previously studied in other mammals to characterize the organization of these organs, providing a simplified estimate of the complexity and self-similarity inherent to these tissues [6].

Here, we present the implementation of a Python algorithm to develop tree-like (Dosel) and Diffusion-Limited Aggregation (DLA) fractals, with the generation of a binary mask that involves controlled interaction between the fractal structures. This procedure permits simulation, visualization, and analysis of tumor growth behavior in lung tissue from morphological and dynamic perspectives.

This study aims to develop an in-silico framework for modeling tumor growth in lung tissue using fractal geometry.

2 Methodology and Tools

Tumor growth in lung tissue was simulated using a computational algorithm implemented in the Python programming language. The design of the implementation allows for dynamic visualization and interaction between morphological patterns, facilitating the understanding of the system's dynamic evolution. The libraries Pygame, NumPy, Matplotlib, and SciPy were used in the development of the algorithm. The SciPy library generates branches from seed points using a stochastic algorithm.

The Matplotlib library is used to generate a fractal tree mask, originating at the base of the window and extending toward the center, with branches forming at each level at an angle of $\pm 30°$, which defines the branching pattern. This creates a fractal influence zone that simulates a biological structure—in this case, the lungs.

Tumoral growth is modeled using a stochastic Diffusion-Limited Aggregation (DLA) process. To implement this computational environment, NumPy defines and manipulates the matrix representing the fractal growth space, while Scipy simulates regions of biological tissue by expanding a binary mask, representing zones of structural influence in the tumor microenvironment.

Finally, when the growth simulation is run and the result is graphically displayed, the fractal tree plot appears in the background, and the fractal growth matrix is superimposed on top. This scenario models a type of biological microenvironment that modulates cell proliferation in complicated systems such as tumor growth.

The algorithm configuration parameters are shown in the Table 1:

Table 1. Algorithm Parameters

Parameter	Variable	Value
Window Size	`ancho, alto`	500×500
Unitary Pixel Size	`tamano_pixel`	2
Initial Fractal Tree Coordinates	`x, y`	250×250
Initial Angle Main Branch	`angulo`	–90
Bifurcation Angle Daughter Branches	`angulo_bifurcacion`	± 30
Branch Initial Longitude	`longitud`	150/120
Tree Depth	`profundidad`	12
Fractal Growth Iterations	`iteraciones`	500
Dilation Iterations	`iterations (binary_dilation)`	3
Accelerated Growth Factor	—	10

The parameters defined in the algorithm determine the behavior of the fractal tumor growth system, along with its interaction with the fractal tree structure that emulates lung tissue. Window and pixel size determine the spatial resolution of the simulation, which affects the fractal's level of detail and computational consumption.

In tomographic image analysis, it has been reported that the mean bronchial bifurcation angle in generations 4–6 is around 40°, and in more distal generations it can reach values close to 48°. Consistently, another morphometric study in adults using CT found that the [24] interbronchial and subcarinal angles averaged between 32° and 38°, values very close to the approximate ±30° commonly used in computational airflow models [9].

Growth iterations control the duration and extent of the fractal expansion process, while dilation iterations expand the tree's area of influence, affecting regions where growth is accelerated. Finally, the accelerated growth factor regulates the acceleration of fractal development in proximity to the tree, simulating non-uniform behavior that seeks to represent tissue-specific propagation in its surroundings.

3 Results

3.1 Interface for Defining Initial Seeds

A 2D binary array of dimensions 250 × 250 is created, derived from a graphic resolution of 500 pixels per side with a unit size of 2 pixels per cell. This array represents the discrete domain of lung tissue, on which the tumor growth is simulated. Each grid cell functions as a spatial unit where cell proliferation may or may not occur.

Fig. 1. Interactive graphical interface designed to facilitate the manual distribution of initial seeds within the simulation matrix.

Through the graphical interface developed with Pygame, the user can manually set the initial conditions by selecting one or more pixels that act as tumor seeds (Fig. 1). These initial points are recorded as active cells with a value of 1 in the matrix, defining the locations where aggregation begins.

3.2 Fractal Tree Generation as Mask

A branched fractal structure of the canopy type is generated, aiming to emulate the hierarchical morphology of the pulmonary bronchial tree. This structure is constructed through a recursive algorithm that replicates the typical symmetry and bifurcation of the airways, as shown in Fig. 2.

Subsequently, the figure is converted into a binary mask that serves as a spatial guide over which the tumor growth pattern is modulated. To expand its area of influence and represent greater interaction between the tumor and the surrounding microenvironment, a morphological dilation is applied to the mask. This allows the simulation of signaling regions or extended structural support within the tissue.

Fig. 2. Fractal tree generated with a depth of 12 and a bifurcation angle of $\pm 30°$.

3.3 Simulation of Fractal Growth with Tree Influence

The simulation of tumor growth is based on an iterative algorithm inspired by the Diffusion-Limited Aggregation (DLA) model, adapted to a two-dimensional discrete domain. In each iteration, a new cell performs an isotropic random walk, moving in the cardinal directions. During this process, at each step, it is evaluated whether the cell lies within the Moore neighborhood [4] of any cell previously adhered to the tumor cluster. Once this neighborhood condition is satisfied, the cell becomes part of the expanding cluster.

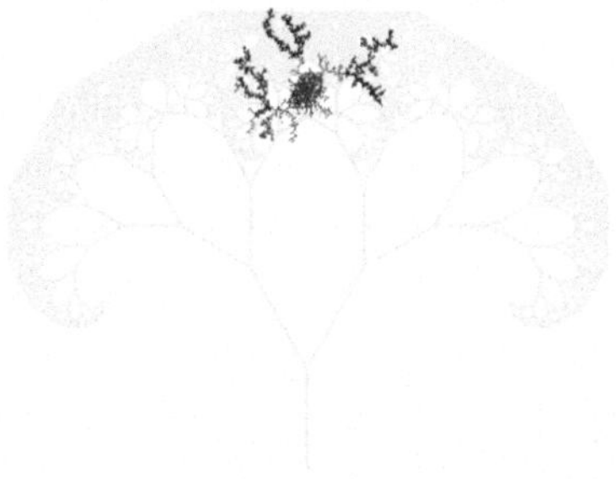

(a) 100 iterations

(b) 300 iterations

Fig. 3. Tumor growth evolution across iterations.

Additionally, if the aggregation point lies within the area of influence defined by the fractal mask, a mechanism of accelerated local growth is activated, in which multiple additional adhesions are generated within the immediate neighborhood. This behavior simulates a more aggressive tumor growth phenomenon associated with the interaction with fractal tissue structures, as might occur in the presence of signaling pathways, vascular networks, or favorable microenvironmental conditions within the bronchial tree.

3.4 Simulation of Fractal Growth

Finally, the simulation is visualized by superimposing the tumor propagation onto the background of the fractal tree structure, enabling a graphical observation of the system's dynamic behavior.

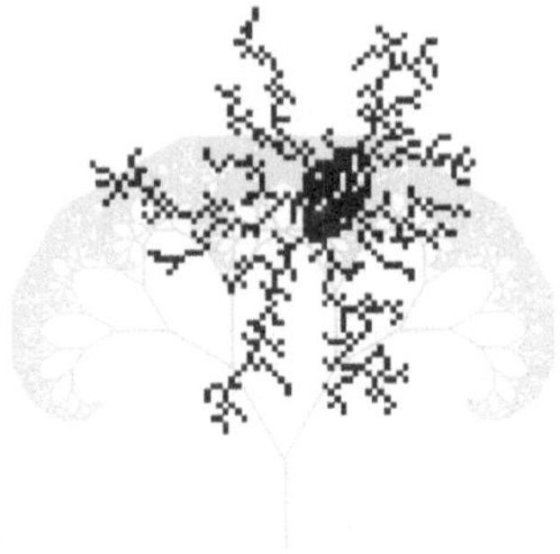

Fig. 4. Temporal evolution of the tumor fractal size modeled through Diffusion-Limited Aggregation (DLA). The graph shows the cumulative number of active pixels (growth zones) over 500 iterations. Regions of accelerated growth are observed, correlated with the contact between the tumor and the environmental fractal structure (tree mask).

This visualization (Fig. 4) clearly highlights the regions where tumor growth is favored by proximity to the branched structure.

It was observed that cellular growth in the simulated model intensifies in areas adjacent to the fractal structure, resulting in more irregular and dense morphological patterns. This finding suggests a possible relationship between the geometry of the environment and the aggressiveness of cellular growth. This observation is consistent with the results reported by [22], who demonstrated that a higher fractal dimension of the tumor border in histological samples of breast cancer is directly correlated with higher histological grades and, therefore, with more aggressive tumor behavior. This reinforces the hypothesis that the geometric characteristics of the microenvironment may act as modulators of tumor progression.

4 Analysis of Results

The results obtained from the computational simulation, based on a Diffusion-Limited Aggregation (DLA) model, show that tumor growth acquires a fractal morphology whose complexity strongly depends on the initial conditions and the structural characteristics of the microenvironment. In particular, it is observed that the initial location of the seed cell critically influences the morphology and expansion of the simulated tumor: if the cell is placed in a topologically favorable region—e.g., with higher connectivity or lower structural density—the growth becomes faster and denser, as shown in Fig. 3a.

This dynamic is consistent with observations in real tumors [11], where it has been demonstrated that the availability of resources such as oxygen, nutrients, extracellular space, and the organization of the extracellular matrix (ECM) modulate both cellular proliferation and tumor aggressiveness [17]. The spatial heterogeneity of the microenvironment can act as a physico-chemical gradient that constrains the evolutionary trajectories of the tumor from early stages.

In the early stages of the model, the simulated growth pattern is relatively isotropic, characterized by homogeneous radial expansion from the point of origin. This behavior has been documented in studies of Diffusion-Limited Aggregation (DLA), where the initial particles tend to distribute symmetrically around the seed due to the uniform probability of free diffusion.

However, when multiple executions of the algorithm are performed using the same fractal seed in a fixed position, it is observed that although the initial shapes are similar, significant differences emerge in the morphology and extent of tumor growth as the system evolves, as shown in Fig. 3b. This divergence is attributed to the inherently stochastic nature of the aggregation process, where small variations in the random walk trajectories of particles lead to divergent growth paths over time [10].

In the biological context, this variability reflects the non-deterministic nature of actual tumor growth, where factors such as fluctuations in nutrient availability, mechanical pressure, and microenvironmental architecture influence diverse morphological patterns, even in tumors with similar genotypes [1].

These results are not only consistent with clinical observations regarding the structural diversity of tumors but also open a pathway for computational exploration of how small perturbations in the microenvironment can be amplified and lead to divergent evolutionary trajectories in tumor progression.

4.1 Limitations

This study provides a qualitative evaluation of the feasibility of integrating two fractal structures to describe lung tumor growth, however, several limitations must be acknowledged.

The growth conditions of the bronchial tree fractal are predefined at the initialization stage and cannot be modified during the simulation, which restricts the adaptability of the model to dynamic physiological conditions. Similarly, tumor growth is currently governed by a stochastic criterion that allows expansion only in cross-shaped directions limiting the spatial realism of the simulated morphology. Additionally, the model assumes accelerated growth under certain local conditions, but this mechanism remains simplified and does not yet capture the complexity of biological triggers that modulate tumor progression.

Another important limitation is that the present implementation is restricted to a single seed point from which tumor expansion originates. This constraint prevents the exploration of scenarios involving multiple initiation marks, which are common in clinical contexts and may significantly alter growth patterns. These limitations highlight the need for further refinements to improve the biological plausibility and predictive power of the model in future work.

5 Result Validation

Although the present work focuses primarily on the computational framework, validation remains an essential step for assessing the reliability of the proposed model. Direct comparison with patient-specific imaging data is beyond the scope of this study, but future work will integrate such datasets to evaluate biological plausibility. As a preliminary consideration, qualitative benchmarking against existing computational tumor growth models could provide insight into the consistency of the simulated patterns. Establishing these validation pathways will be crucial for translating the model into clinically meaningful applications.

6 Conclusions

A computational model was successfully developed and implemented, integrating stochastic simulation techniques with fractal structures to represent tumor growth in lung tissue. The interaction between the tumor and a fractal structure resembling the bronchial tree was shown to directly influence morphology and growth dynamics, leading to more aggressive and directional expansion patterns.

An important research further study involves the calibration of model parameters such as tumor growth factors and lung tissue morphology using experimental datasets and machine learning techniques. As demonstrated in recent studies, this integration would improve precision by reducing the gap between simulation and biological reality. Likewise, the incorporation of statistical metrics related to fractal growth, such as fractal dimension and lacunarity, could provide a deeper and robust insight into the morphological complexity of tumor expansion.

Additionally, extending the framework to three-dimensional representations of lung tissue offers a promising opportunity to capture long-term tumor growth and tissue-level interactions. However, this direction requires rigorous calibration procedures as it was previously mentioned that remain consistent with clinical evidence, particularly through comparisons with diagnostic imaging and histological observations. With this improvement the model may evolve into a robust platform for investigating tumor progression and informing therapeutic strategies.

Acknowledgments. This product is a result of the PIC-CIAS 4241 project, funded by the Universidad Militar Nueva Granada 2025.

Disclosure of Interests. The authors declare that they have no conflict of interest.

References

1. Anderson, A., Quaranta, V.: Integrative mathematical oncology. Nat. Rev. Cancer **8**, 227–234 (2008). https://doi.org/10.1038/nrc2329
2. Angaji, A., Owusu, M., Velling, C., Dick, N., Weghorn, D., Berg, J.: High-density sampling reveals volume growth in human tumours (2024). https://doi.org/10.7554/elife.95338.2

3. Cogno, N., Axenie, C., Bauer, R., Vavourakis, V.: Agent-based modeling in cancer biomedicine: applications and tools for calibration and validation. Cancer Biol. Therapy **25**(1), 2344600 (2024). https://doi.org/10.1080/15384047.2024.2344600. epub 2024 Apr 28
4. Deutsch, A., Dormann, S.: Cellular Automaton Modeling of Biological Pattern Formation, vol. 163. Springer Science+Business Media (2017)
5. Ferlay, J., et al.: Global cancer observatory: Cancer today (2024). https://gco.iarc.who.int/today. lyon, France: International Agency for Research on Cancer. Accessed 05 Nov 2025
6. Glenny, R., M., K., Bauer, C., Beichel, R.R.: The fractal geometry of bronchial trees differs by strain in mice. J. Appli. Physiol. **128**(2), 362–367 (2020) https://doi.org/10.15252/embr.201439246, epub 8 de noviembre de 2014
7. Gu, Q., et al.: Structural and functional alterations of the tracheobronchial tree after left upper pulmonary lobectomy for lung cancer. BioMed. Eng. OnLine **18** (2019). https://doi.org/10.1186/s12938-019-0722-6
8. Lennon, F., et al.: Lung cancer—a fractal viewpoint. Nature reviews. Clini. Oncol. **12** (2015). https://doi.org/10.1038/nrclinonc.2015.108
9. Liu, X., Sun, X., Qi, H., Wang, H.: Morphometric analysis of human tracheobronchial tree based on multi-detector ct scans. Surg. Radiol. Anat. **32**(7), 629–635 (2010). https://doi.org/10.1007/s00276-010-0614-7
10. Lorthois, S., Cassot, F.: Fractal analysis of vascular networks: Insights from morphogenesis. J. Theor. Biol. **262**(4), 614 (2010). https://doi.org/10.1016/j.jtbi.2009.10.037
11. Lu, P., Weaver, V.M., Werb, Z.: The extracellular matrix: a dynamic niche in cancer progression. J. Cell Biol. **196**(4), 395–406 (2012). https://doi.org/10.1083/jcb.201102147
12. Maity, S., Alrubayan, M., Apachigawo, I., Solanki, D., Pradhan, P.: Optical probing of fractal and multifractal connection to the degree of disorder strength in weakly optical disordered media: application to cancer detection (2025) https://doi.org/10.1364/opticaopen.28425095
13. Michallek, F., Huisman, H., Hamm, B., Elezkurtaj, S., Maxeiner, A., Dewey, M.: Prediction of prostate cancer grade using fractal analysis of perfusion mri: retrospective proof-of-principle study. Eur. Radiol. **32**(5), 3236–3247 (2022). https://doi.org/10.1007/s00330-021-08394-8. epub 2021 Dec 16
14. Montana, E., Maina, J.: Fractal analysis of concurrently prepared latex rubber casts of the bronchial and vascular systems of the human lung. Open Biol. **10** (2020). https://doi.org/10.1098/rsob.190249
15. Osada, H., Takahashi, T.: Genetic alterations of multiple tumor suppressors and oncogenes in the carcinogenesis and progression of lung cancer. Oncogene **21** (2002). https://doi.org/10.1038/sj.onc.1205802
16. Paun, M.A., Postolache, P., Nichita, M.V., Paun, V.A., Paun, V.P.: Fractal analysis in pulmonary ct images of covid-19-infected patients. Fractal and Fractional **7** (2023) https://doi.org/10.3390/fractalfract7040285
17. Pickup, M.W., Mouw, J.K., Weaver, V.M.: La matriz extracelular modula las características del cáncer. EMBO Rep. **15**(12), 1243–1253 (2014). https://doi.org/10.15252/embr.201439246
18. Renee Brady, H.E.: Mathematical models of cancer: when to predict novel therapies, and when not to. Bull. Math. Biol. **81**(12), 4367–4382 (2019). https://doi.org/10.1007/s11538-019-00640-x
19. Rockne, R.C., Scott, J.G.: Introduction to mathematical oncology. JCO Clinical Cancer Inform. **3**, 1–4 (2019). https://doi.org/10.1200/CCI.19.00010

20. Sandino Garzon, A.A., Herrera García, R.J.: Detección de candidatos a microcalcificaciones mamarias agrupadas en mamografías. Ingeniería **24**(08), 159–170 (2019). https://doi.org/10.14483/23448393.12512
21. Santos, A., et al.: Complex shape markers can detect alterations in the spatial distribution of cell nuclei in human lung squamous cell carcinoma: a useful tool for automatic analysis? Histochemistry Cell Biol. **163** (2024). https://doi.org/10.1007/s00418-024-02336-3
22. Tambasco, M., Eliasziw, M., Magliocco, A.M.: Complejidad morfológica de la arquitectura epitelial para predecir la supervivencia invasiva del cáncer de mama. J. Transl. Med. **8**, 140 (2010). https://doi.org/10.1186/1479-5876-8-140, publicado el 31 de diciembre de 2010
23. Tanabe, N., Sato, S., Suki, B., Hirai, T.: Fractal analysis of lung structure in chronic obstructive pulmonary disease. Front. Physiol. **11**, 603197 (2020). https://doi.org/10.3389/fphys.2020.603197
24. Tschirren, J., Hoffman, E.A., McLennan, G., Sonka, M.: Intrathoracic airway trees: segmentation and airway morphology analysis from low-dose ct scans. IEEE Trans. Med. Imaging **24**(12), 1529–1539 (2005). https://doi.org/10.1109/TMI.2005.857654
25. Jiao, Y., S.T.: Emergent behaviors from a cellular automaton model for invasive tumor growth in heterogeneous microenvironments. PLoS Comput. Biol. **7**(12) (2011). https://doi.org/10.1371/journal.pcbi.1002314

Simple Transfer Learning Evaluation of Artificial Neural Network Models for Pulse Detection in Bat Echolocation Signals with Different Ultrasound Microphones

Andrés Muñoz[1], Maira Torres[2], Francisco A. Sánchez[2], and Angel Cruz-Roa[1](✉)

[1] GITECX and AdaLab, Universidad de los Llanos, Villavicencio, Colombia
{andres.munoz,aacruz}@unillanos.edu.co
[2] ECOTONOS and MSU, Universidad de los Llanos, Villavicencio, Colombia
{maira.torres,fsanchezbarrera}@unillanos.edu.co

Abstract. Automated analysis of bat echolocation signals is essential for noninvasive acoustic monitoring of neotropical species. This study evaluates the generalizability of artificial neural network models for the detection of echolocation pulses trained using a simple transfer learning strategy. For this study, 18 audio files were obtained with a Song Meter Mini BAT trap microphone, set at a sampling frequency (384 kHz), with 112 pulses manually annotated by expert biologists from *Saccopteryx leptura*. Three architectures were considered: a multilayer perceptron (MLP), a typical 4-layer convolutional neural network (CNN), and a 5-layer CNN with dropout (CNN2), all of these ANNs originally trained with spectrograms derived from ultrasonic recordings captured with a Pettersson M500 microphone at 500 kHz. To evaluate their generalizability, these models were applied to new recordings. The evaluation focused on the ability of models to detect echolocation pulses, taking into account the agreement between predicted pulses and manual annotations, using performance metrics such as precision, recall, F1-score and Dice coefficient. Furthermore, the Sørensen-Dice coefficient was computed to assess the temporal alignment between predicted and annotated events, providing a complementary measure of the quality of event-level detection. Although all three models achieved similar results, the MLP model performed best overall, reaching an average precision of 88.90%, a recall of 82.99% and an F1-score of 84.32%, with a Dice of 76.17%. These preliminary results are promising and have the potential to contribute to temporal dynamic activity monitoring based on artificial intelligence for ecology research.

Keywords: Deep Learning · Echolocation · Transfer Learning · Neotropical Bats · Convolutional Neural Networks

W. J. Sarmiento and E. L. Sierra-Ballén (Eds.): CCC 2025, CCIS 2692, pp. 239–248, 2027.
https://doi.org/10.1007/978-3-032-28447-1_18

1 Introduction

Nowadays, acoustic monitoring is an essential tool in ecology, especially for the study of bats, due to its non-invasive nature and broad spatial and temporal coverage [3]. These animals emit ultrasonic signals that can be recorded for analysis, allowing researchers to assess their diversity, activity levels, and responses to environmental changes [5]. This technique is particularly useful in megadiverse tropical regions, where traditional methods such as mist-netting have logistical and coverage limitations [3]. However, manual analysis of ultrasonic recordings has limitations. Visually identifying calls requires bioacoustics experts and it is highly time-consuming, reducing the scalability of studies in areas with high species diversity [8]. Additionally, intraspecific variability in calls influenced by factors such as habitat or foraging behavior increases the complexity of the process [5]. These limitations highlight the need to develop automated tools to optimize data processing and enable standardized comparisons across regions.

Consequently, the use of artificial neural networks (ANN) and deep learning (DL), such as convolutional neural networks (CNNs), has transformed bioacoustics by enabling systems to learn directly from spectral representations without requiring manual audio transformations [6]. Furthermore, CNNs have proven effective at detecting brief acoustic events even in recordings with variable background noise, improving robustness in sensor networks deployed in natural environments [1]. Numerous studies have demonstrated the effectiveness of CNNs for acoustic detection in various ecological settings. They have been used successfully to automatically identify bird songs in complex recordings [7], classify dolphin clicks in noisy marine environments [4], and detect odontocete clicks through passive monitoring systems [2]. These applications highlight CNNs' adaptability to environmental variations. Although these studies show high performance, they were trained and tested under consistent conditions, leaving their generalizability to varied acquisition parameters largely unexplored. This work addresses that gap, which is particularly relevant in bat echolocation research where hardware changes can alter signal resolution and affect models performance.

The objective of this work was to evaluate the performance of pulse detection models based on neural networks on a small set of recordings obtained with an ultrasound trap microphone at a lower sampling rate than those used to train the models with a larger dataset at higher sampling rate. This variation in the acquisition conditions affects the temporal and spectral resolution of the signals, justifying the use of transfer learning to assess the robustness of the models.

2 Materials and Methods

2.1 Ultrasounds Data Acquisition

For this study, the training data set was obtained from three 5-min audio files of ultrasonic recordings captured with a Pettersson M500 microphone at 500 kHz. Two expert biologists manually annotated a total of 1,662 pulses using Raven software from three bat species: *Saccopteryx leptura*, *Saccopteryx bilineata* and *Noctilio albiventris*.

For the test data set, 18 audio files were used, in which expert biologists manually identified and annotated 112 echolocation pulses from *Saccopteryx leptura*. The recordings were obtained using seven Song Meter Mini BAT trap microphones, installed at different points on the campus of the Universidad de los Llanos, Villavicencio, Colombia. Each file has a maximum duration of 30 s and was recorded at a sampling frequency of 384 kHz with a gain of 12 dB.

2.2 Data Processing

Pre-processing. To prepare the audio data and ensure its compatibility with the expected input of the pre-trained models, a segmentation process was applied using sliding windows of 12 ms with an overlap of 6 ms. Each segment was then transformed into a magnitude spectrogram using the Short-Time Fourier Transform (STFT). The resulting spectrograms were resized to 137×137 pixels and vertically flipped. Finally, for the training phase, a data augmentation process was performed to increase and balance the number of pulse samples per bat species to obtain a total of 4,368 pulse samples from the three bat species.

Spectrograms were used as input instead of traditional acoustic features because the pre-trained models were originally developed and trained with them. In addition, they provide a rich time-frequency representation that preserves the patterns essential for pulse detection in noisy or variable environments.

Post-processing. As part of the post-processing procedure, the predictions generated by the models were filtered based on their spectral content. For each 12 ms window classified as a pulse, the two peak frequencies were extracted from the magnitude spectrogram of the corresponding audio segment. Only detections where at least one of these frequencies fell within the 39–76 kHz range were retained. The application of this filter helped reduce false positives without modifying the model's architecture or parameters.

2.3 ANN and DL Models

The architectures evaluated were chosen for their proven effectiveness in the training phase and for previous experience with their implementation. These factors made them suitable for evaluating performance under different acquisition conditions.

Three pre-trained ANN and DL models were evaluated: a multilayer perceptron (MLP), a typical 4-layer convolutional neural network (CNN), and a 5-layer convolutional network with two layers of dropout (CNN2), as illustrated in Fig. 1. MLP consists of a fully connected (dense) architecture, with a flattening layer followed by two hidden layers of 150 neurons with a ReLU activation function and a sigmoid output layer. CNN includes three convolutional blocks with 32, 64, and 128 filters, each followed by max pooling operations, a dense layer of 100 neurons, and a sigmoid output layer. CNN2 extends this architecture with a fourth convolutional layer of 256 filters, L2 regularization in the dense layer, and two dropout layers with a dropout rate of 0.5.

The above three architectures were trained after using the original pulse vs non-pulse data set from the Petterson M500 microphone divided into three subsets for train (70%), validation (15%) and test (15%) for a training-validation setup and the final evaluation in testing.

The goal of this work is to evaluate the performance of these pretrained models for automatic pulse detection in the test dataset of 18 audio files obtained with a Song Meter Mini BAT trap microphone.

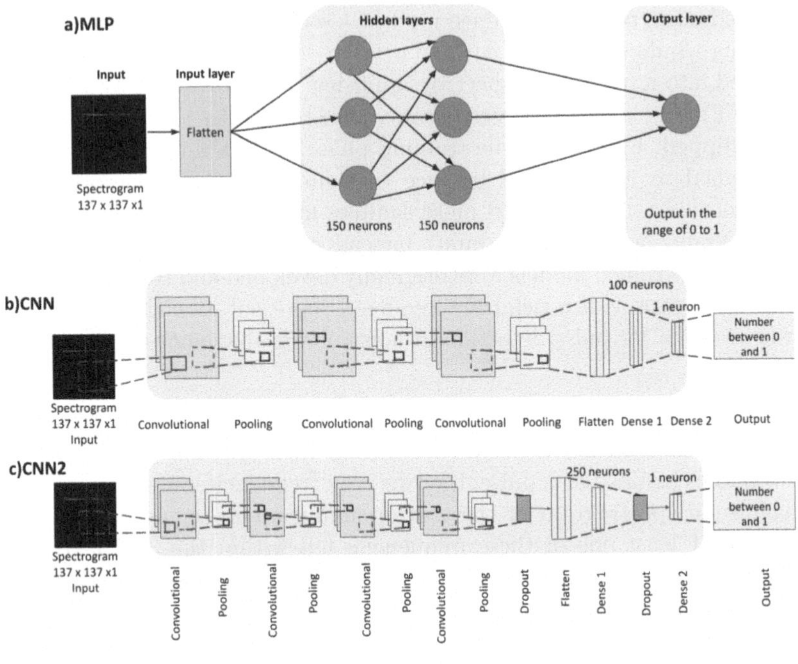

Fig. 1. Schematic representation of the three evaluated ANN and DL models: (a) MLP, (b) CNN y (c) CNN2.

3 Experimental Design

The evaluation of the performance of the models was based on the comparison between predicted pulses and manual annotations. Each audio file was segmented as described in the preprocessing section, generating windows that were predicted as "Pulse" or "No Pulse" by the models.

All pretrained models were used directly, without fine-tuning or additional retraining. This procedure allows us to analyze exclusively their ability to generalize by direct inference under different recording conditions.

In this context, a predicted event is defined as a group of consecutive 12 ms windows predicted as "pulse", which together represent a single echolocation pulse. These predicted events were compared to real events, defined by manually annotated intervals, enabling event-level evaluation.

To quantify performance, confusion matrices were generated for each model and the classification performance metrics of precision, recall and F1-score were computed as follows:

Precision. Precision is the proportion of true positive predictions among all positive predictions made by the model.

$$\text{Precision} = \frac{TP}{TP + FP}$$

where TP is the number of true positives, and FP is the number of false positives.

Recall. Recall is the proportion of true positive predictions among all actual positive instances.

$$\text{Recall} = \frac{TP}{TP + FN}$$

where FN is the number of false negatives.

F-Measure. The F-measure is the harmonic mean of precision and recall. It provides a balance between the two metrics.

$$F_1 = 2 \cdot \frac{\text{Precision} \cdot \text{Recall}}{\text{Precision} + \text{Recall}}$$

All predictions were performed using TensorFlow 2.19 and a NVIDIA GeForce RTX 4060 GPU (8 GB VRAM).

4 Results and Discussion

4.1 Analysis of Performance Metrics and Confusion Matrices

Table 1 presents the precision, recall and F1-score values for both classes. All models showed similar behavior. The MLP model achieved the best overall performance, with the highest precision (88.90%) and the highest average F1-score (84.32%), slightly outperforming the CNN and CNN2 architectures.

Table 1. Summary of classification metrics by class and model. This table supports a detailed comparison of model performance across the Pulses and No Pulses categories.

Model	Class	Precision	Recall	F1-score
MLP	No Pulses	0.9767	0.8047	0.8824
	Pulses	0.5537	0.9265	0.6931
	Average	0.8890	0.8299	0.8432
CNN	No Pulses	0.9754	0.8047	0.8819
	Pulses	0.5526	0.9224	0.6911
	Average	0.8878	0.8291	0.8423
CNN2	No Pulses	0.9754	0.8036	0.8812
	Pulses	0.5512	0.9224	0.6901
	Average	0.8875	0.8283	0.8416

For the "Pulses" class, a high recall value ($\approx$92%) was observed, indicating that the models were able to detect most true pulses, although at the cost of misclassifying some windows without pulses.

All models show a high ability to correctly identify non-pulse segments with precision values above 97%, which is the rejection class, avoiding false negatives. In contrast, the precision for the "Pulses" class was considerably lower ($\approx$ 55%), indicating a notable false positive rate.

For a more detailed analysis of the classification performance of the three models, Fig. 2 shows the normalized confusion matrices for the three models evaluated. The results of Table 1 can be explained because in an audio file, there are more regions of non-pulses than pulses of the same length of 12 ms, which affects the performance evaluation in this test set due to the natural imbalance of samples per class. However, with the normalized confusion matrices, it is easy to see that the three models have around 80% of precision to detect non-pulses against a 92% of precision to detect, a more realistic precision performance for pulses and non-pulses detection.

Although a balanced dataset was created for training the models, the evaluation was performed on fragments of audio recordings without applying any mitigation strategy, in order to reflect real-world conditions, which resulted in a predominance of segments without pulses. This distribution explains the lower performance observed for the minority class (pulses) in terms of precision, as the models tend to produce more false positives, which is reflected in the results obtained.

4.2 Analysis of the Dice Coefficient

To complement the analysis of detection performance, the Sørensen-Dice coefficient (Dice) was calculated for each predicted pulse that overlapped with at least one manually annotated pulse. This metric evaluates the temporal over-

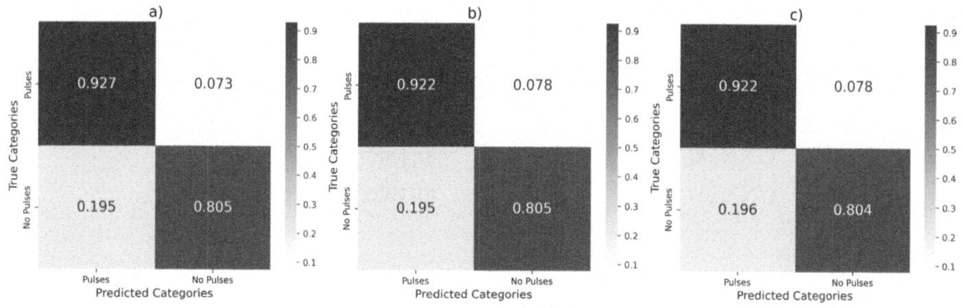

Fig. 2. Normalized confusion matrices for the three models: (a) MLP, (b) CNN and (c) CNN2. These visualizations summarize the effectiveness of each model in distinguishing between pulsed and non-pulsed segments.

lap between two events, considering both duration and alignment. The equation used for each matching pair is as follows:

$$QS = \frac{2 \cdot |A \cap B|}{|A| + |B|} \tag{1}$$

where A and B represent the time intervals (in milliseconds) of a predicted event and its best match among the annotated events, respectively. The intersection $|A \cap B|$ corresponds to the duration of overlap between the two intervals.

Only events with valid overlap were considered for this calculation, as the Dice coefficient is only meaningful when a temporal intersection exists. For the MLP model, 107 pulses were predicted, of which 104 overlapped with at least one annotated pulse. Similarly, both the CNN and CNN2 models predicted 106 pulses, and 104 of them showed valid intersections with actual annotations. Predicted events without overlap were excluded from the analysis.

In addition to the average Dice coefficient value, the standard deviation was computed to assess the variability in detection quality across events.

Table 2 presents the average values of the Dice coefficient and the standard deviation for the three models evaluated. The results indicate that very similar average Dice coefficient values were obtained for all models, suggesting a comparable performance in the temporal alignment of the predicted events with the manual annotations in the test data set of the trap microphone. The MLP model obtained the highest value (0.7617), followed by CNN (0.7578) and CNN2 (0.7567).

In terms of variability, Table 2 also shows that the standard deviation was low for all three models, indicating stable detection between events. The MLP model presented the lowest dispersion (0.1836), followed by CNN (0.1889) and CNN2 (0.1922), although the differences were minimal. These results show that simple architectures such as MLP can maintain robust performance even under acoustic conditions other than training due to a different microphone.

Table 2. Summary of average Dice coefficients and standard deviation for each model.

Model	Average Dice coefficient	Standard Deviation
MLP	0.7617	0.1836
CNN	0.7578	0.1889
CNN2	0.7567	0.1922

4.3 Visual Comparison Between Real and Predicted Pulses (MLP Model)

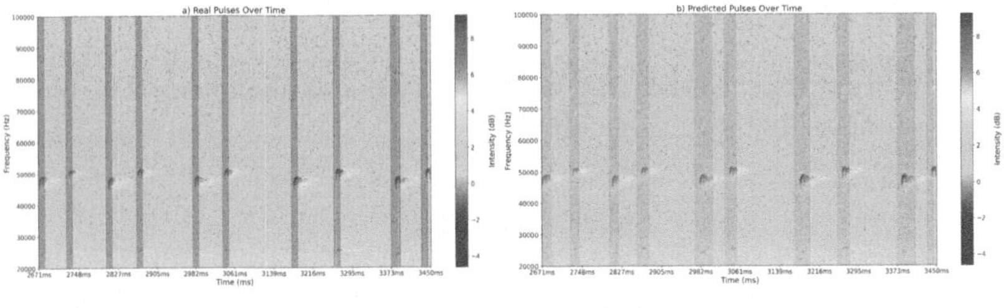

Fig. 3. Spectrograms of *Saccopteryx leptura* pulses from one analyzed recording of the test data set from Song Meter Mini BAT trap microphone: (a) manual annotations by expert biologist and (b) predictions by the MLP model. The figure allows visual comparison of the temporal location of real and predicted pulses.

Figure 3 shows the spectrograms with the actual pulses and those predicted by the MLP model. This model detects most of the pulses, demonstrating its generalizability despite differences in recording conditions. However, the predicted pulses do not always match exactly in location or duration with the true ones, which may be attributed to variations in temporal and spectral resolution introduced by the lower sampling frequency and the characteristics of the new recording device, such as its sensitivity. However, the number of detected pulses is right in most cases, except when there is more than one bat in the same time interval.

5 Conclusions

This study evaluated the generalization capability of three ANN and DL models for the automatic detection of bat echolocation pulses under acquisition conditions different from those of the original training. By applying a simple transfer learning strategy, it was confirmed that the models adapted to recordings from a different microphone and at a lower sampling rate.

Although all three models achieved similar performance metrics, the MLP model obtained the best overall performance, suggesting that, in this context, a simpler architecture may be more effective than more complex ones for bat's pulse detection. However, the models showed lower precision in identifying segments with pulses, highlighting the need to improve discrimination capability for short-duration events under variable acoustic conditions. The high recall rate observed for the "Pulses" class indicates that the models are efficient at detecting most actual events, albeit at the cost of generating some false positives.

Furthermore, the inclusion of the Dice coefficient in the evaluation provided a complementary perspective on the temporal agreement between predicted and annotated pulses. Although the models showed a generally stable overlap at the event level, the temporal localization of individual pulses was not consistently accurate. These results suggest that, while the models can detect most actual events, further refinement is needed to improve their temporal resolution in complex acoustic scenarios.

Nonetheless, the study demonstrates that ANN and DL models can effectively generalize across varying recording conditions, which is particularly valuable for ecological studies conducted in high biodiversity environments. Furthermore, transfer learning and lightweight models could be useful to adapt these automatic approaches to small and portable devices for real-time monitoring.

Future work includes exhaustive parameter exploration and fine-tuning of ANN and DL architectures trained with more and diverse data. In particular, future experiments will involve training the evaluated models with an expanded dataset that integrates echolocation pulses from multiple bat species recorded at various monitoring sites at the Universidad de Los Llanos. This will allow us to evaluate the models' ability to generalize across species and recording conditions. In addition, we plan to implement the models in a web-based acoustic analysis and annotation platform.

Acknowledgments. This work was carried out as part of the project entitled "Dinámica temporal de la actividad de dos especies hermanas de murciélagos insectívoros en el piedemonte llanero colombiano, developed at the Universidad de los Llanos", with the participation and support of the research groups GITECX and ECOTONOS. The authors also acknowledge the specialized software and computing infrastructure provided by the Automatic Data-driven Analytics Laboratory (AdaLab) at Universidad de los Llanos.

Disclosure of Interests. The authors have no competing interests to declare that are relevant to the content of this article.

References

1. Lostanlen, V., Salamon, J., Farnsworth, A., Kelling, S., Bello, J.P.: Robust sound event detection in bioacoustic sensor networks. PLoS ONE **14** (2019). https://doi.org/10.1371/journal.pone.0214168
2. Luo, W., Yang, W., Zhang, Y.: Convolutional neural network for detecting odontocete echolocation clicks. J. Acoust. Soc. Am. **145**, EL7–EL12 (2019). https://doi.org/10.1121/1.5085647

3. López-Baucells, A., Yoh, N., Rocha, R., Bobrowiec, P.E., Palmeirim, J.M., Meyer, C.F.: Optimizing bat bioacoustic surveys in human-modified neotropical landscapes. Ecol. Appl. **31** (2021). https://doi.org/10.1002/eap.2366
4. Seydi, V., et al.: The application of neural networks to classify dolphin echolocation clicks (2022). https://doi.org/10.1101/2022.06.14.496047
5. Stahlschmidt, P., Brühl, C.A.: Bats as bioindicators - the need of a standardized method for acoustic bat activity surveys. Methods Ecol. Evol. **3**, 503–508 (2012). https://doi.org/10.1111/j.2041-210X.2012.00188.x
6. Stowell, D.: Computational bioacoustics with deep learning: a review and roadmap (2022). https://doi.org/10.7717/peerj.13152
7. Stowell, D., Wood, M.D., Pamuła, H., Stylianou, Y., Glotin, H.: Automatic acoustic detection of birds through deep learning: the first bird audio detection challenge. Methods Ecol. Evol. **10**, 368–380 (2019). https://doi.org/10.1111/2041-210X.13103
8. Zamora-Gutierrez, V., et al.: Acoustic identification of Mexican bats based on taxonomic and ecological constraints on call design. Methods Ecol. Evol. **7**, 1082–1091 (2016). https://doi.org/10.1111/2041-210X.12556

Intelligent Generation of Urban Datasets for Road Inspection Using Georeferenced Capture and Visual Enhancement via Deep Learning

Juan Pablo Correa Sandoval, Cristian Acosta, Duber Martinez Torres(✉), and Jairo Antonio Muñoz Arango

Universidad Nacional Abierta y a Distancia - UNAD, Bogotá, Colombia
{duber.martinez,jairo.munoz}@unad.edu.co

Abstract. This paper presents an automated methodology for generating high-quality urban image datasets, aimed at training and validating computer vision models applied to road inspection. The proposed solution combines synchronized video and geolocation data capture from a moving vehicle, followed by an intelligent segmentation process based on spatial displacement. For each identified clip, the frame with the highest visual quality is automatically selected using objective metrics such as sharpness, entropy, and Deep Learning models like HyperIQA. Additionally, the NAFNet model is applied for motion deblurring and overall image enhancement. The results show a reduction of over 95% in the total number of frames while maintaining spatial coverage and improving visual quality by an average of 14.14%. Validation with a downstream defect detection model revealed a 20.93% absolute increase in detection rate, demonstrating that image enhancement directly benefits model performance. This approach represents a significant step forward in the standardization and optimization of data for road defect detection models.

Keywords: Computer vision · pavement inspection · georeferencing · automatic frame selection · video processing

1 Introduction

Automated inspection of road pavement conditions has gained significant relevance in recent years, as it reduces costs and increases safety compared to manual techniques [8,9]. The rise of computer vision and deep learning has enabled the development of systems capable of identifying damage such as cracks and potholes with high accuracy [6,7].

Among the most notable advances are specialized datasets for road inspection, such as RDD2022, which includes over labeled images for multiple types of deterioration [2], and GRDDC, widely used for evaluating detection models

W. J. Sarmiento and E. L. Sierra-Ballén (Eds.): CCC 2025, CCIS 2692, pp. 249–258, 2027.
https://doi.org/10.1007/978-3-032-28447-1_19

based on satellite and panoramic images [1]. These resources have facilitated the training and validation of models such as YOLO, which have demonstrated outstanding performance [11].

Image quality plays a critical role in the performance of these detectors. Image restoration models like NAFNet have set new standards, achieving results of approximately 33.7 dB in deblurring and 40.3 dB in denoising, with high computational efficiency [3]. Extended versions such as NAFSSR apply these principles to stereo-scopic tasks [4]. In various domains, combinations of NAFNet with other architectures have proven robust against noise, motion blur, and adverse conditions [5].

Despite these advances, the use of comprehensive pipelines that cover the entire process, from synchronized video and GPS capture to the intelligent selection of high-quality images for building efficient urban datasets remains limited. Techniques based on spatial segmentation and visual metrics such as sharpness, Shannon entropy, and perceptual quality assessment (HyperIQA) have been explored independently [10], but their integration into a complete pipeline is still emerging and promising.

This work addresses that gap by proposing a complete system that: (i) captures geotagged video, (ii) segments it into clips based on displacement, (iii) automatically selects the highest-quality frame using sharpness and perceptual scores, and (iv) applies restoration using NAFNet. The results show a reduction of up to 95% in data volume without significant loss of quality, with average visual improvements exceeding 14%.

Thus, this study makes a relevant contribution to the efficient generation of urban datasets and demonstrates the feasibility of integrating computer vision and image enhancement techniques for future semi-automated and mobile applications.

2 Methodology

The proposed system was developed following a structured methodology composed of six stages, integrating video processing, georeferenced data, and advanced Deep Learning techniques for the automated generation of high-quality datasets. The methodology can be summarized in the following sequential steps:

- Synchronized data acquisition
- Segment into clips
- Extract one representative frame per clip
- Apply motion deblurring
- Generate initial dataset with the selected processed frames
- Validate results for the obtained dataset

The first stage involves the synchronized capture of video and GPS data from a moving vehicle. The video is recorded using a Samsung Galaxy A35 smartphone equipped with a 50 MP rear camera and digital image stabilization (VDIS), configured to capture footage in Full HD resolution (1920×1080 px) at

60 frames per second (fps). Simultaneously, georeferencing data is collected using the device's internal GNSS receiver, which is compatible with multiple satellite systems (GPS, GLONASS, Galileo, and BDS). A custom Android application stores, for each geolocation point, the following parameters: latitude, longitude, estimated speed, and timestamp in UNIX format.

To ensure synchronization between the video and GPS data, both records use the same timestamp system. During post-processing, linear interpolation of the GPS data is performed to associate each video frame with its exact geographic position at the moment of capture.

In the second stage, the video is segmented into non-redundant clips using a spatial distance threshold as the main criterion. This threshold is set to 1 m to balance the coverage of the inspected area with the reduction of redundant frames. Given the video sampling rate of 60 fps and the low vehicle speeds typical in urban environments (approximately 10 km/h or 2.78 m/s), multiple frames can represent nearly identical scenes. To avoid redundancy, a variable clip size is defined based on the vehicle's speed.

Given:

- V: Vehicle speed in [m/s]
- D: Distance threshold in [m]
- T_m: Sampling period that defines the clip size
- V_c: Video capture rate in [fps]

The time required to reach the minimum displacement defined by the threshold is calculated as:

$$T_m = \frac{D\,(m)}{V\,(m/s)} \tag{1}$$

The compression or reduction is then defined as:

$$Reduction = 100\left(1 - \frac{1}{T_m V_c}\right) \tag{2}$$

This stage ensures that each clip covers a unique section of the road, maintaining spatial continuity without leaving significant gaps uninspected.

In the third stage, the most representative frame within each clip is selected based on objective visual quality metrics. Three independent indicators are used for this purpose:

- **Sharpness**: Calculated using the variance of the Laplacian filter applied to the frame.
- **Entropy**: Measured using Shannon entropy of the luminance channel, which reflects the amount of information in the image.
- **HyperIQA**: A Deep Learning-based metric specifically designed to assess perceptual image quality without requiring a reference.

Each metric is normalized to the [0, 1] range and combined into a weighted score according to the following distribution:

$$Score = 0.4 \times Sharpness + 0.2 \times Entropy + 0.4 \times HyperIQA$$

The weights were assigned considering the relative importance of each metric for road inspection: Sharpness and HyperIQA capture image sharpness and perceptual quality, which are critical for accurate defect detection, while Entropy provides complementary information about the visual content. The weighting values were determined manually through a qualitative analysis based on the visual inspection of test images. For future work, a sensitivity analysis is planned to optimize the coefficients and identify the most appropriate combination of metrics for the automatic selection of high-quality frames.

Finally, in the fourth stage, a Deep Learning model is applied to correct motion-induced distortions in the selected frames. For this task, the NAFNet model is used, downloaded from the official BasicSR repository. This model has demonstrated strong performance in restoring images affected by motion blur. The implementation is carried out using PyTorch with Python 3.10 and CUDA 12.4. Frames are processed individually using an NVIDIA RTX-4070 GPU (8 GB VRAM).

3 Results

To validate the performance of the proposed system, tests were conducted on four urban routes with durations ranging from 7 to 10 min. The routes included mixed segments, such as paved surfaces, visible potholes, and previously repaired areas. This type of scenario allowed for evaluating the system's ability to operate under real-world road inspection conditions.

3.1 Geographic Visualization

As part of the automated process, the georeferencing data collected during each route was used to generate an accurate spatial representation of the trajectory. Figure 1 shows the visualization generated on a map corresponding to the four recorded routes. This functionality allows each selected image to be associated with its exact location, facilitating future geospatial analysis and automated reporting of road anomalies.

3.2 Dataset Reduction

One of the main goals of the system is to reduce the number of redundant frames without compromising spatial coverage. The results show a significant decrease in the total volume of processed images. Table 1 presents a summary comparing the number of original frames with the number of frames selected as

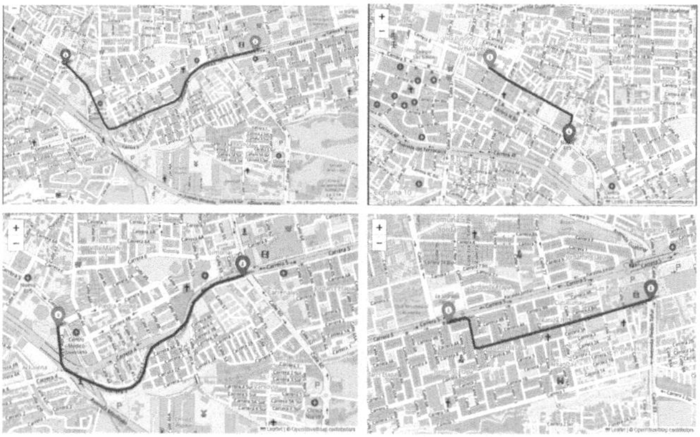

Fig. 1. Visualization of the routes recorded during the data capture stage.

Table 1. Reduction achieved after the automatic selection process.

Route	Original Frame Count	Selected Frame Count	Reduction
1	38100	1560	95.91%
2	23400	630	97.30%
3	29400	1307	95.55%
4	38400	1153	97.00%

representative after applying the segmentation and automatic selection process based on visual quality.

These results indicate that, on average, more than 96% of the original frames are discarded, representing a substantial improvement in terms of storage and post-processing efficiency.

3.3 Visual Enhancement

In addition to dataset reduction, a quantitative evaluation was conducted to assess the improvement in the visual quality of the selected images. For this purpose, objective image quality metrics were calculated before and after the selection and deblurring process using the NAFNet model.

Figure 2 presents examples of frames affected by motion distortion, alongside their corrected versions after applying NAFNet. A clear improvement in sharpness and detail definition can be observed.

Figure 3 shows three curves that illustrate the behavior of the visual quality indicator throughout the different stages of the process:

- **Red curve**: Average visual quality value in each original clip.
- **Blue curve**: Quality of automatically selected frames.
- **Green curve**: Quality of frames after applying motion deblurring.

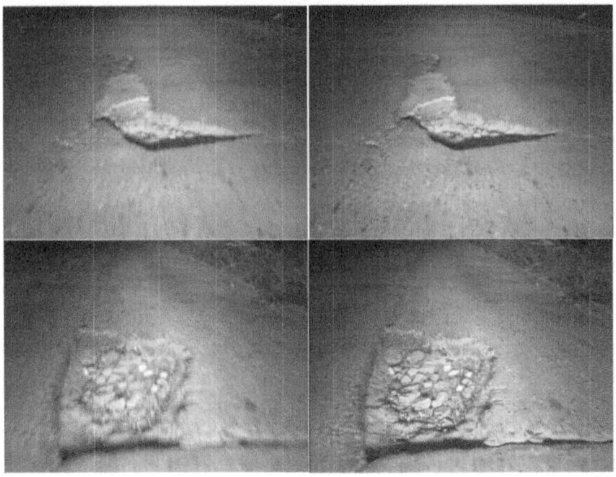

Fig. 2. On the left side, two frames with motion blur are shown. On the right, the corresponding images after applying NAFNet to deblurring.

The average improvement percentages obtained were as follows:

1. Selected image vs. clip average: +**4.86%**.
2. Image with NAFNet vs. clip average: +**14.14%**.
3. Image with NAFNet vs. unprocessed selected image: +**8.88%**.

These values demonstrate the effectiveness of both the selection algorithm and the Deep Learning model in improving the perceptual and quantitative quality of the images.

3.4 Effect of Visual Enhancement in an Automated Road Defect Detection Scenario

To evaluate the effectiveness of the proposed methodology for automated road defect detection, additional experiments were conducted using videos captured from a moving vehicle under nighttime low-light conditions. From these recordings, a dataset was generated following the proposed pipeline, consisting of 681 images containing 1,098 annotated defects of interest. This dataset was divided into three subsets: 70% for training, 10% for validation, and 20% for testing. For the test set, images with the highest degree of motion blur were specifically selected to assess the impact of the correction stage.

Initially, the YOLOv12 model was trained using the original unprocessed images and evaluated on the test set. In a second stage, a duplicate dataset was created with the same structure and images as the original, but with all frames preprocessed using the NAFNet model for motion distortion correction. The model was then retrained with this processed dataset, and the performance on the test set was compared to that of the unprocessed scenario, allowing us to quantify the impact of visual enhancement on road defect detection.

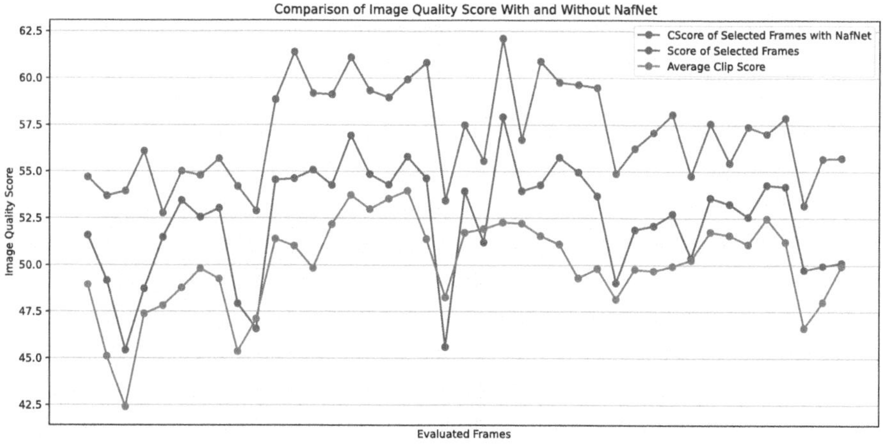

Fig. 3. The vertical axis represents the visual quality indicator value. The horizontal axis displays, in red, the average value of images within each clip; in blue, the value of the selected frames; and in green, the value of the images processed to remove motion blur. (Color figure online)

Table 2 summarizes the results obtained on the test subset considering a Ground Truth of 229 annotated defects. In the first scenario, using the original images without preprocessing, the YOLOv12 model correctly detected 153 defects (66.80%). In the second scenario, where the same images were processed with the NAFNet model for motion deblurring, the number of correctly detected defects increased to 200 (87.73%). This represents an absolute improvement of 20.93%, highlighting the effectiveness of incorporating visual enhancement techniques in the preprocessing stage for automated road defect detection.

It is important to highlight that a very low false positive rate was observed in both scenarios. In the first condition, using the original images without preprocessing, only two false detections were reported in the test set. In contrast, when motion deblurring with NAFNet was applied, no false detections were recorded. Additionally, although a few bounding boxes were generated for regions not included in the original annotations, manual inspection confirmed that these corresponded to actual road defects, indicating the model's capability to detect previously unlabeled anomalies.

Table 2. Detection performance and false positives with and without motion deblurring.

Condition	True Positives	Detection Rate (%)	False Positives
Without Preprocessing	153 / 229	66.80	2
With motion deblurring	200 / 229	87.73	0

Fig. 4. Detection examples. Left: images without Motion deblurring. Right: images after applying deblurring.

The average processing time per image was analyzed for the main stages of the proposed pipeline, as summarized in Table 3. The quality score computation required 0.0254 s per image, and the inference process for defect detection using YOLOv12 averaged 7.3 ms per image. Both stages demonstrate performance suitable for real-time or near real-time implementation. In contrast, the NAFNet-based motion deblurring, with an average of 1.76 s per image, represents the most computationally demanding step, making it more appropriate for offline processing rather than real-time execution (Fig. 4).

Table 3. Average processing time per image for each stage.

Processing Stage	Average Time per Image
Quality score computation	0.0254 s
NAFNet-based motion deblurring	1.76 s
Defect detection inference (YOLOv12)	0.0073 s

3.5 Discussion

The results demonstrate the effectiveness of the proposed system in automatically generating high-quality urban image datasets for road inspection. The methodology significantly reduces the number of processed frames while maintaining spatial coverage and improving visual quality. Video segmentation based on a 1-m geographic threshold reduced over 96% of the total data volume, avoiding redundancy and lowering storage and annotation costs.

Representative frame selection using a weighted combination of sharpness, entropy, and HyperIQA effectively identified the clearest frame in each clip, yielding a 4.86% improvement over the clip average. Applying NAFNet for motion deblurring further enhanced quality, achieving a 14.14% gain over the clip average and 8.88% relative to the original images.

From a practical standpoint, the proposed system significantly reduces manual intervention in dataset construction, minimizing subjective errors and increasing the consistency of the data used for training road defect detection models. Moreover, the integration of georeferencing allows for the precise association of each image with its geographic location, facilitating future applications in automated road anomaly reporting systems.

Additionally, an experimental evaluation was carried out to analyze the impact of motion blur correction on road defect detection performance. Using a test set of 229 annotated defects, the detection rate increased from 66.80% (153 valid detections) with unprocessed images to 87.73% (200 valid detections) after applying NAFNet-based deblurring, resulting in an absolute gain of 20.93%. Furthermore, false positives were reduced from two to zero. These findings confirm that incorporating visual enhancement into the pipeline not only improves image quality but also has a direct and measurable effect on the performance of downstream detection models, especially under challenging conditions such as motion blur and low-light environments.

4 Conclusions and Future Work

The proposed methodology effectively automates the generation of high-quality urban image datasets for training and evaluating computer vision models in road inspection. By integrating video processing, georeferenced data, and Deep Learning, the system reduced frame volume by over 95% while improving visual quality. The combination of sharpness, entropy, and HyperIQA metrics enabled consistent selection of representative frames, and NAFNet-based deblurring enhanced image clarity by 14.14% on average.

The impact on defect detection was significant, with a 20.93% absolute increase in detection rate and reduction of false positives, demonstrating that image enhancement benefits both dataset quality and model performance. Automating the full pipeline—from video capture to frame selection and restoration—reduces time, human effort, and operational costs while improving traceability.

Several lines of improvement are proposed for future work. First, the integration of semantic filters could be explored by incorporating preliminary detection modules for specific features, such as potholes or cracks, to prioritize the selection of relevant frames. Second, semi-automatic labeling strategies based on pretrained models could be implemented to accelerate the dataset annotation process for training neural networks. Finally, a sensitivity analysis could be performed to optimize the weighting distribution among the metrics that compose the image quality score, ensuring a more robust and effective frame selection process.

Acknowledgments. The authors would like to express their gratitude to the Vicerrectoría Académica y de Investigación (VIACI, Vice-Rectory for Academic and Research Affairs), the Sistema de Gestión de la Investigación (SIGI, Research Management System), and the National and Zonal Research Leaders of the Universidad Nacional Abierta y a Distancia (UNAD) for their financial and logistical support in the development of this work.

References

1. Arya, D., et al.: Global road damage detection: state-of-the-art solutions. In: 2020 IEEE International Conference on Big Data (Big Data), pp. 5533–5539. IEEE (2020). https://doi.org/10.1109/BigData50022.2020.9377790
2. Arya, D., Maeda, H., Ghosh, S.K., Toshniwal, D., Sekimoto, Y.: RDD 2022: a multi-national image dataset for automatic road damage detection. Geosci. Data J. **11**(4), 846–862 (2024). https://doi.org/10.6084/m9.figshare.21431547.v1
3. Chen, L., Chu, X., Zhang, X., Sun, J.: Simple baselines for image restoration. In: European Conference on Computer Vision, pp. 17–33. Springer (2022)
4. Chu, J., Chen, L., Yu, W.: NAFSSR: stereo image super-resolution using NAFNet. arXiv (2022). https://doi.org/10.1109/CVPRW56347.2022.00130
5. Esaulov, V., Esfahani, M.M.: A comparative study of NAFNet baselines for image restoration. arXiv preprint arXiv:2506.19845 (2025). https://doi.org/10.48550/arXiv.2506.19845
6. Guan, S., Liu, H., Pourreza, H.R., Mahyar, H.: Deep learning approaches in pavement distress identification: a review. arXiv preprint arXiv:2308.00828 (2023). https://doi.org/10.48550/arXiv.2308.00828
7. Men, T., BaopingWang, Zhang, N., Sun, Q.: Pavement defect detection method based on deep learning. In: International Conference on Computer Engineering and Networks, pp. 494–500. Springer (2023). https://doi.org/10.1007/978-981-99-9243-0_49
8. Nafaa, S., et al.: Automated pavement cracks detection and classification using deep learning. In: 2024 IEEE 3rd International Conference on Computing and Machine Intelligence (ICMI), pp. 1–5. IEEE (2024). https://doi.org/10.1109/ICMI60790.2024.10586098
9. Sami, A.A., Sakib, S., Deb, K., Sarker, I.H.: Improved YOLOv5-based real-time road pavement damage detection in road infrastructure management. Algorithms **16**(9), 452 (2023). https://doi.org/10.3390/a16090452
10. Su, S.: Blindly assess image quality in the wild guided by a self-adaptive hyper network. In: Proceedings of the IEEE/CVF Conference on Computer Vision and Pattern Recognition (CVPR) (2020)
11. Wang, N., Gao, Z., Fang, H., Hu, H., Li, B., Wang, C.: Deep learning network model for road pavement damage detection based on YOLOv8s-GES. Transp. Res. Rec., 03611981241312224 (2025). https://doi.org/10.1177/03611981241312224

Towards Echocardiogram Image Data Augmentation Using GANs

Salomón Hernández-Velandia, Marcela Iregui, and Alexander Cerón-Correa(✉)

ACCEDER, Universidad Militar Nueva Granada, Bogotá, Colombia
alexander.ceron@unimilitar.edu.co

Abstract. Medical imaging datasets are often limited, restricting the training of deep learning models. This work presents a data augmentation approach using a two-stage MedGAN architecture to generate synthetic echocardiographic images, the model combines a pretrained autoencoder for structural reconstruction and adversarial training with an U-Net generator and PatchGAN discriminator, trained on 10,000 grayscale frames from the EchoNet-Dynamic dataset, the model was evaluated after 100 epochs using FID, SSIM, and PSNR, results show basic structural coherence, though visual realism remains limited, highlighting the need for extended training and further optimization.

Keywords: Data Augmentation · Echocardiography · Deep Learning · U-Net · GANs · MedGAN · FID · SSIM · PSNR

1 Introduction

In recent years, deep learning has had a significant impact on medical image processing and analysis, enabling the development of advanced techniques for diagnosing and understanding diseases from images and videos, among the most notable applications are cardiac ultrasound analysis to evaluate heart movement, however, one of the main challenges in training robust and generalizable models in this domain is the scarcity of high-quality labeled data, this is due to patient privacy, high acquisition costs, and the need for medical experts for labeling.

In the medical context, GANs have been successfully used to generate synthetic images in various modalities such as magnetic resonance imaging (MRI), computed tomography (CT), and X-rays [1,2]. Among the most effective architectures for medical image generation tasks is the combination of a U-Net-type generator [3] with a PatchGAN-type discriminator, as proposed by models derived from MedGAN [4], however, for purely generative tasks where no input image is available, approaches such as MedGAN [4] have proven to be more suitable. MedGAN proposes an architecture that integrates a convolutional autoencoder with a generator that maps noise to a latent space, a pre-trained decoder, and refinement techniques, as well as perceptual and feature matching losses, which improves the structural quality of the generated images.

W. J. Sarmiento and E. L. Sierra-Ballén (Eds.): CCC 2025, CCIS 2692, pp. 259–268, 2027.
https://doi.org/10.1007/978-3-032-28447-1_20

This work proposes an adapted MedGAN architecture to generate synthetic echocardiogram images from a real set of 10,000 images collected from various patients by Stanford University, the main objective is to evaluate the feasibility of using synthetic images to increase the available dataset of synthetic echocardiographic images, with the aim of expanding and diversifying the available datasets, the contribution of this work is twofold. (i) The implementation and training of a MedGAN-type architecture focused on echocardiographic images, and (ii) the quantitative evaluation of the quality of images generated using standard metrics such as FID (Frechet Inception Distance), SSIM (Structural Similarity Index), and PSNR (Peak Signal-to-Noise Ratio).

2 State of the Art

2.1 Generative Adversarial Network (GANs)

Since their introduction by Goodfellow [5], Generative Adversarial Networks (GANs) have revolutionized the generation of synthetic data, their structure is based on the simultaneous training of two neural networks: A generator, which attempts to produce fake data that resembles real data, and a discriminator, which attempts to distinguish between real and generated data, this process is modeled as a minimax game, in which both networks improve each other through competition, improvements in training stability and image quality have led to the development of variants such as DCGAN, pix2pix [6], CycleGAN, StyleGAN [7], and MedGAN [4], which have shown great potential in medical imaging.

2.2 GANs in Medical Imaging

The use of GANs in medical imaging has grown rapidly, with key applications such as data augmentation, which improves the performance of classification and segmentation models [8], modality translation such as converting a CT image to MRI [9], reconstruction and resolution enhancement, filling in missing regions or increasing the resolution of medical images [10], and privacy analysis through the generation of synthetic data that does not compromise patient privacy [11].

2.3 MedGAN in the Imaging Domain

Unlike pix2pix, which translates between images conditionally, MedGAN [4] was designed for unconditional generation of medical images, using an architecture based on convolutional autoencoders, in its extended form. MedGAN employs a generator that transforms random noise into a latent space, which is decoded into a synthetic image by a pre-trained decoder, in addition, a refiner network is integrated, and perceptual and feature matching losses are used, which improves the visual and structural quality of the generated images, this architecture has demonstrated greater structural fidelity in medical domains compared to approaches based on U-Net + PatchGAN [12], particularly when conditional input images are not available.

3 Background

3.1 Fundamentals of Generative Adversarial Networks (GANs)

Generative Adversarial Networks (GANs) consist of an unsupervised learning framework where two neural networks are trained simultaneously in a competitive process, these networks are: a Generator (G) that attempts to create synthetic samples from noise or transformed inputs (e.g., images), and a Discriminator (D) that attempts to distinguish between real and generated samples. The goal of the generator is to deceive the discriminator, while the discriminator learns to correctly identify whether a sample is real or synthetic. This zero-sum game is expressed as:

$$\min_G \max_D \mathbb{E}_{x \sim p_{\text{data}}}[\log D(x)] + \mathbb{E}_{z \sim p_z}[\log(1 - D(G(z)))] \tag{1}$$

Since its original proposal, multiple variants have emerged to improve training stability and the quality of results, including improvements in architecture, loss function, and regularization [13].

3.2 MedGAN: Image Generation with Autoencoders and Perceptual Losses

MedGAN, initially proposed by Choi et al. [4], combines a convolutional autoencoder with an MLP-based generator that transforms noise into an intermediate latent space, which is then decoded into an image using a pre-trained convolutional decoder, this architecture was designed for purely generative tasks without input images, and has shown good results in medical images, in addition to standard adversarial loss, MedGAN incorporates two key components to improve the structural quality of the generated images:

- Loss of feature matching, which stabilizes training by minimizing the distance between intermediate activations of the discriminator for real and synthetic images.
- Perceptual loss, based on pre-trained networks (such as VGG), which reinforces the structural similarity between real and generated images.

This design allows the generation of synthetic medical images with high visual fidelity, preserving anatomical structures without requiring encoder-decoder architectures such as U-Net, or local discriminators such as PatchGAN.

4 Dataset and Experiments

4.1 Dataset: EchoNet-Dynamic

For this work, EchoNet-Dynamic was used, a public dataset developed by the Stanford Machine Learning Group, which contains more than 10,000 videos of four-chamber apical echocardiograms (A4C), each video represents a complete cardiac cycle and is associated with a different patient, with clinical metadata such as ejection fraction (EF), age, and sex. The videos have a native resolution of 112×112 pixels and are annotated for tasks such as EF estimation and left ventricular segmentation, however, in this work, only the visual part of the set was used, to transform the video dataset into a set of static images, the first frame of each video was extracted, thus generating a set of 10,000 echocardiographic images, each corresponding to a single patient, this decision was based on the objective of the study: to generate diverse and anatomically plausible synthetic images, rather than temporal sequences.

4.2 Preprocessing

The following preprocessing steps, consistent with the literature on deep learning tasks with medical images, were applied to each extracted image [14]:

1. Conversion to grayscale: the image was reduced to a single channel, which is consistent with the monochromatic nature of ultrasound.
2. Resizing: images were rescaled to a uniform resolution of 128×128 pixels, suitable for training deep convolutional networks while maintaining a balance between structural detail and computational efficiency.
3. Normalization: min-max normalization was applied to bring the pixel values to the range $[-1, 1]$, which improves training stability and is compatible with the *tanh* activation used in the generator output, using the transformation:

$$x_{\text{norm}} = \frac{x - 0.5}{0.5} \tag{2}$$

4.3 Experimental Setup, Adaptation of MedGAN

The training was carried out using Google Colab with GPU acceleration (Tesla T4), under the PyTorch framework, the proposed architecture consists of a U-Net generator and a PatchGAN discriminator, as detailed in the previous section (Table 1).

First, a convolutional autoencoder was trained to reconstruct the real images, this model learns a useful intermediate latent representation to generate structurally coherent images, the encoder and decoder are formed by convolutional and deconvolutional blocks, respectively, with LeakyReLU activations and `BatchNorm`, once trained, the encoder is discarded and the decoder is retained, which will be used in the next stage as part of the MedGAN generator.

As shown in Eq. (3), the MedGAN generator consists of an MLP network that transforms a noise vector $z \sim \mathcal{N}(0, I)$ into a latent vector z', which is then passed to the pre-trained decoder, this design seeks to generate images in the semantic feature space learned by the autoencoder, which improves the anatomical fidelity of the images.

$$z \in \mathbb{R}^{100} \rightarrow \text{MLP} \rightarrow z' \in \mathbb{R}^{512 \times 4 \times 4} \rightarrow \text{Decoder} \rightarrow \hat{x} \tag{3}$$

Table 1. Hyperparameters used during MedGAN training

Hyperparameter	Value
Batch size	16
Number of epochs (autoencoder + GAN)	500 + 500
Optimization	Adam ($\beta_1 = 0.5$, $\beta_2 = 0.999$)
Learning rate	2×10^{-4}
Vector dimension z	100
Generator architecture	MLP + pre-trained decoder
Discriminator architecture	5 blocks Conv + BatchNorm + LeakyReLU
Loss functions	Adversarial, L_1, perceptual, feature matching
Final activation function (generator)	`tanh`
Weight initialization	Normal $\mathcal{N}(0, 0.02)$

A deep convolutional discriminator is used, consisting of 5 `Conv-BatchNorm-LeakyReLU` blocks, which produces a single scalar output per image, unlike PatchGAN-type discriminators, the original MedGAN approach evaluates the authenticity of the entire image at a global level, allowing for greater morphological consistency [4].

During adversarial training, multiple loss functions were used, as proposed in MedGAN:

- **Adversarial loss** with `BCEWithLogitsLoss`
- **Reconstruction loss** type L_1 (between real and generated image)
- **Perceptual loss** based on differences in internal activations of a pre-trained network (VGG)
- **Feature matching loss** measures the distance between internal activations of the discriminator for true and false.

These combined losses help improve both the visual fidelity, the diversity and the perceptual quality of the generated images, which are key aspects in medical applications.

4.4 Evaluation

To quantify the quality and realism of the synthetic images generated by the MedGAN model, three metrics recognized in the scientific literature on image generation were used: FID, SSIM, and PSNR. These metrics offer a complementary perspective on different aspects of the image, from statistical fidelity to structural coherence and pixel-level similarity. Eight real images and their corresponding generated images were randomly selected every 20 epochs. These samples were used to calculate the aforementioned metrics and were also stored for quantitative visual evaluation, each metric allows for the determination and differentiation of a different aspect of the efficiency of the generated images:

1. FID (Frechet Inception Distance) [15]: Evaluate the difference between statistical distributions of deep features extracted with a pre-trained network (*inception-v3*) between real and synthetic images; a low FID indicates greater perceptual and semantic similarity:

$$\text{FID} = \|\mu_r - \mu_g\|^2 + \text{Tr}\left(\Sigma_r + \Sigma_g - 2\left(\Sigma_r \Sigma_g\right)^{\frac{1}{2}}\right) \tag{4}$$

   FID between 50–100 means that it detects learnable patterns but of low quality, between 10–30 are very good, it could be interpreted that the images meet a certain level of realism, and greater than 10 are excellent, almost indistinguishable from the real ones.
2. SSIM (Structural Similarity Index) [16]: It compares two images in terms of structure, luminance, and contrast, modeling human perception, evaluates how similar a generated image is to a real one, and is computed as:

$$\text{SSIM}(x, y) = \frac{(2\mu_x\mu_y + C_1)(2\sigma_{xy} + C_2)}{(\mu_x^2 + \mu_y^2 + C_1)(\sigma_x^2 + \sigma_y^2 + C_2)} \tag{5}$$

   SSIM is calculated between 0–1 dB. Values below 0.8 indicate distortion or loss of details relevant to the diagnosis, values between 0.85 and 0.95 indicate that the essential structures of the image are well preserved, and values above 0.95 are considered excellent, indicating a very high similarity.
3. PSNR (Peak Signal-to-Noise Ratio) [17]: It measures the pixel-by-pixel fidelity between a generated image and its corresponding real image, it is a metric derived from the mean square error (MSE), expressed in decibels (dB). The higher the value, the lower the noise or distortion introduced. It is calculated using:

$$\text{PSNR} = 10 \cdot \log_{10}\left(\frac{\text{MAX}_I^2}{\text{MSE}}\right) \tag{6}$$

   A PSNR greater than 20 dB implies low visual distortion or a lot of noise, between 20–30 it remains low with visible degradation, above 30 it is considered very good, and above 40 it is excellent, almost identical.

5 Results

5.1 Training Progress

The autoencoder training was performed for 500 epochs, reaching a minimum reconstruction loss of 0.033 (see Fig. 1) and a maximum SSIM value of 0.9068, this result suggests that the model is able to reconstruct with an acceptable level of structural fidelity, although without reaching the threshold of excellence required in high-precision medical tasks, there is no evidence of overfitting or collapse, as the metrics stabilize without deterioration.

The analysis of the losses shows an unbalanced behavior: the generator loss ($\mathcal{L}_G$) increases progressively until reaching values above 16, while the discriminator loss ($\mathcal{L}_D$) decreases until stabilizing around 0.16, this pattern suggests

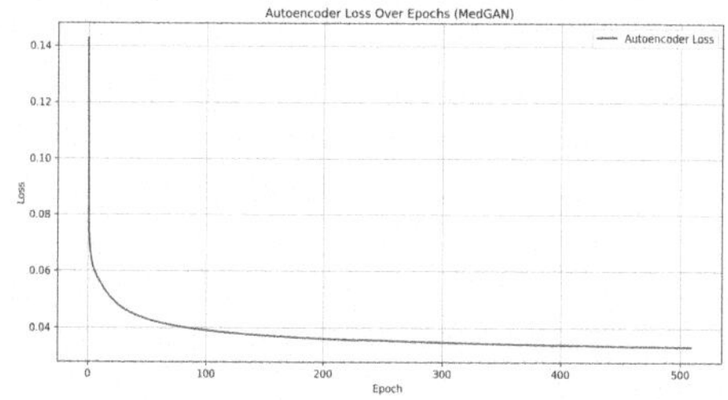

Fig. 1. The behavior of the Autoencoder's loss function progressively decreases during each epoch, as it should.

that the discriminator dominates the generator, limiting its ability to improve and potentially causing a partial collapse of the training, evidenced by the lack of progress in the metrics.

As for adversarial training, it reveals an unbalanced dynamic in adversarial training, with the generator loss ($\mathcal{L}_G$) showing a sustained increase throughout the epochs, reaching values above 16, while the discriminator loss ($\mathcal{L}_D$) decreases rapidly and stabilizes around 0.16, this behavior indicates that the discriminator has gained a significant advantage over the generator, which is often indicative of a partial collapse of adversarial training (Table 2).

Table 2. Generator and discriminator loss functions.

(a) Extremely weak generator.

(b) Very strong discriminator, difficult to fool.

When the discriminator becomes overly competent at distinguishing real images from synthetic ones, the gradient that backpropagates to the generator tends to vanish (a problem known as vanishing gradients), instead of providing useful signals to guide the generator to improve, the discriminator simply classifies the generated images as fake with high confidence, preventing the generator from learning to synthesize more realistic content, this is directly reflected in the metrics (see Table 3): the FID remains high (>180), the SSIM below 0.4, and the PSNR around 13 dB, which shows that the generated images fail to approximate the real ones either visually or statistically.

Furthermore, the constant increase in $\mathcal{L}_G$ indicates that the generator is not only failing to learn how to fool the discriminator, but its performance is also worsening over time (see Table 4), this trend may be due to a combination of factors: lack of an effective adversarial signal, insufficient capacity of the generator to model the distribution of real data, or too weak a reconstruction penalty in the total loss function, the fact that $\mathcal{L}_D$ stabilizes at a very low value also suggests that training has reached a point where the discriminator is no longer challenged, halting meaningful learning in both networks.

Table 3. SSIM and PSNR performance metrics.

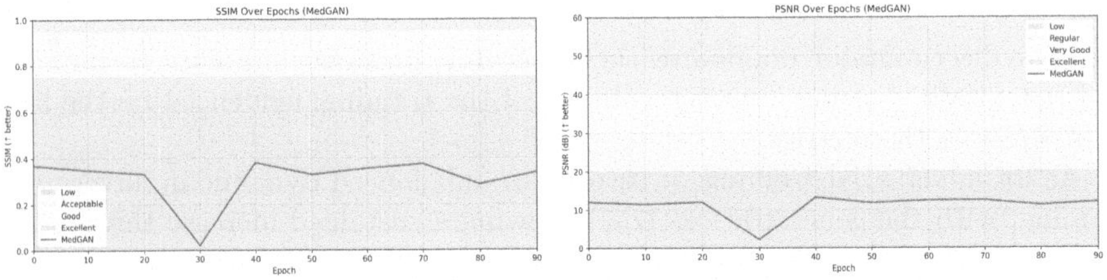

(a) SSIM below 0.4 reflecting low perceptual quality.

(b) Maximum PSNR of 13.19 dB.

Table 4. Results of images generated at different stages as GAN training progresses

epoch 1	epoch 20	epoch 40	epoch 60	epoch 80	epoch 100
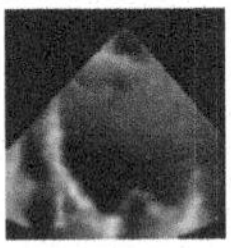	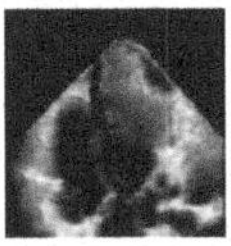		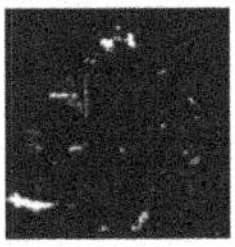	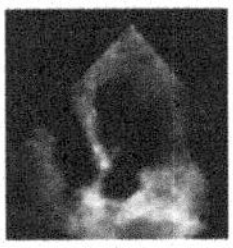	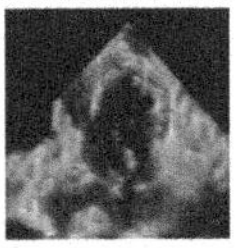

6 Conclusions and Future Work

The architecture implemented, based on MedGAN with staged training, managed to generate echocardiographic images with a coherent basic structure thanks to the autoencoder, although with limitations in visual quality, deficiencies in perceptual quality, and fidelity, after 120 epochs of adversarial training, the metrics obtained (FID $\approx$ 180, SSIM $\approx$ 0.37, PSNR $\approx$ 13.1 dB) indicate that the images still show notable differences from the real ones.

The autoencoder training proved to be effective in capturing the global topology of the images, allowing for a structurally coherent reconstruction, however, the adversarial stage in the loss analysis reveals that the discriminator dominates the generator, which limits effective learning and could be causing stagnation or partial collapse, despite these challenges, the preliminary results are promising in terms of the viability of the approach, to achieve a more complete refinement in the visual and perceptual quality of the images, it will be necessary to significantly extend the training, ideally to 2000 epochs or more, allowing the generator to gradually adapt and close the gap with the discriminator.

Overall, the model shows potential for generating synthetic medical images, although it still requires gradual improvements and more extensive training to reach clinically useful levels.

Future work will focus on evaluating other GAN-type data augmentation architectures.

Acknowledgment. This work was funded by SUE alliance project "Estimation of cardiac work as an index of cardiovascular function in videos echocardiography" with code EXT_ING4073 from Universidad Militar Nueva Granada.

References

1. Frid-Adar, M., Klang, E., Amitai, M., Goldberger, J., Greenspan, H.: Synthetic data augmentation using GAN for improved liver lesion classification. In: 2018 IEEE 15th International Symposium on Biomedical Imaging (ISBI 2018), pp. 289–293 (2018)
2. Salehinejad, H., Colak, E., Dowdell, T., Barfett, J., Valaee, S.: Synthesizing chest X-Ray pathology for training deep convolutional neural networks. IEEE Trans. Med. Imaging **38**(5), 1197–1206 (2019)
3. Ronneberger, O., Fischer, P., Brox, T.: U-Net: convolutional networks for biomedical image segmentation. Arxiv (2015)
4. Choi, E., Biswal, S., Malin, B., Duke, J., Stewart, W.F., Sun, J.: Generating multi-label discrete patient records using generative adversarial networks. In: Doshi-Velez, F., Fackler, J., Kale, D., Ranganath, R., Wallace, B., Wiens, J. (eds.) Proceedings of the 2nd Machine Learning for Healthcare Conference. Proceedings of Machine Learning Research, vol. 68, pp. 286–305. PMLR (2017)
5. Goodfellow, I., et al.: Generative adversarial networks. Commun. ACM **63**(11), 139–144 (2020)
6. Isola, P., Zhu, J.-Y., Zhou, T., Efros, A.A.: Image-to-image translation with conditional adversarial networks. In: 2017 IEEE Conference on Computer Vision and Pattern Recognition (CVPR), pp. 5967–5976 (2017)
7. Karras, T., Laine, S., Aila, T.: A style-based generator architecture for generative adversarial networks. In: CVPR (2019)
8. Frid-Adar, M., Diamant, I., Klang, E., Amitai, M., Goldberger, J., Greenspan, H.: Gan-based synthetic medical image augmentation for increased CNN performance in liver lesion classification. Neurocomputing **321**, 321–331 (2018)
9. Wolterink, J.M., Dinkla, A.M., Savenije, M.H.F., Seevinck, P.R. van den Berg, C.A.T., Išgum, I.: Deep MR to CT synthesis using unpaired data. In: Tsaftaris, S.A., Gooya, A., Frangi, A.F., Prince, J.L. (eds.) Simulation and Synthesis in Medical Imaging. SASHIMI 2017. LNCS, vol. 10557, pp. 14–23. Springer, Cham (2017)

10. Bernard, O., et al.: Deep learning techniques for automatic MRI cardiac multi-structures segmentation and diagnosis: is the problem solved? IEEE Trans. Med. Imaging **37**(11), 2514–2525 (2018)
11. Zhang, X., Xu, C., Yosef, N.: Simulating multiple faceted variability in single cell RNA sequencing. Nat. Commun. **10**, 2611 (2019)
12. Yi, X., Walia, E., Babyn, P.: Generative adversarial network in medical imaging: a review. Med. Image Anal. **58**, 101552 (2019)
13. Arjovsky, M., Chintala, S., Bottou, L.: Wasserstein GAN. In: ICML (2017)
14. Khoshgoftaar, T.M., Shorten, C.: A survey on image data augmentation for deep learning. J. Big Data **06** (2019)
15. Heusel, M., Ramsauer, H., Unterthiner, T., Nessler, B., Hochreiter, S.: GANs trained by a two time-scale update rule converge to a local Nash equilibrium. In: Proceedings of the 31st International Conference on Neural Information Processing Systems, NIPS 2017, Red Hook, NY, USA, pp. 6629–6640. Curran Associates Inc. (2017)
16. Wang, Z., Bovik, A.C., Sheikh, H.R., Simoncelli, E.P.: Image quality assessment: from error visibility to structural similarity. IEEE Trans. Image Process. **13**(4), 600–612 (2004)
17. Horé, A., Ziou, D.: Image quality metrics: PSNR vs. SSIM. In: 2010 20th International Conference on Pattern Recognition, pp. 2366–2369 (2010)

Gender Biases and Vocational Profiles in Systems Engineering

Emilcy Hernández-Leal(✉), Gloria Gasca-Hurtado, and Daniela Higuita-Agudelo

Universidad de Medellín, Carrera 87 #30-65, Medellín, Colombia
ejhernandez@udemedellin.edu.co

Abstract. Choosing an academic program is an important decision in anyone's professional life, but it can be influenced by gender stereotypes that may limit women's aspirations in STEM fields, such as engineering. This study aimed to identify the factors that influence the selection of the Systems Engineering program by students at the Universidad de Medellín. A total of 202 students (76.2% men, 22.3% women) were surveyed to analyze perceptions, cultural factors, family influences, and academic background. The results showed the persistence of significant gender differences. Men were found to be more familiar with digital environments at an early age, as reflected in their frequent use of video games (79% men vs. 35% women) and electronic toys. In addition, a traditional distribution of gender roles in the home was identified, which reinforces stereotypes from an early age and affects self-perception of competence. In conclusion, the gender gap in engineering is multifactorial. Addressing it involves challenging gender stereotypes, strengthening women's self-efficacy and self-confidence, and raising awareness among families and teachers to promote more equitable STEM trajectories.

Keywords: Gender gap · Vocational stereotypes · Systems engineering · Academic selection

1 Introduction

Choosing an academic program in higher education is an important decision in young people's lives, as it largely defines their professional and personal trajectory [1,7]. In this process, the vocational profile plays a central role, as it guides the applicant according to their interests, skills, and expectations [20]. However, these elements are shaped by multiple environmental factors, including gender [21]. Various studies have shown that gender stereotypes can influence how individuals perceive their own competencies, especially in areas traditionally dominated by men, such as engineering and computer science [15,16].

This influence can generate unconscious biases that limit women's aspirations in STEM fields (science, technology, engineering, and mathematics), even when they possess the necessary skills to perform successfully in them [6]. The

W. J. Sarmiento and E. L. Sierra-Ballén (Eds.): CCC 2025, CCIS 2692, pp. 269–278, 2027.
https://doi.org/10.1007/978-3-032-28447-1_21

perception of self-efficacy, for example, can be distorted by cultural, social, and educational factors that reinforce the idea that certain disciplines are more suitable for one gender than another [14,19]. These social constructs affect not only the decision to enter a specific program, but also retention and performance within it, contributing to the persistent gender gap in these areas [17,18].

In Colombia, Systems Engineering is one of the most representative programs within the STEM disciplines and, at the same time, one of those that most clearly reflects the gender gap [5]. According to recent university enrollment data, women's participation in this field continues to be significantly lower than that of men, limiting the diversity of perspectives in a sector that is strategic for the country's technological development and competitiveness. This situation not only highlights a loss of potential talent, but also poses a challenge to equity in access to and retention in programs with high labor demand. Addressing the relationship between program choice and the gender gap in this context is therefore essential to understanding the dynamics that influence the development of engineering careers and to proposing institutional strategies that promote inclusion and equity in the training of professionals in Colombia.

In this context, the present study aims to identify and analyze the factors that influence the selection of the Systems Engineering program by students at the Univeridad de Medellín. To this end, a survey-type instrument is applied to collect information on individual perceptions, cultural factors, family influences, and academic background. Based on this data, a quantitative analysis is performed to examine the role that socioeconomic and gender variables play in young people's vocational decisions, thereby contributing to a better understanding of the mechanisms that perpetuate gender inequality in higher education in STEM fields.

The rest of the article is structured as follows: Sect. 2 presents the state of the art, followed by the motivation for this work; Sect. 3 describes the methodological process through the materials and methods; Sect. 4 introduces the results; and Sect. 5 concludes with the conclusions and future work.

2 State of the Art

The low representation of women in STEM fields is a persistent problem worldwide, especially in disciplines such as engineering and technology, where women's participation is significantly lower [21]. In Mexico, for example, only 8% of people who complete a STEM degree are women, and in computer science, the presence of women barely reaches 15% in university and even lower percentages in vocational training [14]. This gender segregation in higher education and career choice has significant implications for economic and social development, depriving the system of the added value that female talent could bring [15].

Historically, cultural and social stereotypes have associated these careers with masculine skills, discouraging girls from an early age [21]. Phenomena such as "Stereotype Threat," where the fear of confirming a negative stereotype affects performance, and the "Leaky Pipeline," which describes the loss of women throughout their educational and professional careers in STEM, contribute to this disparity [18]. These sexist beliefs and attitudes can be unconsciously transmitted in educational processes, influencing the self-perception of competencies [15]. In fact, boys and girls begin to perceive that boys are more outstanding in intellectual areas from an early age [18].

Self-perception of competence, self-efficacy, and self-confidence play a crucial role in students' vocational decisions [18]. Women tend to underestimate their abilities in mathematics and computer science, even when they perform better academically, while men tend to overestimate their abilities in these subjects [15]. This underestimation often leads women to dismiss participation in these types of studies and professions, assuming that they do not meet the standards, despite having the necessary potential [15]. On the contrary, a growth mindset, which conceives that intellectual skills can be developed through effort, dedication, and motivation, is essential to overcoming these barriers [15].

The immediate environment, including family, peers, and, significantly, teachers, has a decisive influence on the formation of students' perceptions of their abilities [18]. Research shows that teachers' expectations of their students' performance can have a major impact, and that teachers sometimes have lower expectations for girls in mathematics [10]. Although teachers are a source of vocational guidance, students' decisions are more aligned with their personal motivations and interests, and formal recommendations from teachers have little impact on their decisions [21]. Furthermore, the predominance of male teachers in engineering and computer science fields reinforces the masculinization of the perception of these types of careers [19].

Career guidance is an essential tool for providing students with the knowledge to identify their interests and skills, aligning them with available career opportunities and contributing to the reduction of inequalities [7]. However, despite efforts to integrate gender-inclusive practices, stereotypes continue to affect the choice of technical specialties and STEM areas, with segregation between men and women [14,21]. To address these shortcomings, courses have been designed for teachers, seeking to equip them with knowledge about vocational guidance and gender and then develop methodological tools to implement inclusive practices [3,21]. Despite the significant improvement in teachers' skills after these courses, challenges remain in identifying gender stereotypes and biases in vocational guidance toward STEM careers [21].

In the specific case of training programs such as systems engineering, computer engineering, or computer science, these are socially perceived as male-oriented fields [9]. Female students often face stereotypes that suggest they take longer to understand programming or software development topics and are not as skilled [19]. Data reveals that men have greater self-efficacy in computer science and show higher expectations toward activities such as creating video games or repairing computers [14,18]. This perception of women as less competent, even in the professional sphere, is compounded by the lack of role models and support networks, perpetuating inequality in these fields [2].

The persistence of disparities in the choice of STEM careers in countries such as Ecuador, influenced by cultural and social stereotypes, is a reality [21]. These circumstances limit women's aspirations and career options. Paradoxically, it has been observed that countries with lower gender equality indices have a relatively higher proportion of female graduates in STEM disciplines than those with higher indices, suggesting that vocational decisions may be influenced by a strong interaction of individual and contextual factors, beyond equal opportunities [18]. This educational segregation extends into professional life, with women in STEM facing additional challenges in accessing funding or entrepreneurship, often due to discrimination and a lack of extensive networks [13].

In conclusion, the gender gap in engineering and computer science programs is a phenomenon associated with multiple factors that requires comprehensive attention [4]. Addressing this issue involves not only encouraging girls and adolescents to take an interest in STEM, but also challenging gender stereotypes from an early age, strengthening women's self-efficacy and self-confidence in these disciplines, and raising awareness among families and teachers about their role in mitigating biases. The implementation of mentoring programs, awareness workshops for the entire educational community, and the promotion of discrimination-free educational environments are crucial strategies for facilitating more equitable academic and professional trajectories in STEM fields [8,12]. Likewise, research in this field is vital to understanding the evolution of these factors and designing more effective interventions that allow all people to explore their interests and abilities without gender limitations.

3 Materials and Methods

This study is based on a quantitative approach and a non-experimental design. The scope of the research is descriptive, as it seeks to characterize the perceptions, academic conditions, and sociodemographic characteristics of Systems Engineering students in relation to their choice of program. This methodological approach allows us to gain insight into the factors that influence vocational decisions without manipulating the study variables, thus ensuring the validity of the results in the particular context of the participants.

For the development of this study, a survey-type data collection instrument was designed, structured based on a review of literature related to factors that influence the choice of programs in STEM areas, with special emphasis on the gender dimension. This review allowed us to identify recurring variables associated with cultural perceptions, academic conditions, and sociodemographic characteristics that influence vocational decisions. In addition, the particular context of the Universidad de Medellín was considered in order to adapt the instrument to the conditions and experiences of students in the Systems Engineering program.

The questionnaire was organized into three main dimensions: (1) demographic dimension, consisting of six questions about individual characteristics such as age, gender, and parents' occupation; (2) cultural beliefs, consisting of eight questions aimed at understanding family perceptions, gender stereotypes, and parental support; and (3) academic, with eight questions focused on school performance, preferences for areas of knowledge, and professional expectations. The sample consisted of 210 students, representing 45.7% of the total enrollment in the program in the first semester of 2024, ensuring a representative sample. All participants signed an informed consent form, and the data were treated with strict confidentiality and anonymity.

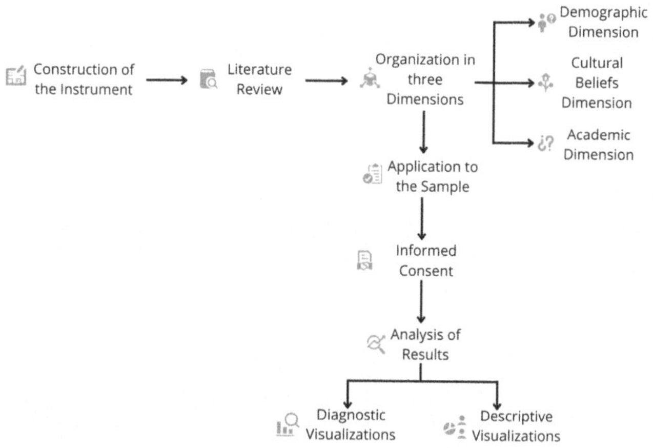

Fig. 1. Methodology used.

El proceso metodológico seguido se resume en la Fig. 1, que ilustra el diseño y ejecución del estudio desde la construcción del instrumento hasta el análisis de resultados. Tras la aplicación de la encuesta, se realizó una fase de análisis que incluyó tanto visualizaciones diagnósticas como descriptivas, permitiendo identificar patrones en las respuestas, diferencias por género, y relaciones entre variables demográficas, culturales y académicas. Esta combinación de análisis cualitativo en la fase de diseño y cuantitativo en la interpretación de los resultados, ofrece una visión de los factores que influyen en la selección de programas de ingeniería, particularmente en un contexto de brecha de género persistente.

4 Results

Of the total number of individuals initially surveyed, 202 students completed the data collection instrument after accepting informed consent, while 8 people abandoned the process before completing it. In terms of gender, the sample consisted of 154 men (76.2%), 45 women (22.3%), and 3 people (1.5%) who chose not to disclose their gender identity. The ages of the participants ranged from 17 to 32 years, reflecting a predominantly young population of undergraduate students in Systems Engineering. This distribution is consistent with the typical demographic characteristics of engineering programs in Colombia [11].

One of the key aspects within the demographic component was the analysis of socioeconomic status, a relevant indicator for contextualizing the respondents' access to and educational conditions. The results show that 76% of participants belong to strata 1, 2, and 3, considered low and middle income in the Colombian economic system, while 20% are in strata 4, 5, and 6 (see Fig. 2). A small remaining percentage preferred not to answer this question. This distribution suggests that most students come from vulnerable or emerging socioeconomic backgrounds, which could influence the type of technological and educational resources available during their previous education.

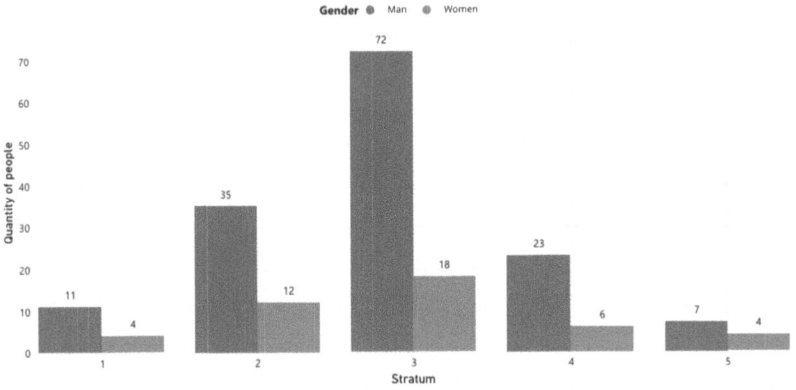

Fig. 2. Relationship between gender and socioeconomic status.

With regard to childhood games, a significant difference was observed in the preference for video games according to gender: 79% of men reported having played video games frequently, in contrast to only 35% of women (see Fig. 3). This finding reinforces the idea of greater early familiarity among men with digital and interactive environments, possibly related to cultural factors, family consumption, and access to technologies. On the other hand, learning or cognitive development games showed a more equitable distribution, with 42% for both men and women, which could indicate a common interest in playful experiences oriented toward intellectual development, without a marked gender distinction.

In addition, men were found to interact more with electronic toys, toy cars, and game consoles, while these objects were less present in women's childhoods (see Fig. 4). This difference reflects not only individual preferences but also possible social and family conditioning that assigns different roles from childhood, reinforcing gender stereotypes associated with the type of play. Such practices could subsequently influence the self-perception of technical or technological skills and, therefore, the choice of programs such as Systems Engineering.

In the academic sphere, the data reveal that both men and women expressed interest in mathematics, with this subject being slightly more preferred by women, which is interesting given the stereotypes that associate mathematics

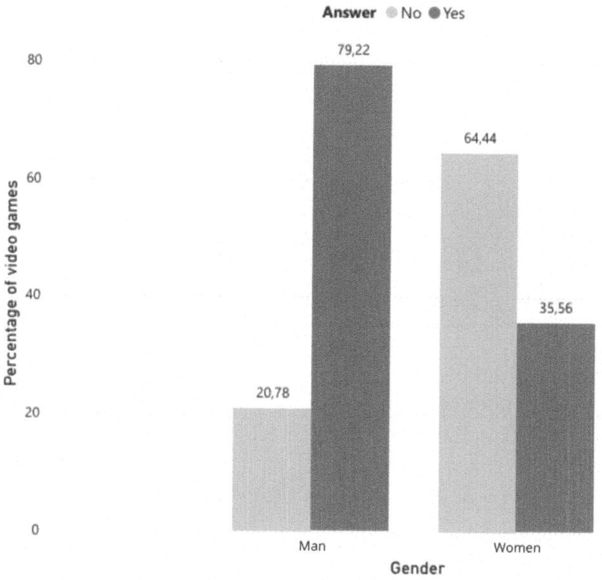

Fig. 3. Relationship between gender and video game use in childhood and adolescence.

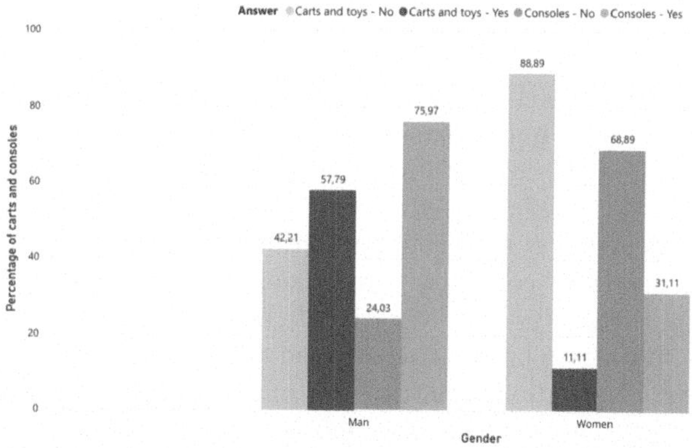

Fig. 4. Relationship between gender and the use of carts and consoles in childhood and adolescence.

with men (see Fig. 5). However, the subject of technology and computer science was more highly valued by men, who showed greater affinity for these fields. This difference may be related to the type of previous experiences, the presence of role models, or the way in which these subjects were promoted or taught in their primary and secondary educational institutions. Taken together, these findings suggest that areas of academic interest continue to exhibit gender patterns, which may directly impact vocational decision-making.

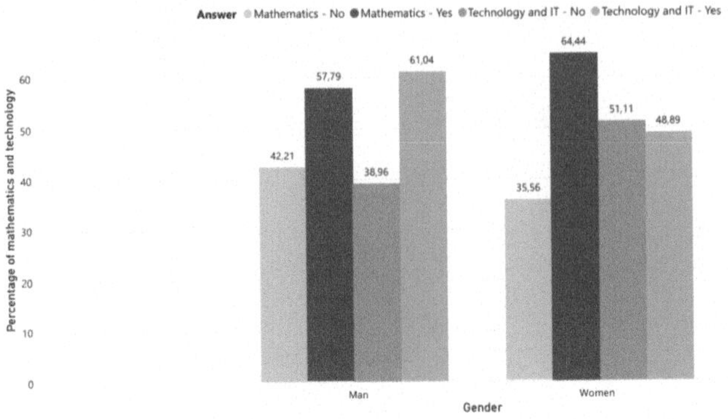

Fig. 5. Relationship between gender and preference in mathematics, technology, and computer science.

5 Conclusions

The results of the study show that significant differences associated with the gender gap persist, influenced by cultural factors and stereotypes that affect vocational guidance and, therefore, the selection of academic programs such as Systems Engineering. This finding is consistent with previous studies that identify early cultural and educational conditioning as determinants of women's lower participation in STEM fields [15,18]. These differences are not only evident at the university decision stage, but originate at an early age, particularly in primary education, where ideas about suitable careers for men and women begin to take shape. The influence of the family environment also plays a decisive role, reinforcing patterns that condition students' aspirations. Similar evidence has been reported in Latin American contexts, where family expectations and the persistence of gender roles strongly shape vocational choices [14,21].

One of the relevant findings relates to the characterization of respondents' households, where a marked traditional distribution of gender roles is observed. Specifically, it was identified that, in most cases, mothers are linked to household activities, caregiving, or administrative jobs, while fathers have occupations in sectors such as engineering, commerce, security, and transportation. This result aligns with the evidence presented by Zambrano et al. [21] and Sainz [15], who highlight that traditional role assignments within the household reinforce stereotypes that limit female self-efficacy in science and technology. This family configuration could be unconsciously reinforcing the perception that certain professions are more aligned with the abilities or responsibilities of a specific gender, thus perpetuating stereotypes that limit female participation in STEM fields.

Based on these results, future work will focus on validating the data collection instrument to ensure its reliability and applicability in similar contexts, using this study as a pilot test of the instrument and subsequent validation by experts. Furthermore, it is proposed to deepen the analysis using data mining and text analysis techniques, which will allow for a more accurate exploration of the narratives and patterns implicit in the students' responses. This methodological expansion will contribute to a clearer identification of the structural and symbolic factors that influence the persistence of the gender gap in engineering programs and, thus, to the generation of more specific recommendations for educational policies with a focus on equity. In this sense, the conclusions reaffirm what has been observed in other contexts [18,19], while providing new evidence for the Colombian case, which underscores the need for institutional strategies that promote inclusive vocational guidance and challenge gender stereotypes from the early stages of education.

Acknowledgments. To Call for Proposals 52 Seed Fund for New Researchers, University of Medellín, which supports the development of the research project entitled "Design of a gamification-based strategy to promote women's participation in science and technology.

References

1. Araya-Pizarro, S.C., Araya-Pizarro, C.R., Rojas-Cabezas, G.M.: Atributos valorados por estudiantes adolescentes en la elección de una institución de educación superior en chile. Formación Universitaria **16**(6), 71–80 (2023)
2. Bermello, D.E.C., Calero, D.C.A., Aguilera, A.G.F., León, C.L.B., Vargas, L.M.R., Centeno, M.V.C.: La equidad de género en la educación stem: estrategias para aumentar la participación femenina. South Florida J. Develop. **5**(10), e4528–e4528 (2024)
3. Briones-Bermeo, M.I., Pinargote-Macías, E.I., et al.: Estrategia metodológica para favorecer la preparación en educación inclusiva del docente de bachillerato. MQRInvestigar **8**(1), 5802–5822 (2024)
4. Card, D., Payne, A.A.: High school choices and the gender gap in stem. Econ. Inq. **59**(1), 9–28 (2021)
5. Barros, C., V.: La construcción del problema público de la brecha de género en ciencia, tecnología, ingeniería y matemáticas-stem en las carreras universitarias colombianas (2021)
6. Dionicio, R.L.C., Luján, R.J.M., Martínez, M.D.C.E.A., Tarazona, M.R.P.: Actitudes hacia los roles de género en la elección de carreras stem: Evaluación del modelo socio cognitivo. Revista de Ciencias Sociales **30**(4), 401–415 (2024)
7. Erazo Guerra, X.F., Rosero Morales, E.d.R.: Orientación vocacional y su influencia en la deserción universitaria. Horizontes Revista de Investigación en Ciencias de la Educación **5**(18), 591–606 (2021)
8. Frigo, L.B., et al.: Mapping women stem initiatives in Latin American countries: Bolivia, Brazil and Peru. In: International Conference on Information Technology & Systems, pp. 401–409. Springer (2024)
9. Guerrero, L.E.M., Ayala, J.A.N., García, C.F.Z.: Estudio comparativo de las competencias de un ingeniero de sistemas y computación entre Colombia Y Latinoamérica. Acad. y Virtualidad **17**(1), 87–98 (2024)
10. Herrera, H., C.A.: Las mujeres stem y sus apreciaciones sobre su transitar por la Carrera universitaria. Nova Sci. **13**(27) (2021)

11. Hernández-Leal, E., Gasca-Hurtado, G., Higuita-Agudelo, D.: Characterization of the female student population in engineering programs in Colombia: registration, admission, enrollment, and graduation (2023)
12. de Paiva Abreu, A.R., Costa, L.F.C., de Lima, Y.O., Nascimento, L.M.A., Barbosa, C.E.: Bridging the gender gap in computing: insights from Latin America with a focus on Brazil. Comput. Sci. Eng. **26** (2024)
13. Piva, E., Rovelli, P.: Mind the gender gap: the impact of university education on the entrepreneurial entry of female and male stem graduates. Small Bus. Econ. **59**(1), 143–161 (2022)
14. Rivera, S.Q., Monroy, V.Y., Rodríguez, L.Á.: Género y steam: Estudio comparativo de las expectativas de niñas y adolescentes en educación básica y media superior. Revista Latinoamericana de Estudios Educativos (México) **55**(2), 1–30 (2025)
15. Sainz, M.: Estereotipos y sesgos de género en la evaluación de las propias competencias científico-tecnológicas. Panorama Soc. **2024**, 39 (2024)
16. Vega, S., Ruiz Alfonso, L.E., Feliciano García, Z., L.: Estereotipos de género y variables vocacionales en alumnas de educación secundaria, Revista de educación (2023)
17. Tandrayen-Ragoobur, V., Gokulsing, D.: Gender gap in stem education and career choices: what matters? J. Appl. Res. Higher Educ. **14**(3), 1021–1040 (2022)
18. Verdugo-Castro, S., García-Holgado, A., Sánchez-Gómez, M.C.: The gender gap in higher stem studies: a systematic literature review. Heliyon **8**(8) (2022)
19. Villaseñor Leal, T.D.J., Macías González, G.G.: Trayectorias de mujeres estudiantes en steam: elección de carrera, percepciones y expectativas profesionales. RIDE Revista Iberoamericana para la Investigación y el. Desarrollo Educativo **15**(30), (2025)
20. Vuyk, A., Montanía, M., Barrios, L., Yubero, A.: Evaluación de un campamento vocacional steam: impacto en la orientación vocacional de estudiantes con altas capacidades en matemáticas. Orientación y Sociedad **24** (2024)
21. Zambrano, D.V., Rodríguez, H.M., Rodríguez, E.: Formación en prácticas inclusivas de género para la orientación vocacional hacia perfiles steam (ciencia, tecnología, ingeniería, arte y matemáticas). Información tecnológica **36**(1), 15–26 (2025)

Master's and Doctoral Symposium

Uncertainty-Informed Framework for Enhancing Decision-Making in Emergency Response Management

Cristian Pulido(✉), Francisco Gómez, and Mario Arrieta-Prieto

Universidad Nacional de Colombia, Bogotá, Colombia
{cpulido,fagomezj,mearrietap}@unal.edu.co

Abstract. Emergency Response Management (ERM) systems increasingly rely on predictive models to support timely and effective decisions in high-stakes contexts such as crime, epidemics, or infrastructure failures. Most of these models produce single-value estimates of future events, which are often treated as exact despite being based on incomplete and noisy data. This neglect of *predictive uncertainty*, i.e., the possibility that real outcomes deviate from predictions, can result in overconfident decisions and suboptimal allocation of limited resources. This research proposes a unified framework that integrates Conformal Prediction (CP), a model-agnostic technique for generating calibrated prediction intervals, with resource allocation strategies in ERM systems. The approach aims not only to quantify uncertainty in spatio-temporal forecasts, but also to incorporate it into prioritization and planning processes using heuristics and optimization-based methods. Partial results on crime data from Chicago show that CP reveals significant variability in forecast confidence and highlights mismatches between prediction reliability and current allocation practices. The methodology is structured in three work packages: uncertainty-aware forecasting, integration into decision models, and empirical evaluation. Ongoing work explores decision-focused learning and robust optimization techniques to improve adaptability in dynamic environments. This paper presents an "in-progress" doctoral research project submitted to the SMD25 symposium, contributing to the development of transparent, uncertainty-informed decision support systems in ERM.

Keywords: Emergency Response · Predictive Modeling · Uncertainty Quantification · Conformal Prediction · Resource Allocation · Decision Support Systems

1 Introduction

Emergency Response Management (ERM) systems are critical to addressing societal threats such as crime, disease outbreaks, natural disasters, and infrastructure failures [10]. These systems guide decision-makers through a structured

W. J. Sarmiento and E. L. Sierra-Ballén (Eds.): CCC 2025, CCIS 2692, pp. 281–287, 2027.
https://doi.org/10.1007/978-3-032-28447-1_22

set of actions, aimed at minimizing the impact of adverse events [7]. In the past decade, ERM systems have increasingly integrated data-driven predictive models that estimate the likelihood, intensity, and location of future events. These forecasts are often used to prioritize intervention zones, allocate emergency resources, and inform real-time planning [10,14].

The central challenge that this research addresses emerges at the intersection of predictive modeling and decision-making under uncertainty. Although predictive models have shown considerable success in anticipating where events are likely to occur, they are typically used in operational settings as if their outputs were exact [11]. Most forecasting systems generate single numerical estimates for each instance, such as forecasting exactly "12 crimes" for a neighborhood on a given day. That results are known as *point predictions*. Often, these outputs are interpreted deterministically, meaning that downstream decisions are made as though the prediction is guaranteed to occur. However, real-world systems are characterized by variability, limited observations, and complex behavioral dynamics [7]. *Predictive uncertainty* refers to the inherent doubt or variability associated with the output of a predictive model: how far the true outcome could be from what the model predicts [2]. Ignoring this uncertainty can lead to overconfident decisions, misallocated resources, and vulnerability to unexpected outcomes, particularly in high-stakes or time-sensitive environments where the cost of error is high [7,11].

This issue becomes more pronounced when point predictions are directly fed into resource allocation processes. Optimization models such as facility location or coverage maximization, commonly used in ERMs, frequently assume input forecasts are accurate [3,5]. This can result in suboptimal allocations, such as overcommitting to stable regions or failing to respond to emerging risks. The consequences include inefficient interventions and erosion of trust in the decision-support system [7]. Despite advances in uncertainty quantification, such as Bayesian inference, quantile regression, and ensemble methods, most ERM systems still lack mechanisms to estimate, represent, or act upon predictive uncertainty [1,10]. The prevailing assumption of certainty in predictions limits the robustness and adaptability of ERM systems. This thesis proposal, submitted to the SMD25 symposium under the "work in progress" modality, aims to fill this gap by developing and evaluating a framework that incorporates uncertainty into both the forecasting and decision-making stages of ERM.

2 Justification

Addressing this problem is relevant both from a methodological and societal perspective. As ERM systems increasingly rely on predictive models to support critical decisions, the ability to account for uncertainty in these predictions becomes essential for operational effectiveness and public trust. In environments where resources are limited and risks are dynamic, such as urban safety or epidemic control, overconfidence in deterministic predictions can lead to misallocated interventions, reduced preparedness, and systemic inefficiencies. While the limitations of point predictions have been widely recognized, few operational systems integrate formal mechanisms to quantify or act upon predictive uncertainty [11]. Previous studies have explored uncertainty estimation in forecasting using approaches like Bayesian inference [1], quantile regression, and ensemble learning [12]. However, these methods often require complex modeling assumptions and rely on computational frameworks or hardware configurations that are difficult to implement in real-time ERM applications. A potential direction that has recently gained traction is *Conformal Prediction* (CP), a model-agnostic framework that produces prediction intervals with formal coverage guarantees [11]. CP is valued for its simplicity, flexibility, and theoretical soundness, and has been applied in domains such as medical diagnosis and natural language processing [2]. However, its integration into ERM systems remains largely unexplored. This research contributes to filling this gap by evaluating CP in real-world ERM scenarios, characterizing its potential to identify overconfident predictions, reveal local variability, and support more cautious and adaptive decisions. In parallel, the challenge of incorporating uncertainty into resource allocation has been approached from the field of optimization under uncertainty. Robust Optimization [3], Stochastic Optimization [6], and Distributionally Robust Optimization (DRO) [8] offer principled frameworks to design decisions that remain effective across a range of plausible scenarios. However, these approaches typically assume that uncertainty is specified *a priori*, often through hand-crafted uncertainty sets or parametric distributions. An emerging alternative is *decision-focused learning*, which aligns the training objective of predictive models with downstream decision performance [16]. This paradigm trains models not only to be accurate, but to produce forecasts that improve final decision outcomes, especially under uncertainty. Despite promising results in areas like scheduling and resource planning, decision-focused learning remains underutilized in ERM systems.

This thesis builds upon these strands of research by proposing an integrated framework that combines Conformal Prediction with uncertainty-aware decision models, and by empirically validating its applicability in realistic ERM settings (Fig. 1). It seeks to move beyond accuracy-focused forecasting, toward systems that acknowledge the limits of model certainty and adapt decisions accordingly. By doing so, the research aligns with growing demands for transparency, accountability, and robustness in data-driven public decision-making [16].

3 Methodology

This research adopts an iterative and modular strategy to design and evaluate an uncertainty-aware decision-support framework for Emergency Response Management (ERM) systems. The central question guiding this work is:

> *How can predictive uncertainty be quantified and integrated into forecasting and resource allocation processes in Emergency Response Management systems, particularly under spatio-temporal dependencies and operational constraints?*

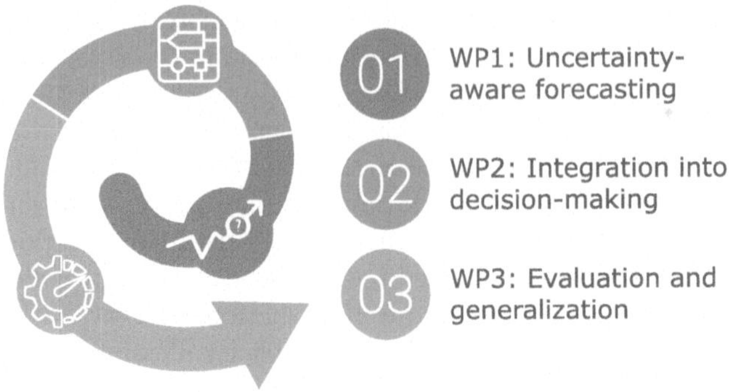

Fig. 1. Modular workflow of the proposed methodology, structured in three interrelated work packages (WPs).

The working hypothesis is that *adapting Conformal Prediction (CP) methods to account for spatial-temporal structure and operational goals will improve the robustness and trustworthiness of decisions in ERM systems.*
The methodology is structured into three interconnected work packages:

- **WP1. Uncertainty-aware forecasting:** Develop CP-based models to generate calibrated prediction intervals over spatio-temporal grids (e.g., crime counts per day and location). The pipeline begins with naïve CP and extends to Adaptive and GeoConformal variants [9].
- **WP2. Integration into decision-making:** Design resource allocation strategies that explicitly use uncertainty estimates such as interval width, local miscoverage, or confidence scores, to support prioritization. Strategies include simple heuristics and robust or scenario-based optimization.
- **WP3. Evaluation and generalization:** Assess both the predictive and decision performance of the framework. Predictive metrics include coverage probability (PICP), interval width, and calibration error. Decision-level metrics cover hotspot recall, allocation efficiency, robustness to variability, and interpretability [4,15].

Experiments will use datasets from urban crime in Chicago, Bogotá, and Buenos Aires, as well as epidemiological risk data from Colombia (e.g., animal mobility and PRRS alerts). This multi-domain design enables testing of the framework's generalizability and operational value. All results will be benchmarked against deterministic baselines that ignore uncertainty.

4 Partial Results

To date, two research fronts have yielded promising results. First, a study on robust prediction under behavioral interventions showed that predictive accuracy is not sufficient to ensure robust decision-making: even models that remain

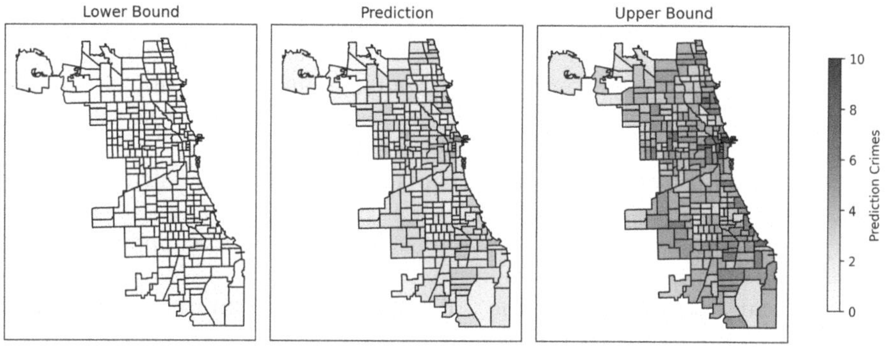

Fig. 2. Example of prediction intervals generated using Conformal Prediction on crime data from Chicago. The maps show the lower bound, point prediction, and upper bound of daily crime counts per cell. Spatial heterogeneity in prediction confidence is clearly visible.

stable under distributional shifts may fail when used within classical optimization strategies like P-median. This points to the need for tighter integration between prediction and allocation stages. Second, a model-agnostic framework was developed to quantify uncertainty in crime forecasting using Conformal Prediction (CP). Applied to crime data from Chicago, the framework generated calibrated prediction intervals and spatially explicit confidence scores. As shown in Fig. 2, CP produces calibrated prediction intervals that reflect spatial heterogeneity in model confidence. Furthermore, the results show that static models tend to be overconfident in high-crime areas and that confidence scores derived from interval width align with areas of operational importance. Preliminary evaluations achieved near-nominal coverage (95%) and revealed substantial spatial heterogeneity in prediction reliability. These results demonstrate the feasibility and value of integrating CP into real-world ERM scenarios. Technical tools and reproducible pipelines have been implemented using the MAPIE library and are publicly available in a dedicated repository. The outputs support the design of the uncertainty-aware strategies proposed in WP1 and WP2, and provide the foundation for future work on decision-focused integration.

Expected Contributions and Related Work

This research aims to contribute both theoretically and operationally to the field of uncertainty-aware decision support in ERM systems. The main expected contributions are as follows:

- **A unified framework** for incorporating predictive uncertainty into ERM decision making. This framework combines model-agnostic uncertainty quantification (via Conformal Prediction) with risk-aware resource allocation strategies. It offers a modular design that can be adapted to diverse domains.

- **Empirical evidence** that predictive uncertainty is not evenly distributed in space or time, and that overconfidence is particularly problematic in high-risk zones. These insights are grounded in real crime data from Chicago and will be extended to other contexts such as epidemiological risk monitoring in Colombia.
- **Operational metrics and visual tools** for interpreting prediction reliability, including spatial confidence scores, miscoverage rates, and their relationship with hotspot alignment. These tools are designed to support decision-makers in prioritization under uncertainty.
- **A critical assessment of the interaction between prediction and decision models**. Results from synthetic and real data show that even robust predictors may fail to produce robust allocations when used naively in optimization pipelines. This underscores the need for integrated approaches like decision-focused learning or uncertainty-aware optimization.

In relation to existing work, previous studies in crime forecasting and predictive policing have focused largely on improving point prediction accuracy [10,14]. Some recent works have recognized the role of uncertainty [7,13], yet formal integration into decision-making remains limited. Conformal Conformal Prediction [2,11] has shown strong performance in diverse fields such as medical diagnosis, finance, and natural language processing (NLP), but its application in spatio-temporal ERM systems is still in early stages. By bridging the gap between uncertainty quantification and operational decision-making, this research complements and extends current work in robust optimization [3], decision-focused learning [16], and risk-sensitive ERM. Ultimately, it aims to enable more adaptive, interpretable, and resilient decisions in public safety, health, and crisis management contexts.

Competing Interests. The authors declare no competing interests related to this work.

References

1. Abdar, M., et al.: A review of uncertainty quantification in deep learning: techniques, applications and challenges. Info. Fus. **76**, 243–297 (2021). https://doi.org/10.1016/j.inffus.2021.05.008
2. Angelopoulos, A.N., Bates, S.: Conformal prediction: a gentle introduction. Found. Trends Mach. Learn. **16**(4), 494–591 (2023). https://doi.org/10.1561/2200000101
3. Bertsimas, D., Thiele, A.: A robust optimization approach to supply chain management. In: Integer Programming and Combinatorial Optimization: 10th International IPCO Conference, New York, NY, USA, June 7-11, 2004. Proceedings 10, pp. 86–100. Springer (2004). https://doi.org/10.1007/978-3-540-25960-2_7
4. Cordier, T., Blot, V., Lacombe, L., Morzadec, T., Capitaine, A., Brunel, N.: Flexible and systematic uncertainty estimation with conformal prediction via the MAPIE library. In: Conformal and Probabilistic Prediction with Applications (2023)

5. D'Angelo, N., Payares, D., Adelfio, G., Mateu, J.: Self-exciting point process modelling of crimes on linear networks. Stat. Model. **24**(2), 139–168 (2024). https://doi.org/10.1177/1471082X221094146
6. Fouskakis, D., Draper, D.: Stochastic optimization: a review. Int. Stat. Rev. **70**(3), 315–349 (2002). https://doi.org/10.1111/j.1751-5823.2002.tb00174.x
7. Hardyns, W., Rummens, A.: Predictive policing as a new tool for law enforcement? Recent developments and challenges. Eur. J. Crim. Policy Res. **24**, 201–218 (2018). https://doi.org/10.1007/s10610-017-9361-2
8. Lin, F., Fang, X., Gao, Z.: Distributionally robust optimization: a review on theory and applications. Num. Algeb. Control Optimiz. **12**(1), 159–212 (2022). https://doi.org/10.3934/naco.2021057
9. Lou, X., Luo, P., Meng, L.: Geoconformal prediction: a model-agnostic framework of measuring the uncertainty of spatial prediction. arXiv preprint arXiv:2412.08661 (2024). https://doi.org/10.48550/arXiv.2412.08661
10. Mukhopadhyay, A., et al.: A review of incident prediction, resource allocation, and dispatch models for emergency management. Accident Anal. Prevent. **163**, 106234 (2021). https://doi.org/10.1016/j.aap.2021.106234
11. Vovk, V., Gammerman, A., Shafer, G.: Algorithmic learning in a random world. Springer Sci. Bus. Media (2005). https://doi.org/10.1007/b106715
12. Wan, S., Sinclair, R.C., Coveney, P.V.: Uncertainty quantification in classical molecular dynamics. Phil. Trans. R. Soc. A **379**(2197), 20200082 (2021). https://doi.org/10.1098/rsta.2020.0082
13. Wang, Z., et al.: Uncertainty-aware crime prediction with spatial temporal multivariate graph neural networks. In: ICASSP 2025-2025 IEEE International Conference on Acoustics, Speech and Signal Processing (ICASSP), pp. 1–5. IEEE (2025). https://doi.org/10.1109/ICASSP49660.2025.10889685
14. Weisburd, D.: Hot spots of crime and place-based prevention. Criminol. Public Policy **17**(1), 5–25 (2018). https://doi.org/10.1111/1745-9133.12350
15. White, V.M., Hunt, J., Green, B.: A discussion of current crime forecasting indices and an improvement to the prediction efficiency index for applications. Secur. J., 1–18 (2023). https://doi.org/10.1057/s41284-023-00367-4
16. Wilder, B., Dilkina, B., Tambe, M.: Melding the data-decisions pipeline: decision-focused learning for combinatorial optimization. In: Proceedings of the AAAI Conference on Artificial Intelligence, vol. 33, pp. 1658–1665 (2019). https://doi.org/10.1609/aaai.v33i01.33011658

Towards an Artificial Intelligence Model for Latency Prediction in Cloud-Based Microservices Architectures

Juan David Ramos-López(✉) and Arles Rodríguez

Universidad Nacional de Colombia, Bogotá, Colombia
{jdramosl,aerodriguezp}@unal.edu.co

Abstract. Microservices architectures offer scalability and flexibility but introduce challenges in maintaining consistent performance due to the complex interdependencies between services. One of the key issues is the uncertainty associated with latency, which can significantly affect the quality of service. This study proposes an artificial intelligence-based approach to predict latency in cloud-based microservices environments using publicly available runtime datasets. The methodology follows the CRISP-DM framework, including data preprocessing, graph construction, and the computation of structural metrics from microservice call graphs. A parallel processing strategy was implemented to merge resource usage data with service interaction logs, resulting in an enriched dataset of over seven million records. Graph-based features were extracted to support future training and evaluation of predictive models. This work-in-progress aims to assess whether the inclusion of graph-derived metrics can enhance latency prediction and thereby assist in informed resource provisioning decisions.

Keywords: Microservices · Latency Prediction · Graph Analysis · Cloud Computing · CRISP-DM · Performance Optimization

1 Introduction

A microservices-based application architecture consists of a collection of small, independent services that can be deployed autonomously [4]. This architectural style has gained widespread adoption in recent years, largely due to the flexibility and efficiency offered by cloud-based platforms, which facilitate resource management and scalability [3].

One of the primary advantages of the microservices architecture is the ability to scale individual services independently. This enables deploying more instances of services with high request volumes while limiting instances of those with lower demand [13]. However, its inherently decentralized nature presents challenges in maintaining acceptable response times when handling a sequence of interdependent service calls triggered by a single user request [14]. Depending on the load

W. J. Sarmiento and E. L. Sierra-Ballén (Eds.): CCC 2025, CCIS 2692, pp. 288–294, 2027.
https://doi.org/10.1007/978-3-032-28447-1_23

of a specific microservice at any given time, the total response time may increase unpredictably, potentially degrading the overall quality of service.

Ensuring high system quality in this context is particularly complex, as it requires balancing multiple quality attributes, such as performance and scalability. For example, prioritizing scalability may negatively affect performance [13]. An increase in the number of microservices typically leads to a higher number of inter-service calls, which can degrade system performance. Conversely, reducing the number of services may improve performance but can limit scalability, as each service becomes responsible for a broader range of requests.

Latency, defined as the time elapsed between a user request and the system's response [12], is a critical metric for quality of service, as it directly influences users' perception of performance. High latency can result in a poor user experience. One way to mitigate latency is by provisioning additional resources to the services with the highest demand; however, this must consider dynamic factors such as time of day and request type. Since overprovisioning leads to increased operational costs, it is essential to reduce resource usage when demand is low. Predictive models that estimate latency in advance can support informed resource allocation decisions by providing actionable insights to system administrators.

In the proposal, the Alibaba Cluster Data [7] is utilized. This dataset comprises runtime metrics from Alibaba production clusters, spanning ten thousand nodes over twelve hours. Some previous works, such as those by [17] and [5], have used the same dataset to predict latency. In [17], the construction of PERT graphs is proposed to capture temporal dependencies and time execution dynamics that a microservice call graph cannot capture. A PERT graph is a directed acyclic graph that represents requests and responses, illustrating parallel and sequential calls. From the dataset, node resource statistics (CPU, Memory), API calls, timestamps, and request/response information are used. However, its implementation poses significant computational challenges, requiring a hardware setup with 8 RTX 3090 GPUs and 1 TB of memory, as well as approximately 10 h of training. These constraints highlight the need for exploring more computationally efficient alternatives. In [5], only topological information derived from the microservice call graph is extracted, and latency prediction is modeled as a classification problem focused on determining whether latency remained constant or fluctuated.

This document presents a proposal for the development of a model aimed at predicting latency in a cloud-based microservices environment using the Alibaba Cluster dataset. In contrast to the works by [5] and [17], this approach combines data from the microservices call graph and resource utilization, matching request times. Additionaly, enriches the dataset with structural information from the microservice call graph at both node-level (via metrics like Betweenness, Closeness, Degree, Eigenvector, PageRank, and community detection) and graph-level (via metrics like diameter, radius, and eccentricity) to predict latency. In this paper, the focus is on outlining the motivation behind the approach and detailing the data preprocessing steps required to support future model implementation and evaluation.

1.1 Research Objective

The primary objective of this research is to build an artificial intelligence-based model aimed at predicting latency in a cloud-based microservices environment dataset. By developing this model, it can optimize resource provisioning, enhance the end-user experience, and leverage system monitoring logs more effectively.

Without tools like this, the opportunity to analyze large volumes of information automatically would be lost. In this context, the research seeks to answer the following question:

> *How can an artificial intelligence-based model be developed to support decision-making by predicting latency based on data from a microservices architecture?*

2 Methodological Strategy

This research follows a quantitative experimental design, structured around the CRISP-DM methodology to predict latency in microservices. The methodological framework is composed of the following phases: Data Understanding, Data Preparation, Modeling, Evaluation.

2.1 Data Understanding

The two datasets used to build the data are available in [7]. Those traces contain the detailed runtime metrics of nearly twenty thousand microservices. They are collected from Alibaba production clusters of over ten thousand *bare-metal nodes* during twelve hours in 2021.

The datasets selected were *MS CallGraph Table* and *MS Resource Table*. The former contains information about microservice call graphs. Due to the large volume of data, a sampling process was applied to the call graphs with a rate of 0.5%. It includes over twenty million call graphs involving more than 20,000 microservices. The second dataset, *MS Resource Table*, contains MS runtime information. It records CPU and memory utilization of 90000+ containers for 1300+ MSs in the same production cluster. The response time of the call (rt) refers to the time elapsed from when the UM microservice sends the request to the DM microservice until the request is completed.

The dataset shows a mostly uniform distribution of the *timestamp* values, with two noticeable data peaks. CPU usage is concentrated between 0% and 40%, while memory usage is higher, typically between 50% and 90%. Response time values are mostly centered around zero, requiring a logarithmic scale to visualize extreme values (Fig. 1). Positive values represent the duration from when UM sends a request to its completion, while negative values represent the duration from when DM receives a request to its response [7].

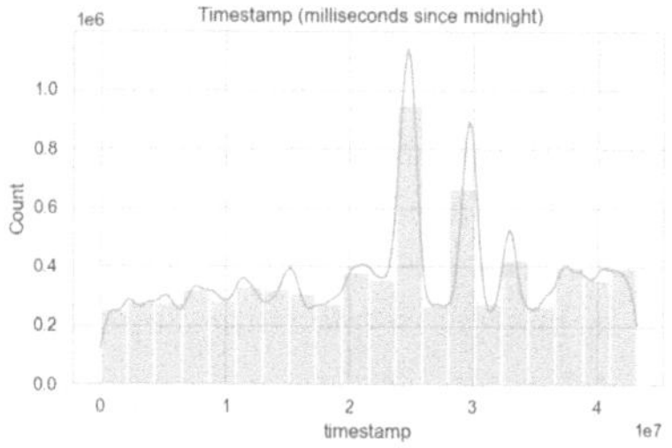

(a) Timestamp

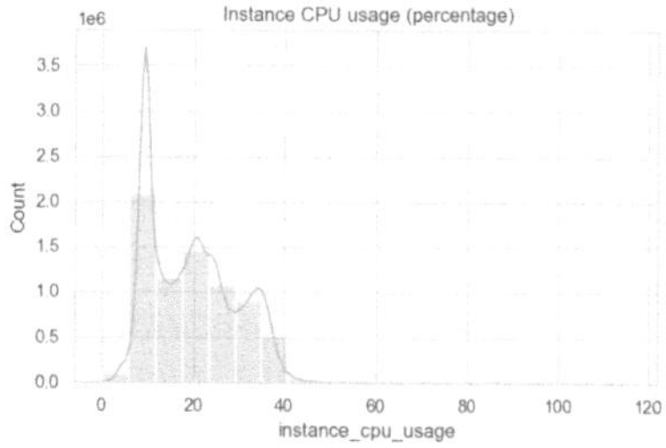

(b) CPU Usage

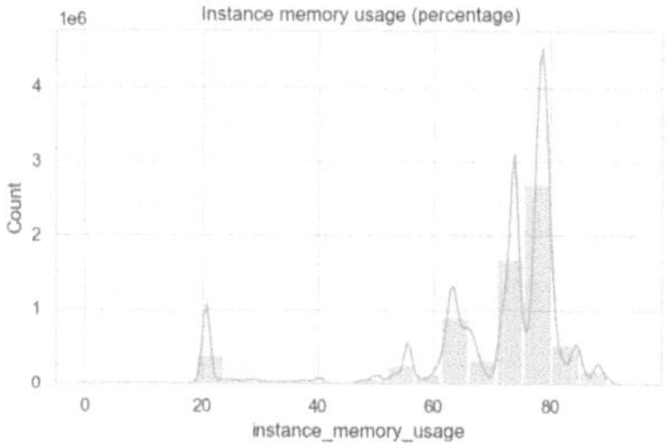

(c) Memory Usage

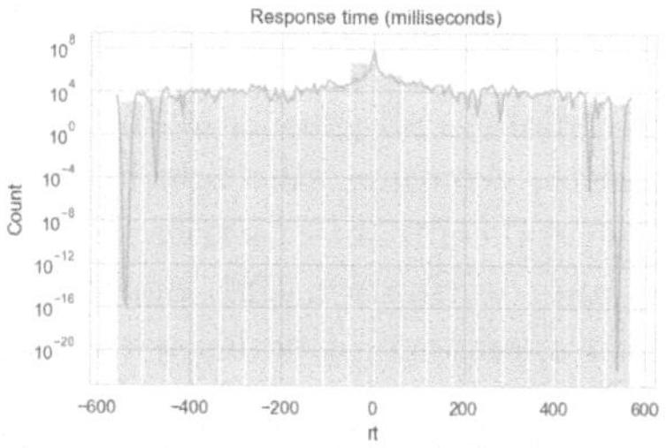

(d) Response Time (log scale)

Fig. 1. Distributions of key variables in the dataset.

2.2 Data Preparation

Since the target variable to be predicted is the response time (`rt`), and this requires information like the CPU and memory usage, it was necessary to combine data from both datasets. The columns `timestamp` (common to both datasets), `msname` (from the *MS Resource Table*), and `um` (from the *MS Call-Graph Table*) were used for the join. Records without matching entries were discarded. As the *MS CallGraph Table* consists of 145 CSV files and the *MS Resource Table* contains 12 files, parallel processing was required. The resulting dataset contained 7,389,113 records.

Parallel Execution. The computer used for processing had an 8-core processor, the workload was divided into 8 batches. Each of the 145 files in the *MS CallGraph Table* dataset had to be combined with each of the 12 CSV files from the *MS Resource Table.* Within each thread, a temporary file was generated by performing an inner join using the fields `um` and `timestamp` from the first dataset with `msname` and `timestamp` from the second dataset.

After completing the 8 processes, all temporary files were merged into a single dataset. Since the `timestamp` column in the *MS Resource Table* has intervals of 60 s, a large portion of the data was discarded, as the `timestamp` in the *MS CallGraph Table* had shorter intervals.

Graph Construction. To build the graphs, the data was processed using the implementation provided in the article *Network Analysis of Microservices: A Case Study on Alibaba Production Clusters* [10], available on GitHub [11]. Several modifications were made to reconstruct the call graph data, compute graph metrics, and relate them to latency. The original code was designed for the dataset `cluster-trace-microservices-v2022`, which is hosted in the same repository as the datasets used in the current work.

Since the dataset structures are similar, the first step was to adapt the column renaming process. The changes were the following: `rpcid` to `service`, as each call has a unique identifier `rpcid`. The next fields renamed were `um` to `source` and `dm` to `target`. The fields `traceid` and `rt` remained unchanged.

After renaming the columns, the next step was to allow processing all of the available call graphs in the dataset. This step generates a CSV file for each `service`, grouping all related data for a given call into one file.

Several centrality metrics were computed for each node, from both the source (`UM`) and destination (`DM`) perspectives. These metrics help characterize the relative importance of nodes within the network:

- **Betweenness centrality (UM/DM)**: Measures how often a node lies on the shortest paths between other node pairs [6]. Higher values indicate strategic positions for communication. Quantitatively, $B_i = \sum_{1 \leq j < k \leq n} \frac{n_{jk}^i}{n_{jk}}$, where n_{jk}^i is the number of shortest paths between nodes j and k passing through i, and n_{jk} is the total number of shortest paths between j and k [8].
- **Closeness Centrality (UM/DM)**: Is the inverse of the sum of shortest path distances from a node to all others [1]. Nodes with high closeness can quickly access or spread information throughout the network. It is $C_i = \frac{n}{\sum_{j \in [n]} dist(i,j)}$, where n is the total number of nodes and $dist(i, j)$ is the shortest distance between i and j [8].
- **Degree Centrality (UM/DM)**: Counts the number of direct connections (edges) a node has [9]. A higher degree suggests greater activity or visibility. This is given by $D_i = \#(\{j \in V_G | i, j \in E_G\})$, where V_G is the set of nodes and E_G is the set of edges in graph G.
- **Eigenvector Centrality (UM/DM)**: Assigns higher values to nodes connected to other highly connected nodes, reflecting their structural influence [15]. A node's importance here stems from its connections to other important nodes. The formula is $E_i = \frac{1}{\lambda} \sum_{j \in neigh_i} E_j$, where λ is the largest eigenvalue and $neigh_i$ represents the neighbors of node i.
- **PageRank (UM/DM)**: Adapts Google's ranking algorithm to measure a node's influence based on the quality and quantity of its connections and the importance of linked nodes. It is a variation of eigenvector centrality, incorporating a damping factor and outbound edges, L_j [16]. $PR_i = \sum_{j \in neigh_i} \frac{PR_j}{L_j}$, where PR_j is the PageRank of node j and L_j is its number of outbound links.

- **Community (UM/DM)**: Identifies the community affiliation for each node using the Louvain method [2]. This algorithm groups densely connected nodes by optimizing modularity, a measure of community partition quality.

In addition, global graph-level metrics were calculated to characterize the overall network structure:

- **Number of Nodes**: total number of vertices in the graph.
- **Number of Edges**: total number of connections (links) between nodes.
- **Density**: proportion of existing edges relative to the maximum possible. Indicates how interconnected the graph is.
- **Average Clustering Coefficient**: measures the tendency of nodes to form cliques or triangles, reflecting local cohesion.
- **Average Degree**: average number of connections per node, obtained by dividing twice the number of edges by the number of nodes.
- **Diameter**: length of the longest shortest path between any two nodes in the graph.
- **Radius**: minimum eccentricity across all nodes, representing the shortest maximum distance from any node.
- **Eccentricity**: for each node, it is the longest shortest path to any other node. It indicates how far a node can be from the rest.

Once these metrics were calculated for each call (identified by its `rpcid`), they were added to the original dataset as twenty new columns. The final dataset contains the twenty computed metrics along with the original columns.

3 Results

As mentioned earlier, the current results include the enriched dataset with graph-related data, which will be used to train models in order to evaluate the usefulness of this additional information. To assess this, the same models will be trained both with and without the enriched data to determine whether the inclusion of graph-based data leads to improved performance. This comparison will help validate the effectiveness of the preprocessing and feature engineering steps.

References

1. Bavelas, A.: Communication patterns in task-oriented groups. Acoustical Soc. America J. **22**(6), 725 (1950). https://doi.org/10.1121/1.1906679
2. Blondel, V.D., Guillaume, J.L., Lambiotte, R., Lefebvre, E.: Fast unfolding of communities in large networks. J. Stat. Mech: Theory Exp. **2008**(10), P10008 (2008). https://doi.org/10.1088/1742-5468/2008/10/p10008
3. Cloud, H.: The NIST definition of cloud computing. National Inst. Sci. Technol. Special Public. **800**(2011), 145 (2011)
4. Dragoni, N., et al.: MicroServices: yesterday, today, and tomorrow. In: Present and Ulterior Software Engineering, pp. 195–216 (2017). https://doi.org/10.1007/978-3-319-67425-4_12

5. Fındık, B.: Using topological features of microservice call graphs to predict the response time variation. Master's Thesis, Middle East Technical University (2024). https://open.metu.edu.tr/handle/11511/108868
6. Freeman, L.C.: A set of measures of centrality based on betweenness. Sociometry **40**(1), 35–41 (1977). http://www.jstor.org/stable/3033543
7. Group, A.: Cluster data - microservices trace dataset (v2021) (2021). https://github.com/alibaba/clusterdata/tree/master/cluster-trace-microservices-v2021, consultado el 20 de Mayo de 2025
8. Hofstad, R.V.D.: Random Graphs and Complex Networks. Cambridge Series in Statistical and Probabilistic Mathematics, Cambridge University Press (2016). https://doi.org/10.1017/9781316779422
9. Jahoda, M., Deutsch, M., Cook, S.W.: Research Methods in Social Relations: With Especial Reference to Prejudice: Part Two: Selected Techniques, Dryden Press (1951)
10. Khodabandeh, G., Ezaz, A., Ezzati-Jivan, N.: Network analysis of microservices: a case study on Alibaba production clusters. In: Companion of the 15th ACM/SPEC International Conference on Performance Engineering, pp. 67–71. (2024)
11. Khodabandelou, G., Committee, I.: ICPE 2024 data challenge - microservice tracing dataset (2024). https://github.com/ghazalkhb/ICPE2024_DataChallenge, consultado el 20 de Mayo de 2025
12. Learn, M.: Monitor a cloud environment - cloud adoption framework (2025). https://learn.microsoft.com/en-us/azure/cloud-adoption-framework/manage/monitor/monitoring, consultado el 14 de febrero de 2025
13. Li, S., et al.: Understanding and addressing quality attributes of microservices architecture: a systematic literature review. Info. Softw. Technol. **131** (2021). https://doi.org/10.1016/j.infsof.2020.106449
14. Luo, S., et al.: The power of prediction: microservice auto scaling via workload learning. In: SoCC 2022 - Proceedings of the 13th Symposium on Cloud Computing, pp. 355–369 (2022). https://doi.org/10.1145/3542929.3563477
15. Negre, C.F.A., et al.: Eigenvector centrality for characterization of protein allosteric pathways. Proc. National Acad. Sci. **115**(52), E12201–E12208 (2018). https://doi.org/10.1073/pnas.1810452115
16. Sayama, H.: Introduction to the Modeling and Analysis of Complex Systems. Open SUNY Textbooks (2015)
17. Tam, D.S.H., Liu, Y., Xu, H., Xie, S., Lau, W.C.: PERT-GNN: latency prediction for microservice-based cloud-native applications via graph neural networks, pp. 2155–2165. ACM (2023). https://doi.org/10.1145/3580305.3599465

A Model-Driven and Data-Driven Framework for Automatic Needle Trajectory Assessment in XR-Based Lumbar Puncture Simulation

Byron Perez-Gutierrez[1](✉), Oscar I. Caldas[1], and Carlos Rodriguez-Guerrero[2]

[1] Universidad Militar Nueva Granada, Bogota D.C., Colombia
byron.perez@unimilitar.edu.co
[2] KU Leuven, Leuven, Belgium

Abstract. Lumbar puncture (LP) simulation using Extended Reality (XR) poses significant engineering challenges, particularly in modeling tissue biomechanics and accurately replicating needle–tissue interactions. This research proposes a framework integrating model-driven biomechanical modeling with data-driven algorithms to enable automated, timely needle trajectory assessment in XR-based LP training. Preliminary developments validate the technical feasibility of the approach, and this paper expands on the methodological details, user-centered requirements, and planned validation strategies to ensure scalability, robustness, and seamless integration into advanced medical simulation systems.

Keywords: Extended Reality · Lumbar Puncture Simulation · Needle Trajectory Assessment

1 Introduction

Lumbar puncture (LP) is a key clinical procedure to diagnose central nervous system conditions by collecting a sample of cerebrospinal fluid (CSF) through needle insertion in the lower back to access the spinal column. It remains a fundamental procedure that internal medicine physicians, anesthesiologists, and neurologists are expected to understand and perform competently [2]. Traditional learning methods for LP, such as lectures, instructional videos, and the "see one, do one, teach one" approach, have become less acceptable due to concerns about patient safety [1]. As a result, medical simulation has emerged as an alternative learning tool, allowing trainees to practice in a safe and controlled environment before performing the procedure in a clinical setting [8].

Conventional simulation tools, such as part-task trainers, are widely used for procedural training; however, they often fall short in offering anatomical variability, exhibit limited physiological realism, and depend on in-person supervision for feedback and assessment, thereby limiting both scalability and overall training effectiveness [7].

W. J. Sarmiento and E. L. Sierra-Ballén (Eds.): CCC 2025, CCIS 2692, pp. 295–301, 2027.
https://doi.org/10.1007/978-3-032-28447-1_24

Extended Reality (XR) simulators, including Virtual, Augmented, and Mixed Reality, offer promising solutions to the limitations of conventional training by enabling dynamic and customizable anatomical models, patient-specific variability, and an essentially unlimited range of training scenarios [3–6]. XR technologies also support the integration of multimodal sensory interfaces, enhancing learners' perceptual engagement and improving the understanding and retention of procedural steps in lumbar puncture.

Despite growing interest in XR-based medical simulation, several research gaps remain unaddressed in the context of LP training. A comprehensive review of the literature reveals the following key limitations:

- Limited integration of immersive technologies such as XR into validated LP training frameworks.
- Lack of real-time or near-instantaneous needle tracking for procedural guidance.
- Absence of standardized assessment tools for evaluating procedural performance within XR environments.
- Insufficient implementation of dynamic haptic feedback to represent tissue resistance accurately.
- Incomplete simulation of procedural steps based on established LP curricula.
- Limited incorporation of ultrasound imaging in LP simulation scenarios.

To address the limitations identified in the literature, such as the lack of automated assessment tools, limited integration of immersive technologies, and insufficient representation of procedural complexity, this research proposes a novel approach that integrates both model-driven and data-driven techniques within an XR-based simulation framework. The model-driven approach relies on predefined kinematic and kinetic parameters, including insertion angles, depth thresholds, and biomechanical constraints derived from anatomical models of the lumbar region. This method is particularly effective for well-defined tasks, where expert knowledge can be formalized into deterministic rules, boundaries, and reference trajectories for lumbar puncture. These parameters may also incorporate patient-specific biomechanical data to simulate realistic anatomical conditions.

In contrast, the data-driven approach involves analyzing and learning from recorded expert performances to train algorithms, often through machine learning, that can recognize patterns of correct and incorrect needle trajectories. This approach is better suited for modeling complex, variable, and dynamic movements, which are difficult to capture through static rules alone. By combining these two paradigms, the proposed hybrid framework seeks to provide a more robust, adaptable, and accurate solution for the automatic assessment of needle trajectories during XR-based lumbar puncture simulations. The following sections present the central research question, outline the general and specific objectives of the study, and state the working hypothesis that guides the development and evaluation of the proposed solution.

1.1 Research Question

How effective is a framework combining model-driven and data-driven approaches aimed at providing timely and accurate automatic assessment of needle trajectory during XR-based lumbar puncture simulations?

1.2 Objectives

General objective:

To propose a hybrid framework that integrates model-driven and data-driven approaches for enabling timely and accurate automatic assessment of needle trajectory in XR-based lumbar puncture simulations.

Specific objectives:

1. To identify the requirements for combining model-driven and data-driven approaches for needle trajectory assessment.
2. To propose a model-driven and data-driven model hybrid framework within a lumbar puncture simulation in XR environment.
3. To implement the proposed framework using metrics and algorithms that quantify needle trajectory accuracy and procedural adherence.
4. To validate the performance of the proposed framework against conventional evaluation methods based on checklist.

1.3 Hypothesis

A framework that combines model-driven and data-driven approaches provides significantly more accurate and timely automatic assessments of needle trajectory during XR-based lumbar puncture simulations compared to traditional check-list assessment methods.

2 Methodology

The methodology for this research is structured around seven key stages that guide the development and evaluation of the proposed hybrid framework for automatic needle trajectory assessment in XR-based LP simulation. The process begins with a requirements analysis, which involves a thorough literature review, semi-structured interviews with domain experts (anesthesiologists and internal medicine physicians) to elicit user needs, and the identification of clinical, technical, and computational constraints relevant to the LP procedure. Building on these findings, the framework's design incorporates three complementary components. First, a model-driven module establishes reference trajectories, insertion angles, and depth limits based on clinical guidelines and biomechanical properties. Second, a data-driven module uses supervised learning on annotated expert-performed procedures to assess needle trajectories. For this purpose, candidate algorithms include recurrent neural networks like Long Short-Term Memory (LSTM) and Gated Recurrent Unit (GRU), which are well-suited for the sequential and time-dependent nature of the needle's path. Third, a middleware to fuse deterministic and learned predictions.

Subsequently, the framework integration within a XR LP simulation system with force and tactile feedback. The methodology also includes a dedicated phase for metric design and algorithm development, where both performance metrics (e.g., trajectory deviation, insertion force, target reach time) and timeliness metrics (e.g., latency from event to feedback (< 100 ms), update rate ($\geq 10\,\text{Hz}$)) are defined and computed. In the experimental design stage, a randomized controlled trial within-subjects will be conducted comparing the XR hybrid simulator with a baseline checklist-based assessment. The study will include 40 medical residents, a sample size determined to provide adequate statistical power for detecting moderate effect sizes (Cohen's d ≈ 0.5) with an alpha of 0.05 and power of 0.8, while remaining feasible within the available recruitment pool. Inclusion criteria require completion of a basic LP module, and participants with prior advanced LP experience will be excluded. The primary outcome will be needle trajectory accuracy, while secondary outcomes will include procedural time and user confidence. Data analysis and validation will employ quantitative methods and hypothesis testing to assess the effectiveness and accuracy of the framework.

Finally, the optimization and final validation phase will involve iterative refinement of the algorithms and a conclusive testing cycle to ensure the system's robustness and readiness for broader application.

1. Requirements Analysis
   - Literature review
   - Semi-structured interviews domain experts
   - Identification of biomechanical, technical, clinical, and computational constraints.
2. Framework Design
   - Model-Driven Component
   - Data-Driven Component
3. Framework implementation
   - Implement the middleware that fuses outputs of biomechanical model and data driven module for decision support
   - Integration with the XR environment
4. Metric design and algorithm development
   - Performances metrics: Trajectory deviation, insertion force, target reach time
   - Timeliness metrics
5. Experimental design for evaluation
   - Experimental protocol design, XR prototype vs baseline system (traditional checklist simulator)
6. Data analysis and system validation
   - Quantitative method analysis
   - Hypothesis testing
7. Optimization and final validation
   - Refine algorithms
   - Final testing

2.1 Value Proposition

This research proposes a novel assessment framework that combines model-driven biomechanics with data-driven learning to enable timely, and automated evaluation of needle trajectories in XR-based LP simulation. By integrating deterministic models of insertion kinematics with machine learning algorithms trained on expert performance data, the framework addresses key limitations of current LP simulators, such as the lack of timely assessment, limited personalization, and reliance on manual evaluation. To the best of our knowledge, no existing LP simulation system has successfully fused both paradigms in a unified XR environment for needle trajectory evaluation, making this research a significant contribution to the field of simulation-based medical education

3 Preliminary Results

As part of the initial development phase, several foundational components of the proposed XR-based LP simulation framework have been implemented and tested to demonstrate technical feasibility and support the planned research objectives.

First, a functional virtual environment has been developed in Unity 3D, simulating an emergency room setting in which the patient is positioned in the seated posture commonly used for LP procedures. The environment provides a contextual and immersive backdrop for procedural simulation, including elements of clinical realism and interaction possibilities with the virtual patient.

Second, anatomically accurate 3D models of patients were reconstructed from anonymized CT scan data to provide realistic volumetric representations of lumbar anatomy. These models serve as the anatomical foundation for simulating tissue structures and interactions during needle insertion, and will support both model-driven and data-driven assessment layers in future iterations.

Third, a vibrotactile prototype was developed to explore a low-fidelity approach to tactile feedback. Instead of simulating precise resistance forces, the device uses vibration cues to create awareness of tissue transitions (e.g., skin, ligamentum flavum, dura mater) during needle advancement. This approach offers a cost-effective alternative for procedural awareness training and has shown promising perceptual differentiation during early testing.

Additionally, an initial haptic model with force feedback was integrated into the Unity environment using a custom-built simulator. This prototype applies variable resistance forces corresponding to different tissue layers and provides the groundwork for validating force-based feedback alongside vibrotactile cues, as illustrated in Fig. 1. Together, these developments demonstrate the viability of combining immersive environments, biomechanical modeling, and tactile feedback mechanisms to support enhanced procedural training in LP simulation.

4 Expected Contributions

This doctoral research is expected to produce both theoretical advancements and practical outcomes in the field of XR-based simulation systems engineering. From

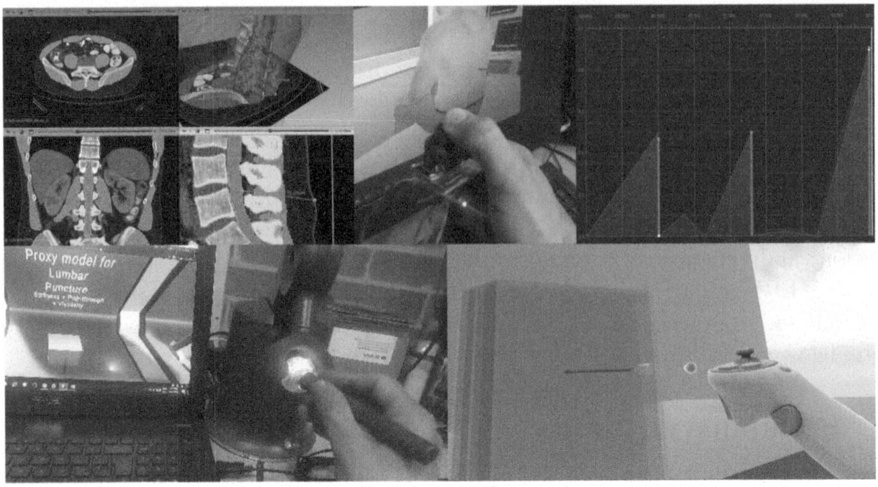

Fig. 1. Preliminary results: CT-based anatomy reconstruction, virtual patient model, and tactile/force feedback prototypes for needle–tissue interaction.

a scientific perspective, it will advance the understanding of how model-driven biomechanics and data-driven algorithms can be seamlessly integrated into a unified computational framework for real-time, automated assessment of complex procedural tasks—specifically, needle trajectory during lumbar puncture. The project will deliver a hybrid architecture that introduces a novel engineering methodology for combining deterministic biomechanical modeling with machine learning–based trajectory analysis, enabling scalable and adaptive simulation platforms

On the practical side, the research will deliver a functional hybrid assessment framework embedded within an XR-based training environment. This system will include validated performance metrics, algorithms for trajectory classification, and tactile feedback mechanisms, both vibrotactile and force-based, designed to enhance procedural awareness. The resulting prototype is intended to support scalable, feedback-rich training for novice practitioners and to reduce reliance on continuous expert supervision.

5 Conclusion

This research proposes and demonstrates the feasibility of a hybrid model-driven and data-driven framework for automatic needle trajectory assessment in XR-based lumbar puncture simulation. By integrating deterministic biomechanical parameters with machine learning algorithms trained on expert data, the framework aims to overcome key limitations of traditional simulation systems, including lack of timely feedback, limited anatomical variability, and dependence on manual evaluation. Preliminary developments—comprising anatomically accurate 3D models, multimodal tactile feedback prototypes, and a functional XR environment—establish a robust foundation for the system's full implementation and experimental validation. The expected outcome is a scalable simulation architecture delivering timely, objective feedback to advance XR system engineering.

Acknowledgments. This research was conducted with the support of the Research Division of Universidad Militar Nueva Granada (UMNG) through grant INV-ING-4047/2024. B. Perez-Gutierrez serves as Assistant Professor of Mechatronics Engineering at UMNG and is currently pursuing doctoral studies. The doctoral work is supervised by Dr. O.I. Caldas, Associate Professor of Mechatronics Engineering at UMNG, who acts as the primary PhD advisor, and co-supervised by Dr. C. Rodriguez-Guerrero, Associate Professor at KU Leuven and member of the Robot Assisted Surgery research group.

References

1. Ayub, S.M.: "See one, do one, teach one": balancing patient care and surgical training in an emergency trauma department. J. Global Health **12** (2022). https://doi.org/10.7189/jogh.12.03051
2. Brydges, R., Stroud, L., Wong, B.M., Holmboe, E.S., Imrie, K., Hatala, R.: Core competencies or a competent core? A scoping review and realist synthesis of invasive bedside procedural skills training in internal medicine. Acad. Med. J. Assoc. Am. Med. Coll. **92**(11), 1632–1643 (2017). https://doi.org/10.1097/ACM.0000000000001726
3. Huang, X., et al.: A mixed-reality stimulator for lumbar puncture training: a pilot study. BMC Med. Educ. **23**(1) (2023).https://doi.org/10.1186/s12909-023-04173-9
4. Lee, D., Bathish, M., Nelson, J.: Transforming nursing education: developing augmented reality procedural training. Cyberpsychol. Behav. Social Network. **27**(6), 372–378 (2024). https://doi.org/10.1089/cyber.2023.0403
5. Meliagros, P., Diener-Brazelle, J., Garber, A.: Using mixed reality to teach medical students lumbar punctures. Am. J. Med. **137**(12) (2024). https://doi.org/10.1016/j.amjmed.2024.08.003
6. Rios-Hernández, M., Jacinto-Villegas, J., Zemiti, N., Vilchis-González, A., Padilla-Castañeda, M., Debien, B.: Development of a lumbar puncture virtual simulator for medical students training: a preliminary evaluation. Int. J. Med. Rob. Comput. Assist. Surg. **19**(6) (2023). https://doi.org/10.1002/rcs.2572
7. Schaff, J., Russell, C.: Mannequin-based simulators and part-task trainers. In: Mahoney, B., Minehart, R.D., Pian-Smith, M.C.M. (eds.) Comprehensive Healthcare Simulation: Anesthesiology. CHS, pp. 107–115. Springer, Cham (2020). https://doi.org/10.1007/978-3-030-26849-7_11
8. Yee, J., San Miguel, C., Khandelwal, S., Way, D., Panchal, A.: Procedural curriculum to verify intern competence prior to patient care. West. J. Emerg. Med. **24**(1), 8–14 (2023). https://doi.org/10.5811/westjem.2022.11.58057

Exploring Surgeon-Radiologist Communication Using Extended Reality During Preoperative Surgical Planning. A User-Centered and Ethnographical Approach

Carlos J. Latorre-Rojas(✉) and Wilson J. Sarmiento

Grupo de Investigación en Multimedia, Universidad Militar Nueva Granada, Bogotá, Colombia
{est.carlos.latorre,gim}@unimilitar.edu.co

Abstract. Recent applications of XR technologies in medical settings, including 3D diagnostic image data analysis, have shown potential benefits. However, there is a lack of association with the design process and performance of medical staff. This lack can impact the intended use of the application and the cost-benefit of including these technologies massively. The present proposal pretends to contribute to the establishment of standards for the design and evaluate XR technologies applied in preoperative SP, considering the context radiologist-surgeon multidisciplinary work.

Keywords: Human Centered Computing · Extended Reality · Medical imaging

1 Introduction

Currently, 3D diagnostic image visualization is integral to treating patients with structural diseases. However, for complex malformations surgeons may struggle to manage the radiologist's anatomical information due to the specific skills required for 3D image interpretation, creating a communication gap [3,6].

eXtended Reality (XR), which includes technologies for 3D interaction and immersion, has been used to enhance the treatment of fields like oncology, gynecology, and cardiovascular surgery [4,5]. While XR applications have shown promise in improving safety and precision in preoperative surgical planning (SP), many studies lack detailed technical and methodological information about device design and user performance. The FDA highlighted the need for specific evaluation standards for XR technologies in medical applications during their 2020 workshop [1]. The study aims to provide a comprehensive framework for designing and evaluating XR solutions for medical image visualization in multidisciplinary surgical planning dialogues between surgeons and radiologists applying a CSCW approach.

W. J. Sarmiento and E. L. Sierra-Ballén (Eds.): CCC 2025, CCIS 2692, pp. 302–307, 2027.
https://doi.org/10.1007/978-3-032-28447-1_25

2 Research Questions

Four research questions guide the investigation of interaction and communication enhancements between surgeons and radiologists during Extended Reality (XR)-supported preoperative surgical planning (SP):

- RQ1: What are the interaction and communication requirements between the surgeon and radiologist during preoperative planning of complex malformations supported by Extended Reality?
- RQ2: How does the current state-of-the-art technology align with the identified interaction requirements?
- RQ3: How can the interaction and communication requirements be effectively fulfilled?
- RQ4: What is the effect of employing user-centered design approaches in XR solutions for preoperative surgical planning?

3 Methodological Approach

- *Ethnography:* The ethnographical approach covered open-ended interviews and fieldwork, capturing data through annotations, photographs, and recordings. The information was curated to develop an affinity diagram that explained the surgeon and radiologist profiles, preferences, tasks, and interaction requirements.
- *Technical Review:* A structured review of current state-of-the-art XR technologies was conducted, guided by the clinical interaction requirements and tasks previously identified. This helped to evaluate technological alignment with clinician needs.
- *Mental Modeling and Interaction Design:* Indirect mental modeling (analysis of recordings and interviewing) methods were applied to capture clinicians' perceptions of the system. Additionally, eight interaction models, oriented equally to the surgeon and the radiologist, were conceptualized and prototyped through video demonstrations emphasizing natural interactions. The clinicians subsequently participated in the design by providing feedback for correction and enhancement of errors.
- *Literature Review and Expert Interviews:* A comprehensive literature review and expert interviews were performed to assess a framework that allow evaluating effects of user-centered XR systems for effective interdisciplinary communication during preoperative surgical planning.

4 Partial Results

The work in progress is presented in response to each research question.

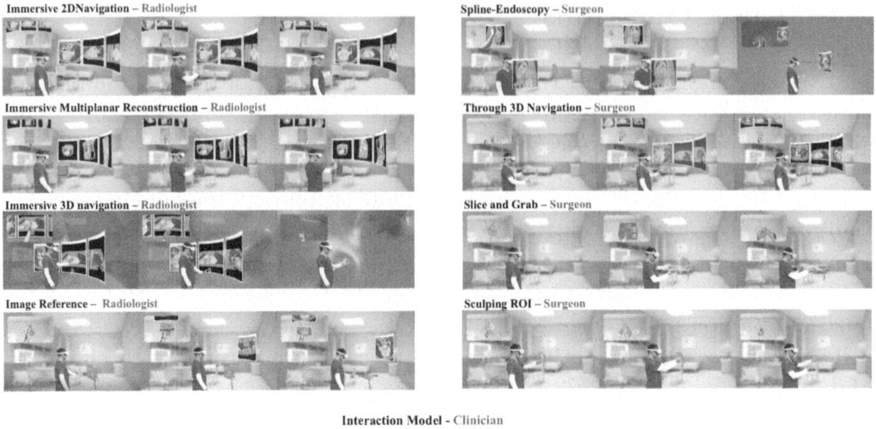

Fig. 1. Illustrative sequence for interaction model prototypes. This interaction models were presented to the radiologist and surgeons whose adjust and enhance them. The process decrease developing efforts.

4.1 RQ1:What are the Interaction and Communication Requirements Between Surgeons and Radiologists During Preoperative Planning of Complex Malformations Supported by Extended Reality?

Two experienced healthcare professionals—a radiologist with 24 years of experience and a surgeon with 47 years of experience in general and cardiovascular surgery—utilized an XR development tool for preoperative planning in cases of congenital heart disease. Over a three-year period, promising clinical outcomes were observed. During this time, a total of 16 preoperative surgical plans were developed, with three of them being recorded for further analysis. Additionally, three open-ended interviews were conducted before, during, and at the end of the ethnography. The first and second interviews facilitated the creation of an affinity diagram, which highlighted the surgeons' past experiences as well as variations in their interaction preferences and requirements [7].

4.2 RQ2: How Does the Current State-of-the-Art Technology Align with the Identified Interaction Requirements?

This unique technical review, based on clinical interaction preferences and requirements performed in 2023, demonstrated significant gaps between existing XR technologies and clinicians' needs [7]. Findings emphasize the necessity of incorporating awareness components into XR systems to enhance intuitive interaction, particularly for surgeons, aiming to support natural interactions that reflect both surgeons' and radiologists' requirements.

4.3 RQ3: How can the Interaction and Communication Requirements be Effectively Fulfilled in the Context of This Research?

Conceptual and causal-loop mental models specified the required interaction mechanisms and communication flows. Guided by these models, we produced eight *video prototypes*—four for surgeons and four for radiologists—to elicit clinical judgment without building functional systems. This early–stage approach enabled validation while minimizing development effort and cost and accommodating clinicians' limited availability. The prototypes were reviewed with clinicians and iteratively refined following a user-centered design process; Fig. 1 shows representative still frames (photograms) from the final set.

4.4 RQ4: How Does the Design of an XR System for Preoperative SP, Based on the Mental Model Approach, Influence Communication and Collaboration Between Surgeons and Radiologists?

The process of eliciting mental models involves considering various human, clinical, technical, and logistical factors and challenges when evaluating an interdisciplinary clinical collaboration heavily reliant on technology. While evaluating medical XR, in general, involves considering optics, ergonomics, and usability, additional evaluation is needed for communication and collaboration, given that radiology, as a medical specialty, aims to communicate information to other specialties and has been studied as a communication phenomenon by previous authors.

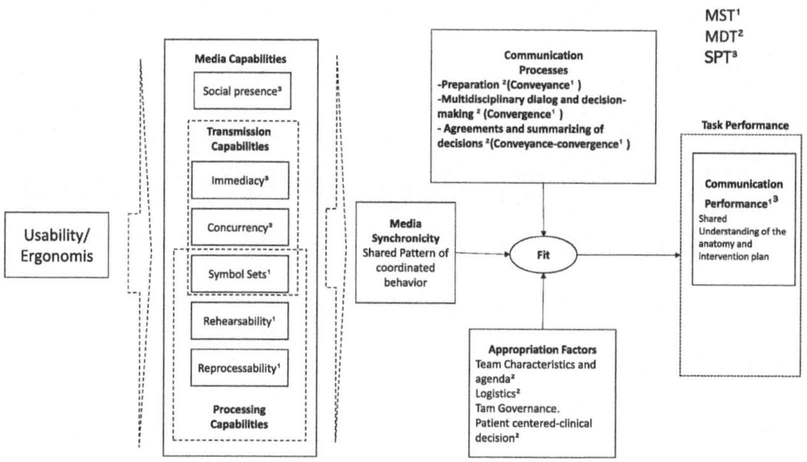

Fig. 2. Research model for evaluating human, clinical, technical and logistical factors for SP with and without the assistance of XR.

4.5 RQ4: What is the Impact of Employing User-Centered Design Approaches in XR Solutions for Preoperative Surgical Planning?

The extensive contact with the radiologist and the surgeon allowed to identify the need to develop a communication-based evaluation approach. A comprehensive framework for evaluating communication effectiveness was formulated. An initial approach using Media Synchronicity Theory (MST) helped to understand both synchronous and asynchronous processes [6]. MST helps to categorize communication features mediated by technology, including transmission velocity, rehearsability, reprocessability, and symbol set [2].

Social presence was also incorporated as a media capability considering the strong body of state of the art that study the immersive environment's presence (Social Presence Theory SPT) [9]. Furthermore, the complexity of human and logistical factors during multidisciplinary(MDT) treatment of patients with complex pathologies has helped to outline strategies to measure the differences between clinical specialties and patient safety [8]. As a Result, Fig. 2 presents a holistic integration of previous theoretical models aiming to consider most of the factors present in a clinical setting.

5 Expected Contributions

The current doctoral research aims to strengthen communication between surgeons and radiologists by establishing design and evaluation methods for medical extended reality (XR) in preoperative surgical planning. Specifically, it proposes (i) a guideline for technical reviews of XR software, (ii) methods for innovating XR interaction techniques, and (iii) a comprehensive framework for evaluating multidisciplinary, technology-mediated communication in healthcare.

The ethnographic approach enabled deep understanding of the role of communication in this clinical context—insights unlikely to emerge from surveys, brief interviews, or external observation alone—thereby positioning communication as a central object of study.

A key gap in the literature is that very few experimental studies explicitly treat VR/AR/MR as a communication medium for surgeon–radiologist coordination (beyond visualization). A forthcoming experiment will operationalize XR-mediated communication and evaluate its effects across preparation, real-time convergence, and agreement/decision phases, generating evidence on when—and for whom—XR improves communicative effectiveness.

Current findings are context-specific and not yet directly replicable; however, they have yielded clear, testable hypotheses to be examined in future experimental settings.

A major outcome to date is a methodology for designing interaction models that align users' mental models with technical requirements. The work further indicates that perceived "naturalness" of interaction varies with clinicians' subjective experience and professional background, underscoring the need for role- and experience-sensitive design.

Acknowledgments. This work results from research project IMP-ING-3407, funded by the Chancellor for Research at the *Universidad Militar Nueva Granada*. Carlos J. Latorre acknowledges the academic mentorship received during his Ph.D. studies under the supervision of Wilson J. Sarmiento.

References

1. Beams, R.: Evaluation challenges for the application of extended reality devices in medicine. J. Digit. Imaging **35**(5), 1409–1418 (2022). https://doi.org/10.1007/s10278-022-00622-x
2. Dennis, A.R., Fuller, R.M., Valacich, J.S.: Media, tasks, and communication processes: a theory of media synchronicity. MIS Q. 575–600 (2008). https://doi.org/10.2307/25148857
3. England, J.R., Cheng, P.M., Romero, M.: Qualitative reporting of lesion number: do radiologists and referring physicians understand each other? J. Am. Coll. Radiol. **15**(8), 1178–1181 (2018). https://doi.org/10.1016/j.jacr.2018.03.049. https://www.jacr.org/article/S1546-1440(18)30392-2/abstract
4. González Izard, S., Juanes Méndez, J.A., Ruisoto Palomera, P., García-Peñalvo, F.J.: Applications of virtual and augmented reality in biomedical imaging. J. Med. Syst. **43**(4), 1–5 (2019). https://doi.org/10.1007/s10916-019-1239-z
5. Jing, W., Masoud, G.: Using virtual reality to facilitate communication in the AEC domain: a systematic review. Constr. Innov. **20**(3), 509–542 (2020). https://doi.org/10.1108/CI-11-2019-0122
6. Larson, D.B., Froehle, C.M., Johnson, N.D., Towbin, A.J.: Communication in diagnostic radiology: meeting the challenges of complexity. Am. J. Roentgenol. **203**(5), 957–964 (2014). https://doi.org/10.2214/AJR.14.12949
7. Latorre-Rojas, C.J., Luzon, J.A., Rozo-Torres, A., Sarmiento, W.J.: Immersive visualization for surgical planning through mental model elicitation: work in progress. In: 2024 IEEE Conference on Virtual Reality and 3D User Interfaces Abstracts and Workshops (VRW), pp. 685–686 (2024). https://doi.org/10.1109/VRW62533.2024.00139
8. Shang, C., Feng, L., Gu, Y., Hong, H., Hong, L., Hou, J.: Impact of multidisciplinary team management on the survival rate of head and neck cancer patients: a cohort study meta-analysis. Front. Oncol. **11** (2021). https://doi.org/10.3389/fonc.2021.630906
9. Tang, F., Wang, X., Norman, C.S.: An investigation of the impact of media capabilities and extraversion on social presence and user satisfaction. Behav. Inf. Technol. **32**(10), 1060–1073 (2013). https://doi.org/10.1080/0144929X.2013.830335

Author Index

W. J. Sarmiento and E. L. Sierra-Ballén (Eds.): CCC 2025, CCIS 2692, pp. 309–310, 2027.
https://doi.org/10.1007/978-3-032-28447-1